Handbook of Complementary, Alternative, and Integrative Medicine

Volume 1 focuses on complementary, alternative, and integrative medicine (CAM) education. Its 20 chapters cover CAM education history, needed competencies, and curriculum reform, among other topics. It is Volume 1 of 6 that describes the education, practice and research-related issues and the efficacy and safety of CAM in treating various conditions. The purpose of these six volumes (sold individually or as a set) is to explain how complementary, alternative, and integrative medicine is practiced around the world; to share the best practices/experiences in terms of education, practice and research; and identify the challenges and suggest recommendations to overcome the identified challenges.

Key Features

- Addresses worldwide issues of *education, training, assessment, and accreditation in complementary and alternative medicine*
- Deals with such hot topics as access/equitable access, online education, and quality and accreditation
- Serves as part of a six-volume comprehensive treatment of complementary, alternative, and integrative medicine as practiced around the world

Prof. Yaser Mohammed Al-Worafi is a Professor of Clinical Pharmacy at the College of Pharmacy, University of Science and Technology of Fujairah, UAE (Previously known as Ajman University). He graduated with a bachelor's degree in Pharmacy (BPharm) from Sana'a University, Yemen, and earned his Master's and PhD degrees in Clinical Pharmacy from the Universiti Sains Malaysia (USM), Malaysia. Prof. Yaser has many postgraduate program certificates including Training to Teach in Medicine 2023 from Harvard Medical School, Harvard University; Contemporary Approaches to University Teaching from The Council of Australasian University Leaders in Learning and Teaching (CAULLT). He has more than 20 years of experience in education, practice, and research in Yemen, Saudi Arabia, the United Arab Emirates, and Malaysia. He has held various academic and professional positions including Deputy Dean and acting dean for Medical Sciences College and Pharmacy College; PharmD program director, Head of Clinical Pharmacy/Pharmacy Practice Department; Head of Teaching & Learning Committee, Head of Training Committee, Head of the Curriculum Committee and other committees. He has authored over 100 peer-reviewed papers in international journals, editing/authoring more than 30 books and 800 book chapters by Springer, Elsevier, Taylor & Francis, USA. Prof. Yaser has supervised/co-supervised many PhD, Master's, PharmD, and BPharm students. He is a reviewer for eight recognized international peer-reviewed journals. Prof. Yaser taught, prepared, designed, and wrote many pharmacy programs for many universities including the Master of Clinical Pharmacy/Pharmacy Practice program; PharmD program, and BPharm program; internship/clerkships for Master's, PharmD and BPharm programs; more than 30 courses related to Clinical Pharmacy, Pharmacy Practice, Social Pharmacy, and Patient Care.

Handbook of Complementary, Alternative, and Integrative Medicine

Education, Practice, and Research

Volume 1: Education, Training, Assessment, and Accreditation

Yaser Mohammed Al-Worafi

CRC Press
Taylor & Francis Group
Boca Raton London New York

CRC Press is an imprint of the
Taylor & Francis Group, an **informa** business

First edition published 2025
by CRC Press
2385 NW Executive Center Drive, Suite 320, Boca Raton FL 33431

and by CRC Press
4 Park Square, Milton Park, Abingdon, Oxon, OX14 4RN

CRC Press is an imprint of Taylor & Francis Group, LLC

© 2025 Yaser Mohammed Al-Worafi

Library of Congress Cataloging-in-Publication Data
Names: Al-Worafi, Yaser, author.
Title: Handbook of complementary, alternative, and integrative medicine /
Yaser Mohammed Al-Worafi.
Description: First edition. | Boca Raton, FL : CRC Press, 2025- |
Includes bibliographical references and index.
Identifiers: LCCN 2024013531 | ISBN 9781032346823 (hardback ; v. 1) |
ISBN 9781032355115 (paperback ; v. 1) | ISBN 9781003327202 (ebook ; v. 1)
Subjects: MESH: Complementary Therapies | Integrative Medicine
Classification: LCC R733 | NLM WB 890 | DDC 610–dc23/eng/20240329
LC record available at https://lccn.loc.gov/2024013531

ISBN: 978-1-032-34682-3 (hbk)
ISBN: 978-1-032-35511-5 (pbk)
ISBN: 978-1-003-32720-2 (ebk)

DOI: 10.1201/9781003327202

Typeset in Times
by codeMantra

To
My wife

Contents

Preface

Volume 1 will focus on complementary, alternative, and integrative medicine education. Further, it includes 20 chapters about complementary, alternative, and integrative medicine education-related issues. It describes how the history of complementary, alternative, and integrative medicine (CAM) education began in Chapter 1. The required and necessary competencies of CAM education are detailed in Chapter 2. In Chapters 3 and 4, the curriculum and curriculum reform in CAM programs, and admission criteria will be described. Teaching strategies and assessment methods will be presented in Chapters 5 and 6. Chapter 7 covers online education-related issues. Training during CAM and integrative medicine education and internship will be presented in Chapter 8. Quality of education and program accreditation will be discussed in Chapter 9. Access/equitable access to education, continuous medical education, cost of education and training, and research and graduation projects will be presented in Chapters 10–13. Chapter 14 focuses on technology and its impact on CAM and integrative medicine education and training. The importance of community/public services, and the contribution of CAM schools/programs in the community and public services, and recommendations for effective community services will be discussed in Chapter 15. Library and educational resources are very important for students and educators, this will be discussed in Chapter 16. Chapter 17 will describe the interprofessional education-related issues. The perspective and practice of CAM education and training will be discussed in Chapter 18. Achievements in CAM education and training, challenges, and the future will be discussed in Chapters 19 and 20.

Key Features:

- Describes the complementary, alternative, and integrative medicine education-related issues.
- Describes the complementary and alternative medicine programs, competencies, curriculums, teaching, assessment methods, and training.
- Explains the quality of complementary and alternative medicine education as well as the accreditation of CAM programs.
- Delves into complementary and alternative medicine online education, facilitators, and barriers to online CAM education.
- Covers technology-related issues in complementary and alternative medicine education, library, and educational resources.
- Describes the challenges of complementary and alternative medicine education and suggests recommendations to overcome them.

Yaser Mohammed Al-Worafi

Acknowledgments

It would have been difficult to write such a book without the help of:

My wife for providing me the time and support to work on the book and spend less time with the family.

Stephen Zollo for his valuable guidance, advice, and help during the writing of this book.

Laura Piedrahita, Randy Brehm, Tom Connelly, Sathya Devi and the production team for their great efforts.

1 History of Complementary and Alternative Medicine (CAM) Education

BACKGROUND

Complementary and alternative medicine (CAM) is a term that encompasses a wide range of health-care practices, including herbal medicine, acupuncture, chiropractic, and other non-conventional therapies. The origins of CAM can be traced back to ancient civilizations, where traditional healing practices were the norm. The use of CAM has grown in popularity over the years, with more people seeking alternatives to conventional medicine. CAM education has also evolved over time, with many institutions now offering courses in various CAM practices. Herewith are the summaries of the history of CAM education, from its early beginnings to its current state (Al-Worafi, 2020a,b, 2023, 2024a,b; Bodeker, 2005; Hasan et al., 2019).

ANCIENT PRACTICES

The origins of CAM can be traced back to ancient civilizations such as China, India, and Egypt. These cultures relied heavily on traditional healing practices, which included the use of herbs, acupuncture, and massage. In ancient China, for example, traditional medicine was practiced along-side Confucianism and Taoism. The Huangdi Neijing, a Chinese medical text that dates back to 2600 BC, outlines the principles of traditional Chinese medicine (TCM). This system of medicine emphasizes the importance of maintaining balance and harmony within the body, using herbs, acupuncture, and other non-conventional therapies to achieve this.

In India, Ayurvedic medicine was the predominant form of healthcare. Ayurveda emphasizes the use of natural remedies and the importance of maintaining balance in the body, mind, and spirit. This holistic approach to healthcare involves diet, exercise, herbal remedies, and other therapies.

In Egypt, healers used a combination of herbal remedies, prayer, and other non-conventional therapies. Egyptian medical texts, such as the Ebers Papyrus, contain information about the use of herbs and other remedies to treat a variety of ailments.

THE RISE OF MODERN MEDICINE

With the advent of modern medicine in the 19th century, traditional healing practices were largely replaced by conventional medical treatments. However, the use of CAM persisted in some parts of the world, and interest in non-conventional therapies began to grow in the West.

In the 1960s and 1970s, the counterculture movement in the United States sparked an interest in alternative forms of healthcare. People began to question the effectiveness of conventional medicine and turned to non-conventional therapies such as acupuncture, chiropractic, and herbal medicine. This led to the development of CAM education programs in the United States.

EARLY CAM EDUCATION PROGRAMS

The first CAM education programs in the United States were established in the 1970s. The National College of Naturopathic Medicine (NCNM), founded in 1956, was one of the first institutions to offer a degree program in naturopathic medicine. In 1973, the New England School of Acupuncture (NESA) was established, becoming the first acupuncture school in the United States.

In 1975, the American Holistic Medical Association (AHMA) was founded to promote the integration of conventional and non-conventional therapies. The AHMA advocated for the inclusion of CAM therapies in medical education programs, and it helped to establish the first CAM departments at medical schools.

In 1978, the National Institutes of Health (NIH) established the Office of Alternative Medicine (OAM), which was later renamed the National Center for Complementary and Alternative Medicine (NCCAM). The NCCAM was created to research the safety and effectiveness of CAM therapies and to provide information to the public and healthcare providers.

CAM EDUCATION IN THE 1980S AND 1990S

In the 1980s and 1990s, interest in CAM continued to grow, and more CAM education programs were established. In 1989, the Council on Naturopathic Medical Education (CNME) was founded to accredit naturopathic medicine.

Here is a timeline of key events in the history of CAM education:

- **2600 BC**: The Huangdi Neijing, a Chinese medical text, outlines the principles of TCM.
- **1550 BC**: The Ebers Papyrus, an Egyptian medical text, contains information about the use of herbs and other remedies to treat a variety of ailments.
- **1500 BC**: The Ayurvedic medical system is established in India.
- **1956**: NCNM is founded in the United States.
- **1973**: NESA is established, becoming the first acupuncture school in the United States.
- **1975**: AHMA is founded to promote the integration of conventional and non-conventional therapies.
- **1978**: NIH establishes the OAM, later renamed as NCCAM.
- **1989**: CNME is founded to accredit naturopathic programs.
- **1990**: The Alternative Medicine Program is established at Columbia University, becoming the first program of its kind at a major medical school in the United States.
- **1992**: The American Board of Holistic Medicine is established to certify physicians in holistic medicine.
- **1993**: The White House Commission on Complementary and Alternative Medicine Policy is established to advise the President and Congress on CAM issues.
- **1998**: NCCAM is elevated to become one of the NIH.
- **2000**: The Consortium of Academic Health Centers for Integrative Medicine is established to promote CAM education and research.
- **2005**: ACAOM is recognized by the United States Department of Education as a specialized accrediting agency.
- **2007**: The Samueli Institute is established to advance integrative health and wellness through research and education.
- **2010**: The Integrative Health Policy Consortium is established to promote the integration of conventional and CAM therapies in healthcare.
- **2014**: NCCIH is established to reflect a shift in focus toward a more integrated approach to healthcare.
- **2016**: The American Association of Naturopathic Physicians releases a position paper on CAM education, calling for increased integration of CAM therapies in medical education.

HISTORY OF COMPLEMENTARY AND ALTERNATIVE MEDICINE (CAM) EDUCATION IN SELECTED COUNTRIES

HISTORY OF COMPLEMENTARY AND ALTERNATIVE MEDICINE (CAM) EDUCATION IN CHINA

CAM has a long history in China, where TCM has been practiced for over 2,500 years. TCM encompasses a range of modalities, including acupuncture, herbal medicine, moxibustion, and massage, and is based on the belief that health is maintained by the balance of yin and yang and the flow of Qi (energy) throughout the body.

In the early 20th century, the Chinese government began to promote Western medicine as the primary healthcare system, and TCM fell out of favor. However, in the 1950s, the government began to recognize the value of TCM and began to integrate it into the healthcare system. Today, TCM is widely practiced in China, and many hospitals have TCM departments alongside Western medicine departments.

In terms of education, TCM schools and universities have been established in China since the 1950s. The first TCM university, the Beijing College of Traditional Chinese Medicine, was founded in 1956. Today, there are over 40 TCM universities and colleges in China, offering undergraduate and graduate programs in TCM.

In addition to TCM, other forms of CAM, such as qigong and tai chi, are also widely practiced in China. These practices are often taught through community centers and schools rather than through formal education institutions.

Overall, CAM education in China has a long history and continues to be an important part of the healthcare system and culture. While Western medicine is also widely practiced in China, TCM and other forms of CAM remain popular and widely used.

HISTORY OF COMPLEMENTARY AND ALTERNATIVE MEDICINE (CAM) EDUCATION IN INDIA

CAM has a long and rich history in India, dating back thousands of years. In India, CAM practices are often considered part of traditional medicine, which is referred to as Ayurveda, Yoga, and Naturopathy. These practices have been passed down through generations and are deeply ingrained in Indian culture and belief systems.

The formal education of CAM practices in India began in the 1920s with the establishment of institutions such as the Central Council of Indian Medicine and the Central Council of Homeopathy. These organizations were responsible for regulating the practice of Ayurveda, Yoga, Naturopathy, Unani, and Homeopathy.

In 1970, the Indian government established the Department of Indian Systems of Medicine and Homeopathy, which was later renamed the Department of Ayurveda, Yoga & Naturopathy, Unani, Siddha and Homoeopathy (AYUSH). The AYUSH department is responsible for the development and promotion of Ayurveda, Yoga, Naturopathy, Unani, Siddha, and Homeopathy in India. The department also oversees the education and training of practitioners in these fields.

Today, there are many institutions in India that offer formal education and training in CAM practices. These institutions include universities, colleges, and specialized institutes that offer undergraduate and postgraduate programs in Ayurveda, Yoga, Naturopathy, Unani, Siddha, and Homeopathy. Some of the leading institutions include the National Institute of Ayurveda, the Morarji Desai National Institute of Yoga, and the National Institute of Naturopathy.

In recent years, the Indian government has taken steps to promote the integration of CAM practices with modern medicine. The government has also initiated programs to encourage research in CAM practices and to develop evidence-based guidelines for their use.

Overall, the education and practice of CAM in India continue to be an important part of the country's healthcare system, and the government is committed to promoting and developing these practices for the benefit of its citizens.

History of Complementary and Alternative Medicine (CAM) Education in Egypt

CAM has a long history in Egypt, dating back thousands of years to the time of the ancient Egyptians. The ancient Egyptians used a variety of natural remedies, such as herbs and minerals, to treat a range of illnesses and conditions.

In modern times, CAM education in Egypt has been influenced by the country's traditional practices and the global interest in holistic health. In 1960, the Ministry of Health established the Institute of Folk Medicine to promote the use of traditional medicine in the country.

In the 1990s, the government created NCCAM to oversee the development of CAM education and research in the country. The center offered training programs and courses for healthcare professionals in various fields, including acupuncture, herbal medicine, and massage therapy.

In 2000, the government passed a law to regulate the practice of alternative medicine in the country. The law established a licensing system for practitioners of CAM and required them to meet certain standards of education and training.

Since then, there has been a growing interest in CAM education and research in Egypt. Several universities and institutions now offer degree programs and courses in various CAM modalities, including acupuncture, herbal medicine, homeopathy, and naturopathy.

However, CAM education and practice in Egypt are still facing challenges, including limited access to resources and funding, a lack of standardization in education and training, and a lack of integration with conventional medicine. Despite these challenges, the interest in CAM education and practice in Egypt is growing, and there is a growing recognition of the importance of integrating traditional and complementary practices with conventional medicine to improve patient outcomes.

History of Complementary and Alternative Medicine (CAM) Education in Europe

CAM education in Europe has a complex and diverse history, with significant variations across different countries and regions. While the use of CAM practices has a long and rich history in many parts of Europe, the formalization of CAM education and training has been a more recent development.

One of the earliest initiatives to establish CAM education in Europe was the founding of the British College of Naturopathy and Osteopathy in 1936. This institution offered courses in osteopathy, chiropractic, and naturopathy, and played a significant role in the development of these fields in the UK and other parts of Europe.

During the latter half of the 20th century, interest in CAM practices grew rapidly in many European countries, leading to the establishment of a range of new educational institutions and programs. In some cases, these programs were established within existing universities and medical schools, while in other cases, they were created as stand-alone institutions.

One important milestone in the history of CAM education in Europe was the establishment of the European Congress for Integrative Medicine in 1993. This organization brought together practitioners, educators, and researchers from across Europe to share knowledge and best practices in the field of CAM.

Today, CAM education in Europe continues to evolve, with a growing emphasis on evidence-based practice and interdisciplinary collaboration. Many countries have established regulatory bodies to oversee CAM education and practice, and some CAM therapies have been integrated into the national healthcare systems of certain European countries.

However, there is also ongoing debate and controversy around the role of CAM in healthcare, with some critics raising concerns about the safety, efficacy, and scientific validity of certain CAM practices. As a result, the future of CAM education and practice in Europe is likely to remain a subject of debate and discussion for many years to come.

HISTORY OF COMPLEMENTARY AND ALTERNATIVE MEDICINE (CAM) EDUCATION IN THE UNITED STATES

CAM refers to a diverse set of healthcare practices and products that are not considered part of conventional Western medicine. CAM includes practices such as acupuncture, chiropractic, herbal medicine, homeopathy, and naturopathy. The history of CAM education in the United States dates back to the early 20th century when interest in non-conventional healthcare practices started to gain popularity.

In the 1920s, chiropractic education emerged as a separate profession from medicine, and chiropractic schools began to appear across the country. Naturopathic medicine also gained popularity during this time and naturopathic schools were established, but they faced opposition from the medical establishment, which led to many of these schools closing down.

During the 1960s and 1970s, interest in alternative medicine grew, and a number of schools were established to teach a variety of CAM practices. Many of these schools were founded by practitioners who wanted to create a formal education system for their respective fields.

In the 1990s, NIH established the OAM, which later became NCCAM in 1998. This signaled a shift in the perception of CAM from being considered fringe medicine to being recognized as a legitimate field of study.

As interest in CAM education continued to grow, many traditional medical schools began to integrate CAM into their curriculum. Some medical schools established separate CAM departments, while others offered elective courses in various CAM practices. In addition, many colleges and universities began to offer degree programs in CAM, including acupuncture, chiropractic, and naturopathy.

Currently, the field of CAM education is still evolving, with some states offering licensing and certification for CAM practitioners, and others still in the process of developing regulatory frameworks. CAM education has become more mainstream in the United States, and it is likely that the demand for CAM practitioners will continue to grow in the future.

HISTORY OF COMPLEMENTARY AND ALTERNATIVE MEDICINE (CAM) EDUCATION IN CANADA

CAM education in Canada has a relatively short history compared to conventional medical education. CAM practices, such as TCM, naturopathy, and chiropractic, have been used in Canada for centuries, but formal education in these fields did not start until the latter half of the 20th century.

One of the earliest CAM educational programs in Canada was established in 1945 when the Canadian Memorial Chiropractic College (CMCC) was founded in Toronto. The CMCC was the first chiropractic college in Canada and offered a four-year program that led to a Doctor of Chiropractic degree. In 1978, the CMCC became the first Canadian chiropractic college to receive accreditation from the Council on Chiropractic Education.

In the 1970s, naturopathic medicine gained recognition as a distinct healthcare profession in Canada. In 1978, the Ontario College of Naturopathic Medicine (now the Canadian College of Naturopathic Medicine) was founded in Toronto. The college offered a four-year program leading to a Doctor of Naturopathic Medicine degree. Today, there are two accredited naturopathic colleges in Canada: the Canadian College of Naturopathic Medicine in Toronto and the Boucher Institute of Naturopathic Medicine in New Westminster, British Columbia.

Acupuncture and TCM education also started to gain momentum in Canada in the 1970s. In 1974, the first TCM program was established at the British Columbia Institute of Technology. Today, there are several accredited acupuncture and TCM programs across Canada, including the International College of Traditional Chinese Medicine in Vancouver, the Alberta College of Acupuncture and Traditional Chinese Medicine in Calgary, and the Ontario College of Traditional Chinese Medicine in Toronto.

In the 1990s, massage therapy education became more formalized in Canada. The College of Massage Therapists of Ontario was established in 1991, and the College of Massage Therapists of British Columbia was established in 1994. Today, there are many accredited massage therapy programs in Canada, including those at community colleges and private institutions.

In conclusion, CAM education in Canada has evolved over the past few decades to include a range of healthcare professions. Chiropractic, naturopathic medicine, acupuncture and TCM, and massage therapy are among the most established CAM professions in Canada. These professions have developed their own educational standards and accreditation processes, and graduates are licensed by their respective regulatory bodies.

HISTORY OF COMPLEMENTARY AND ALTERNATIVE MEDICINE (CAM) EDUCATION IN ARAB COUNTRIES

CAM education in Arab countries has a long history, as many traditional healing practices have been passed down through generations. However, formal education and training in CAM practices in Arab countries have been established more recently.

One of the earliest CAM educational programs in Arab countries was the establishment of the Ibn Sina National College for Medical Studies in Jeddah, Saudi Arabia in 2003. The college offered courses in traditional medicine and complementary and alternative medicine, including acupuncture, homeopathy, and herbal medicine.

In 2005, the United Arab Emirates established the Dubai Healthcare City, which included the Dubai School of Alternative Medicine (DSAM). DSAM offered courses in acupuncture, aromatherapy, herbal medicine, and reflexology, among others.

In 2009, the Egyptian Ministry of Health and Population established the Complementary and Alternative Medicine Unit, which aimed to integrate CAM practices into the healthcare system in Egypt. The unit provides training and education for healthcare professionals in traditional Arabic medicine and other CAM practices.

In 2012, the Moroccan Ministry of Health established the Moroccan Institute for Traditional Medicine, which aimed to preserve and promote traditional medicine in Morocco. The institute offers training programs for healthcare professionals in traditional Moroccan medicine and other CAM practices.

In 2016, the Gulf Cooperation Council (GCC) established the Gulf Health Council for Traditional and Alternative Medicine, which aims to promote and integrate traditional and alternative medicine into the healthcare systems of GCC member states.

Overall, CAM education in Arab countries is still in its early stages, and the level of education and training varies greatly among different countries. However, there is a growing interest in integrating CAM practices into the healthcare system, and many universities and institutions are beginning to offer courses and training programs in CAM practices.

CONCLUSION

In conclusion, the history of CAM education is closely linked to the evolution of traditional healing practices and the rise of modern medicine. From its ancient roots in China, India, and Egypt, CAM has evolved to become a diverse range of healthcare practices, including herbal medicine, acupuncture, chiropractic, and other non-conventional therapies. In the United States, interest in CAM grew in the 1960s and 1970s, leading to the establishment of the first CAM education programs. The 1980s and 1990s saw the continued growth of CAM education, with more institutions offering programs in various CAM practices. Today, CAM education continues to evolve and expand, with increasing recognition of the importance of a more integrated approach to healthcare. The establishment of organizations such as the Consortium of Academic Health Centers for Integrative Medicine and the Integrative Health Policy Consortium reflects a growing commitment to promoting CAM education and research, and to advancing a more holistic approach to healthcare. As the field of CAM education continues to evolve, it will be important to ensure that education programs are evidence-based, that practitioners are properly trained and licensed, and that patients have access to accurate information about CAM therapies. By working together to advance CAM education and research, we can help promote a more integrated and effective approach to healthcare that meets the diverse needs of patients and communities around the world.

REFERENCES

Al-Worafi, Y.M. (Ed.). (2020a). *Drug Safety in Developing Countries: Achievements and Challenges*. Academic Press.

Al-Worafi, Y.M. (2020b). Herbal medicines safety issues. In: *Drug Safety in Developing Countries* (pp. 163–178). Academic Press.

Al-Worafi, Y.M. (2023). *Patient Safety in Developing Countries: Education, Research, Case Studies*. CRC Press.

Al-Worafi, Y.M. (Ed.). (2024a). *Handbook of Medical and Health Sciences in Developing Countries*. Springer, Cham.

Al-Worafi, Y.M. (2024b). Complementary and alternative medicine (CAM) in developing countries. In: Al-Worafi, Y.M. (ed), *Handbook of Medical and Health Sciences in Developing Countries*. Springer, Cham. https://doi.org/10.1007/978-3-030-74786-2_301-1

Bodeker, G. (2005). *WHO Global Atlas of Traditional, Complementary and Alternative Medicine* (Vol. 1). World Health Organization.

Hasan, S., Al-Omar, M.J., AlZubaidy, H., and Al-Worafi, Y.M. (2019). Use of medications in Arab countries. In: Laher, I. (ed), *Handbook of Healthcare in the Arab World* (p. 42). Springer, Cham.

2 Competencies in Complementary and Alternative Medicine (CAM) Education

BACKGROUND

Literature defined competency-based education(CBE) as "a data-based, adaptive, performance-oriented set of integrated processes that facilitate, measure, record, and certify within the context of flexible time parameters the demonstration of known, explicitly stated, and agreed upon learning outcomes that reflect successful functioning in life roles" (Spady, 1977), or "A form of education that derives curriculum from an analysis of a prospective or actual role in modern society and that attempts to certify student progress on the basis of demonstrated performance in some or all aspects of that role" (Riesman, 1979). The background and history of competencies in complementary and alternative medicine (CAM) education reflect a growing recognition and integration of CAM practices into mainstream healthcare. Here's an overview.

EARLY STAGES AND RECOGNITION

1. **Rise of CAM Popularity**: In the latter part of the 20th century, there was a significant increase in the public's interest in and the use of CAM therapies. This trend was driven by a desire for more holistic and natural health approaches.
2. **Initial Lack of Standardization**: Initially, CAM practices were diverse, with varying degrees of formal training and standardization. This diversity ranged from long-standing traditional practices like traditional Chinese medicine to newer modalities like energy healing.

DEVELOPMENT OF FORMAL EDUCATION AND STANDARDS

1. **Inclusion in Medical Education**: Recognizing the public's growing use of CAM, medical schools and healthcare institutions began to incorporate CAM topics into their curricula. This was partly in response to a need for practitioners to be informed about therapies their patients were using.
2. **Research and Evidence-Based Focus**: The National Institutes of Health in the United States, particularly through its National Center for Complementary and Integrative Health, played a significant role in funding research to evaluate the efficacy and safety of CAM practices. This research base helped in developing educational standards.
3. **Establishment of Competencies**: Professional organizations and educational institutions began to develop specific competencies for CAM education. These competencies were designed to ensure that healthcare providers could competently and safely incorporate CAM into their practice.

DOI: 10.1201/9781003327202-2

INTEGRATION AND COLLABORATION

1. **Interdisciplinary Collaboration**: There was an increasing emphasis on interdisciplinary collaboration between conventional medical practitioners and CAM professionals. This led to a more integrative approach to healthcare, where different modalities are used in a complementary manner.
2. **Global Influences**: The World Health Organization and other international bodies also recognized the importance of traditional and complementary medicine. They advocated for the integration of these practices into national health systems and for the establishment of regulatory frameworks.

CURRENT TRENDS

1. **Increasing Formalization and Regulation**: Today, many CAM practices are becoming more formalized, with standardized curricula and certification processes. This includes accreditation for educational programs and licensing for practitioners in certain CAM modalities.
2. **Focus on Patient-Centered Care**: CAM education now emphasizes patient-centered care, recognizing the need to respect patient choices and to integrate CAM practices into conventional treatment plans based on individual patient needs.
3. **Continued Research and Development**: Ongoing research into CAM therapies continues to shape educational competencies, ensuring that they are based on the latest evidence and best practices.

In summary, the development of competencies in CAM education reflects a broader trend toward a more integrative, holistic, and patient-centered approach to healthcare. It underscores the importance of ensuring safe and effective practice in the field of CAM, guided by ongoing research and interdisciplinary collaboration.

HISTORY OF COMPETENCY-BASED EDUCATION

Historical accounts in higher education trace the use of CBE back to the Morrill Act of 1862. The Morrill Land-Grant Acts "provided the basis for an applied education oriented to the needs of farm and townspeople who could not attend the more exclusive and prestigious universities and colleges of the eastern United States" (Clark, 1976). In medical education, the Accreditation Council for Graduate Medical Education (ACGME) has been the major driver of the competency movement due to its ability to set regulatory and certification standards for medical education (ACGME, 2020). CBE in medical education shares the same roots as CBE in higher education, competency-based medical education diverged in the 1970s and 1980s when it began to adopt competencies in education to address public demand for accountability in medical education (Carraccio et al., 2002). However, the new emphasis on competencies in medical education failed to take hold until 1999. At that time, ACGME and the American Board of Medical Specialties endorsed six "core competencies" that medical professionals should demonstrate upon graduation from their residency program. The six competencies are patient care; medical knowledge; practice-based learning and improvement; systems-based practice; and professionalism and interpersonal skills communication (Al-Worafi, 2022a–c).

TERMINOLOGIES

There are many terminologies related to the competencies and CBE as detailed below (ACGME, 2020; WHO, 2022; Englander et al., 2013, 2017; Frank et al., 2010a; O'Grady and Jadad, 2010; Gronlund and Brookhart, 2009; UNESCO, 2013; Van Melle et al., 2019; Al-Worafi, 2022a–c).

COMPETENCIES

The abilities of a person to integrate knowledge, skills, and attitudes in their performance of tasks in a given context. Competencies are durable, trainable, and, through the expression of behaviors, measurable.

COMPETENCE

The state of proficiency of a person to perform the required practice activities to the defined standard. This incorporates having the requisite competencies to do this in a given context. Competence is multidimensional and dynamic. It changes with time, experience, and setting.

COMPETENCY-BASED EDUCATION

An approach to preparing [health workers] for practice that is fundamentally oriented to outcome abilities and organized according to competencies. It de-emphasizes time-based training and facilitates greater accountability, flexibility, and learner-centeredness.

COMPETENCY-BASED CURRICULUM

A curriculum that emphasizes the complex outcomes of learning rather than mainly focusing on what learners are expected to learn about in terms of traditionally defined subject content. In principle, such a curriculum is learner-centered and adaptive to the changing needs of students, teachers, and society. It implies that learning activities and environments are chosen so that learners can acquire and apply the knowledge, skills, and attitudes to situations they encounter in work environments.

COMPETENCY FRAMEWORK

An organized and structured representation of a set of interrelated and purposeful competencies.

COMPETENT

Descriptive of a person who has the ability to perform the designated practice activities to the defined standard. This equates to having the requisite competencies.

KNOWLEDGE

The recall of specifics and universals, the recall of methods and processes, and/or the recall of a pattern, structure, or setting.

ATTITUDE

A person's feelings, values, and beliefs, which influence their behavior and the performance of tasks.

BEHAVIOR

Observable conduct toward other people or tasks that express a competency. Behaviors are measurable in the performance of tasks.

SKILL

A specific cognitive or motor ability that is typically developed through training and practice, and is not context specific.

PRACTICE ACTIVITY

A core function of health practice comprising a group of related tasks. Practice activities are time-limited, trainable, and, through the performance of tasks, measurable. Individuals may be certified to perform practice activities.

PROFICIENCY

A person's level of performance (for example, novice or expert).

PERFORMANCE (INDIVIDUAL WORK PERFORMANCE)

What the organization hires one to do and do well. Performance is a function of competence, motivation, and opportunity to participate or contribute. Where competence reflects what a health worker can do, performance is what a health worker does do.

DOMAIN

A broad, distinguishable area of content; domains, in aggregate, constitute a general descriptive framework.

CURRICULUM

The totality of organized educational activities and environments that are designed to achieve specific learning goals. The curriculum encompasses the content of learning; the organization and sequencing of content; the learning experiences; teaching methods; the formats of assessment; and quality improvement and programmatic evaluation.

INTERPROFESSIONAL EDUCATION

A situation in which learners from two or more occupations learn about, from, and with each other.

STANDARD

The level of required proficiency.

TASK

Observable unit of work within a practice activity that draws on knowledge, skills, and attitudes. Tasks are time-limited, trainable, and measurable.

COLLABORATIVE PRACTICE

A process by which multiple health workers from different professional backgrounds work together with individuals, caregivers, families, and communities to deliver the highest quality of care. It allows health workers to engage any individual whose skills can help achieve local health goals.

COLLABORATIVE DECISION-MAKING

A process of engagement in which health workers and individuals, caregivers, families, and communities work together to understand health issues and determine the best course of action, beyond the two-way knowledge exchange of shared decision-making.

LEARNING OUTCOMES

Learning outcomes describe what learners should know, be able to do, and value as a result of integrating knowledge, skills, and attitudes learned throughout the course. They are stated in measurable terms.

GOALS

Course goals or learning goals are the broad desired results of a course. Goals reflect the purpose of the course and may be derived from a program of study. They are what you want students to learn or get out of your course.

OBJECTIVES

Learning objectives describe the intended result of a learning experience. They are stated in measurable terms. Learning objectives identify discrete aspects of a learning outcome or goal. Collectively, they roll up to meet learning outcomes or goals.

MILESTONES

The ACGME defines milestones as "competency-based developmental outcomes (e.g., knowledge, skills, attitudes, and performance) that can be demonstrated progressively by residents/fellows from the beginning of their education through graduation to the unsupervised practice of their specialties" (ACGME, 2020). Englander et al. (2017) further expanded on this definition by defining milestones as observable markers of an individual's abilities.

COMPETENCY-BASED EDUCATION PRINCIPLES

CBE has the following principles (Frank et al., 2010b):

FOCUSING ON OUTCOMES

To ensure that all graduates are competent in all essential domains. Prepare the graduates for the practice and to be good professionals able to provide effective patient care services.

EMPHASIZING ABILITIES

Medical curricula must emphasize the abilities to be acquired. An emphasis on the abilities of learners should be derived from the needs of those served by graduates (i.e., societal needs).

DE-EMPHASIZING TIME-BASED TRAINING

Medical education must shift from a focus on the time a learner spends on an educational unit to a focus on the learning attained. Greater emphasis should be placed on the developmental progression of abilities and on measures of performance.

PROMOTING GREATER LEARNER-CENTEREDNESS

Medical education must promote greater learner engagement in training. A curriculum of competencies provides clear goals for learners.

CHARACTERISTICS OF LEARNING OUTCOMES

Effective learning outcomes are (Gronlund and Brookhart, 2009):

Clear statements, containing a verb and an object of the verb, of what students are expected to know or do
Action-oriented
Free of ambiguous words and phrases
Learner-centered—written from the perspective of what the learner does
Clearly aligned with the course goals: each learning outcome will support a course goal
Aligned with the course content, including assessments
Realistic and achievable such that the audience must be able to achieve the learning outcome within the logistics of the course such as time, environment, and others
Appropriate for the level of the learner

COMPLEMENTARY AND ALTERNATIVE MEDICINE (CAM) COMPETENCIES: RATIONALITY AND IMPORTANCE

The rationality and importance of developing competencies in CAM can be understood from multiple perspectives, including patient safety, healthcare integration, and the evolving landscape of healthcare needs and preferences.

RATIONALITY

1. **Meeting Patient Demand**: As more patients turn to CAM therapies, healthcare providers need to be knowledgeable about these treatments. This knowledge allows them to guide patients safely and effectively, respecting patient preferences while ensuring informed decision-making.
2. **Ensuring Safety and Efficacy**: Establishing competencies in CAM ensures that practitioners are trained to provide therapies safely and effectively. This is crucial since CAM practices vary widely in technique and approach, and not all are supported by robust scientific evidence.
3. **Evidence-Based Approach**: CAM competencies promote an evidence-based approach to these therapies, integrating scientific research and clinical expertise. This is important for discerning which therapies are beneficial and which may be ineffective or potentially harmful.
4. **Interprofessional Education**: As healthcare becomes more interdisciplinary, CAM competencies facilitate better communication and collaboration between conventional and CAM practitioners, leading to more comprehensive patient care.

IMPORTANCE

1. **Quality of Care**: Competencies ensure that practitioners have a solid foundation in both the theory and practice of CAM, leading to higher quality care for patients who seek these treatments.

2. **Ethical Practice**: By establishing clear competencies, practitioners are better equipped to navigate the ethical considerations involved in CAM, such as informed consent, understanding placebo effects, and managing expectations.

3. **Regulatory Compliance and Professionalism**: Defined competencies help in aligning CAM practices with regulatory standards, enhancing the professionalism and legitimacy of CAM practitioners.

4. **Patient-Centered Care**: Competencies in CAM support a more holistic, patient-centered approach to health, recognizing the importance of treating the whole person rather than just symptoms.

5. **Public Health and Prevention**: Many CAM practices emphasize wellness and prevention. Competencies in these areas can contribute to broader public health goals by encouraging healthy lifestyles and preventive care.

6. **Cultural Competence**: Given that many CAM practices are rooted in specific cultural traditions, competencies include respect for and understanding of cultural differences, enhancing the cultural competence of healthcare providers.

7. **Integration of Traditional Knowledge**: Competencies help bridge traditional knowledge systems with modern healthcare practices, preserving valuable traditional practices and integrating them into contemporary medical care.

In summary, the development of CAM competencies is rational and important for ensuring that CAM therapies are used safely, effectively, and ethically. It supports a more integrative, evidence-based, and patient-centered approach to healthcare, respecting patient preferences and cultural diversity, and contributing to the overall improvement of health outcomes.

COMPETENCIES IN COMPLEMENTARY AND ALTERNATIVE MEDICINE (CAM) EDUCATION

Medical and Health Sciences schools around developing countries adopted the program's competencies from the international recommendations in addition to the recommendations of the local and international accreditation agencies in order to provide graduate students with the necessary knowledge and skills related to CAM to be competent in the health care system and able to provide the most effective patient care services. Competencies in CAM education are essential to ensure that healthcare providers are adequately prepared to integrate CAM practices into patient care safely and effectively. The suggested competencies based on literature review experts' and stakeholders' opinions are as follows (Al-Worafi, 2020a, b, 2023, 2024a, b; Bodeker, 2005; Hasan et al., 2019). These competencies typically encompass a broad range of skills, knowledge, and attitudes, and they are crucial for the responsible and informed application of CAM therapies. Key competencies in CAM education include:

1. **Understanding of CAM Modalities**: Proficiency in various CAM therapies such as acupuncture, herbal medicine, homeopathy, naturopathy, chiropractic, mind-body practices, and others. This includes an understanding of their historical, cultural, and theoretical foundations, as well as their mechanisms of action.

2. **Clinical Skills in CAM**: Developing practical skills for specific CAM therapies, including hands-on techniques and patient management strategies tailored to each modality.

3. **Evidence-Based Practice**: The ability to critically evaluate the scientific evidence supporting various CAM modalities. This includes understanding research methodologies, interpreting research findings, and integrating evidence-based CAM therapies into patient care.

4. **Safety and Efficacy Knowledge**: Understanding the safety profiles, potential side effects, contraindications, and interactions of CAM therapies, particularly in relation to conventional treatments.

5. **Ethical and Legal Considerations**: Knowledge of the ethical and legal implications of using CAM, including informed consent, patient confidentiality, and regulatory issues surrounding CAM practices.
6. **Cultural Competence**: The ability to provide culturally sensitive care, acknowledging the diverse cultural backgrounds and health beliefs of patients, which can influence their approach to and use of CAM therapies.
7. **Interdisciplinary Collaboration**: Skills in collaborating with other healthcare professionals, both within CAM specialties and in conventional medicine, for comprehensive patient care.
8. **Patient Assessment and Integrative Planning**: Competency in conducting thorough patient assessments, understanding the patient's overall health, and integrating CAM therapies into conventional treatment plans as appropriate.
9. **Communication Skills**: Effective communication with patients about CAM therapies, including explaining treatments, managing expectations, and discussing potential risks and benefits.
10. **Professionalism and Quality of Care**: Adhering to high standards of professionalism, including maintaining patient confidentiality, demonstrating empathy, and committing to providing the highest quality of care.
11. **Lifelong Learning and Self-Reflection**: Engaging in continuous learning to stay current with the latest developments in CAM and reflecting on one's own practice to continually improve patient care.
12. **Holistic Health Perspective**: A fundamental competency in CAM is understanding and applying a holistic approach to health. This involves considering the physical, emotional, mental, social, and spiritual aspects of a patient's well-being, rather than just focusing on specific symptoms or diseases.
13. **Patient-Centered Care**: Emphasizing the importance of viewing patients as active participants in their healthcare. This includes respecting patient choices, preferences, and values regarding health and wellness, and involving them in decision-making processes.
14. **Integrative Health Knowledge**: Understanding how to combine CAM practices with conventional medicine to provide integrative care. This requires knowledge of when and how different therapies can be effectively and safely combined or used in sequence.
15. **Research Literacy**: Developing skills not only in interpreting research but also in understanding the limitations and gaps in CAM research. This includes recognizing the challenges in studying CAM therapies and being able to communicate these issues effectively to patients and colleagues.
16. **Regulatory Awareness**: Keeping abreast of the changing landscape of regulations and guidelines governing the practice of various CAM modalities, both at the local and national levels.
17. **Health Promotion and Disease Prevention**: Skills in using CAM approaches for health promotion and disease prevention, including lifestyle counseling, nutrition advice, stress management techniques, and other preventive measures.
18. **Reflective Practice and Self-Care**: Recognizing the importance of self-care for healthcare providers and engaging in regular reflective practice to enhance personal well-being and professional performance. This is particularly important in CAM, where the practitioner's well-being can directly impact the therapeutic relationship and effectiveness.
19. **Networking and Community Involvement**: Building networks with other CAM practitioners and involvement in professional communities to share knowledge, stay informed about best practices, and support the development of the field.
20. **Advocacy and Education**: Engaging in advocacy and educational activities to promote wider understanding and acceptance of CAM practices among the public and within the healthcare system.

21. **Business and Practice Management**: For those in private practice, competencies in business management, including marketing, finance, and operations, are important to sustain a successful CAM practice.
22. **Technology and Digital Health Literacy**: As technology becomes more integrated into healthcare, CAM practitioners should be competent in using digital health tools. This includes telehealth platforms, electronic health records, and digital resources for patient education and health monitoring.
23. **Environmental Health Awareness**: Understanding the impact of environmental factors on health and how various CAM practices can address these issues. This includes knowledge of how environmental toxins, lifestyle, and diet can affect health, and using CAM approaches to support patients in managing these influences.
24. **Public Health Integration**: Skills in understanding and applying CAM in the context of public health, including knowledge of how CAM can contribute to addressing community health issues, health disparities, and global health challenges.
25. **Critical Analysis of Traditional Practices**: While respecting traditional CAM practices, practitioners should also be able to critically analyze and modernize these practices where necessary, ensuring they meet contemporary health and safety standards.
26. **Ethnobotany and Herbal Medicine**: For those specializing in herbal medicine, deep knowledge of ethnobotany, pharmacognosy, and the therapeutic uses of plants is essential. This includes understanding herb sourcing, preparation, dosing, and potential interactions with pharmaceuticals.
27. **Mind-Body Connection**: Competency in understanding and applying the principles of the mind-body connection in health and healing. This includes practices like meditation, yoga, tai chi, and other techniques that integrate mental and physical health.
28. **Nutritional Competence**: Understanding the role of nutrition in health and disease, including knowledge of dietary supplements, nutritional counseling, and the use of food as medicine.
29. **Specialized Population Care**: Skills in providing care to specialized populations such as pediatrics, geriatrics, or those with chronic illnesses. This requires an understanding of how CAM modalities can be safely and effectively tailored to meet the unique needs of these groups.
30. **Crisis and Trauma-Informed Care**: Competence in recognizing and addressing the impacts of crisis, trauma, and mental health issues, and how CAM practices can be integrated into trauma-informed care models.
31. **Quality Improvement and Patient Safety**: Skills in implementing quality improvement initiatives and patient safety protocols in CAM practice. This includes tracking outcomes and continually striving to enhance the safety and effectiveness of CAM therapies.
32. **Global Health Perspectives**: Understanding global health trends and practices, and how CAM fits into various healthcare systems around the world. This includes awareness of international variations in the regulation, practice, and integration of CAM.
33. **Ethical Marketing and Promotion**: For those in private practice, ethical considerations in marketing and promoting CAM services, ensuring accuracy, honesty, and clarity in communicating the benefits and limitations of CAM therapies are important.
34. **Pain Management**: Competency in using CAM modalities for pain management, understanding the mechanisms of pain, and how different CAM practices can alleviate acute and chronic pain conditions.
35. **Genomics and Personalized Medicine**: Understanding the emerging field of genomics and its implications for personalized medicine, including how individual genetic profiles might influence the effectiveness of certain CAM therapies.
36. **Integrative Oncology**: Knowledge of how CAM can be used in oncology, including understanding the specific needs of cancer patients and how various CAM practices can support conventional cancer treatments, manage side effects, and improve quality of life.

37. **Pedagogical Skills for CAM Education**: For those involved in teaching CAM, skills in educational methodologies, curriculum development, and student assessment are important. This includes the ability to convey complex CAM concepts effectively to students or trainees.

38. **Sports Medicine and Physical Rehabilitation**: Understanding how CAM practices can be applied in sports medicine and physical rehabilitation, including injury prevention, recovery, and enhancing athletic performance.

39. **Geriatric CAM Practice**: Specialized knowledge in applying CAM therapies to address the health concerns of older adults, taking into account age-related physiological changes and the prevalence of chronic conditions in this population.

40. **Mental Health and Emotional Wellness**: Competence in addressing mental health and emotional wellness through CAM modalities, such as using mind-body practices for stress reduction, anxiety, and depression, and enhancing overall emotional resilience.

41. **Healthcare Systems and Policy**: Understanding the broader healthcare system and policies that affect CAM practice, including insurance coverage, reimbursement issues, and advocating for policy changes to support CAM integration.

42. **Sustainability and Ethical Sourcing**: Awareness of environmental sustainability, especially in practices like herbal medicine, ensuring that resources are ethically and sustainably sourced, and promoting environmental stewardship in healthcare.

43. **Collaborative Research Skills**: Competence in engaging in collaborative research efforts, including interdisciplinary studies that bridge CAM and conventional medicine, to contribute to the evidence base for CAM practices.

44. **Disaster and Emergency Response**: Understanding the role of CAM in disaster and emergency response, including how CAM practices can support physical and psychological recovery in crisis situations.

45. **Spiritual and Existential Aspects of Health**: Recognizing and addressing the spiritual and existential dimensions of health, and how CAM practices can support patients in these areas, particularly in palliative and end-of-life care.

46. **Health Literacy and Community Education**: Skills in enhancing health literacy among patients and in the community, including the ability to communicate complex health information in an accessible manner.

47. **Neuroscience and CAM**: Understanding the neuroscience behind CAM practices, especially how certain therapies like meditation, acupuncture, or yoga can affect the nervous system and brain function.

48. **Integrative Mental Health**: Developing skills in integrating CAM therapies into mental health care, understanding the interactions between physical, psychological, and social factors in mental wellness.

49. **Phytotherapy and Pharmacognosy**: Advanced knowledge in the field of phytotherapy (herbal medicine) and pharmacognosy, including the bioactive compounds in medicinal plants, their extraction, and clinical applications.

50. **Bioethics in CAM**: Understanding and addressing bioethical issues in CAM, including patient autonomy, beneficence, non-maleficence, and justice, especially as they pertain to the use of traditional and alternative therapies.

51. **Healthcare Informatics**: Competence in healthcare informatics and the use of data in CAM practice. This includes understanding how to collect, analyze, and apply data for better patient outcomes and research purposes.

52. **Traditional Healing Practices**: In-depth knowledge of various traditional healing practices from around the world, respecting and integrating indigenous and historical healing arts into modern practice.

53. **Cosmetic and Aesthetic CAM Therapies**: For practitioners involved in aesthetic treatments, understanding the role of CAM in cosmetic and aesthetic applications, including natural and non-invasive therapies.

54. **Pediatric CAM Practices**: Specialized skills in applying CAM therapies to children, understanding the unique considerations, safety profiles, and developmental aspects of treating pediatric patients with CAM.
55. **Women's Health and CAM**: Expertise in addressing women's health issues through CAM, including reproductive health, pregnancy, menopause, and conditions like polycystic ovary syndrome (PCOS) and endometriosis.
56. **Addiction and Recovery Support**: Using CAM modalities to support addiction recovery and manage substance abuse issues, including understanding the psychological and physiological aspects of addiction.
57. **Veterinary CAM**: For practitioners working with animals, competencies in veterinary CAM, understanding how various CAM modalities can be safely and effectively applied to animal health.
58. **Health Coaching and Lifestyle Counseling**: Skills in health coaching and lifestyle counseling, empowering patients to make informed health choices and changes that align with their overall wellness goals.
59. **Global and Cross-Cultural Health Practices**: Understanding and respecting global health practices, and being able to adapt CAM modalities to suit different cultural contexts and health belief systems.
60. **Advanced Diagnostic Techniques in CAM**: Proficiency in advanced diagnostic techniques specific to certain CAM modalities, such as pulse diagnosis in Traditional Chinese Medicine or iridology.

These competencies are integral to ensuring that healthcare practitioners who incorporate CAM into their practice do so responsibly, ethically, and effectively, contributing to the overall well-being and satisfaction of their patients.

CONCLUSION

In conclusion, the competencies required for effective and responsible practice in CAM are both diverse and comprehensive, reflecting the multifaceted nature of this field. These competencies span a wide range of knowledge areas, skills, and attitudes, emphasizing the importance of a holistic, evidence-based, and patient-centered approach to healthcare. From understanding various CAM modalities and their historical, cultural, and theoretical foundations to developing specific clinical skills; from critically appraising research evidence to ensuring safety and ethical practice; CAM education encompasses a broad spectrum of competencies. These include effective communication, cultural competence, interdisciplinary collaboration, legal and ethical considerations, continuous professional development, and more. Furthermore, CAM competencies extend into specialized areas such as mental health, neurology, women's health, pediatrics, geriatrics, and even veterinary practice, demonstrating the versatility and adaptability of CAM in addressing a wide range of health issues. The development and continuous refinement of these competencies are crucial for ensuring that CAM practitioners are well-equipped to provide safe, effective, and ethical care. They also play a vital role in integrating CAM practices into mainstream healthcare, fostering a more inclusive and comprehensive healthcare system that acknowledges and utilizes the strengths of both conventional medicine and CAM for the benefit of patient health and well-being. Ultimately, the competencies in CAM education underscore a commitment to lifelong learning, professional excellence, and a deep understanding of the dynamic interplay between different health modalities, patient needs, and the evolving landscape of healthcare.

REFERENCES

Accreditation Council for Graduate Medical Education (ACGME). (2020). Frequently asked questions: Milestones. Retrieved December 12, 2020, from https://www.acgme.org/portals/0/milestonesfaq.pdf

Al-Worafi, Y.M. (Ed.). (2020a). *Drug Safety in Developing Countries: Achievements and Challenges.* Academic Press.

Al-Worafi, Y.M. (2020b). Herbal medicines safety issues. In: Al-Worafi, Y.M. (ed), *Drug Safety in Developing Countries* (pp. 163–178). Academic Press.

Al-Worafi, Y.M. (2022a). *A Guide to Online Pharmacy Education: Teaching Strategies and Assessment Methods.* CRC Press.

Al-Worafi, Y.M. (2022b). History and importance. In: Al-Worafi, Y.M. (ed), *A Guide to Online Pharmacy Education: Teaching Strategies and Assessment Methods.* CRC Press.

Al-Worafi, Y.M. (2022c). Competencies and learning outcomes. In: Al-Worafi, Y.M. (ed), *A Guide to Online Pharmacy Education: Teaching Strategies and Assessment Methods.* CRC Press.

Al-Worafi, Y.M. (2023). *Patient Safety in Developing Countries: Education, Research, Case Studies.* CRC Press.

Al-Worafi, Y.M. (Ed.). (2024a). *Handbook of Medical and Health Sciences in Developing Countries.* Springer, Cham.

Al-Worafi, Y.M. (2024b). Complementary and alternative medicine (CAM) in developing countries. In: Al-Worafi, Y.M. (ed), *Handbook of Medical and Health Sciences in Developing Countries.* Springer, Cham. https://doi.org/10.1007/978-3-030-74786-2_301-1

Bodeker, G. (2005). *WHO Global Atlas of Traditional, Complementary and Alternative Medicine* (Vol. 1). World Health Organization.

Carraccio, C., Wolfsthal, S.D., Englander, R., Ferentz, K., and Martin, C. (2002). Shifting paradigms: From flexner to competencies. *Academic Medicine*, 77, 361–367.

Clark, F.W. (1976). Characteristics of the competency-based curriculum. In: Arkava, M.L., and Brennen, E.C. (eds), *Competency-Based Education for Social Work: Evaluation and Curriculum.*

Englander, R., Cameron, T., Ballard, A.J., Dodge, J., Bull, J., and Aschenbrener, C.A. (2013). Toward a common taxonomy of competency domains for the health professions and competencies for physicians. *Academic Medicine*, 88(8), 1088–1094.

Englander, R., Frank, J.R., Carraccio, C., Sherbino, J., Ross, S., Snell, L., and ICBME Collaborators. (2017). Toward a shared language for competency-based medical education. *Medical Teacher*, 39(6), 582–587.

Frank, J.R., Mungroo, R., Ahmad, Y., Wang, M., De Rossi, S., and Horsley, T. (2010a). Toward a definition of competency-based education in medicine: A systematic review of published definitions. *Medical Teacher*, 32(8), 631–637.

Frank, J.R., Snell, L.S., Cate, O.T., Holmboe, E.S., Carraccio, C., Swing, S.R., Harris P., Glasgow, N.J., Campbell, C., Dath, D., Harden, R.M., Lobst, W., Long, D.M., Mungroo, R., Richardson, D.L., Sherbino, J., Silver, I., Taber, S., Talbot, M., and Harris, K.A. (2010b). Competency-based medical education: Theory to practice. *Medical Teacher*, 32(8), 638–645.

Gronlund, N.E., and Brookhart, S.M. (2009). *Writing Instructional Objectives* (8th Edition). Pearson Education Inc., Upper Saddle River, NJ.

Hasan, S., Al-Omar, M.J., AlZubaidy, H., and Al-Worafi, Y.M. (2019). Use of medications in Arab countries. In: Laher, I. (ed), *Handbook of Healthcare in the Arab World* (p. 42). Springer, Cham.

O'Grady, L., and Jadad, A. (2010). Shifting from shared to collaborative decision making: A change in thinking and doing. *Journal of Participatory Medicine*, 2(13), 1–6.

Riesman, D. (1979). Society's demands for competence. In: Grant, G., Elbow, P., Ewens, T., Gamson, Z., Kohli, W., Neumann, W., Olesen, V., and Riesman, D. (eds), *On Competence: A Critical Analysis of Competence-Based Reforms in Higher Education* (pp. 18–65). Jossey-Bass Inc., San Francisco, CA.

Spady, W.G. (1977). Competency based education: A bandwagon in search of a definition. *Educational Researcher*, 6(1), 9–14.

UNESCO International Bureau of Education. (2013). *IBE Glossary of Curriculum Terminology.* UNESCO, Geneva.

Van Melle, E., Frank, J.R., Holmboe, E.S., Dagnone, D., Stockley, D., Sherbino, J., and International Competency-Based Medical Education Collaborators. (2019). A core components framework for evaluating implementation of competency-based medical education programs. *Academic Medicine*, 94(7), 1002–1009.

World Health Organization. (2022). *Global Competency and Outcomes Framework for Universal Health Coverage.*

3 Curriculum and Curriculum Reform in Complementary and Alternative Medicine (CAM) Programs

INTRODUCTION

Complementary and alternative medicine (CAM) represents a diverse group of medical and health-care systems, practices, and products that are not generally considered part of conventional medicine. The curriculum and curriculum reform in CAM programs have become increasingly important as the demand for such practices grows. These programs aim to provide comprehensive education and training in various CAM modalities, ensuring practitioners are well-equipped to meet the needs of patients seeking alternative or supplementary healthcare. In the context of CAM programs, the curriculum is typically designed to offer a broad understanding of holistic health principles while focusing on specific areas such as herbal medicine, acupuncture, homeopathy, naturopathy, and mind-body therapies. An essential aspect of these programs is to provide a foundational understanding of the human body from both conventional and alternative perspectives. This includes a deep dive into anatomy, physiology, pathology, and pharmacology, juxtaposed with training in CAM-specific theories such as the meridian system in acupuncture or the concept of vitalism in naturopathy. One of the major challenges in CAM education is integrating these modalities into a framework that is both scientifically sound and true to the traditional roots of each practice. Curriculum reform efforts often focus on bridging this gap, emphasizing evidence-based approaches while respecting historical and cultural origins. As a result, modern CAM curricula are increasingly incorporating research methodology and critical thinking skills, enabling practitioners to evaluate the efficacy and safety of CAM therapies and to integrate this knowledge into their practice.

Clinical training is another critical component of CAM programs. Students typically undergo extensive training in clinical settings, which may include on-campus clinics, externships, or internships at healthcare facilities that offer CAM services. This hands-on experience is vital for developing practical skills and understanding the nuances of patient care in a CAM context. It also provides opportunities for students to learn how to communicate effectively with patients and other healthcare professionals, particularly in integrative healthcare settings where CAM is used alongside conventional medicine. The ethical and legal aspects of CAM practice are also emphasized in the curriculum. This includes understanding the scope of practice, informed consent, confidentiality, and the legalities surrounding the use of various CAM therapies. As the regulatory landscape for CAM continues to evolve, it's crucial for practitioners to stay informed about laws and regulations affecting their practice. Curriculum reform in CAM education also reflects a growing emphasis on integrative medicine, which combines conventional and CAM therapies to optimize patient health. This approach requires a thorough understanding of when and how to integrate different modalities safely and effectively. As such, CAM programs are increasingly offering courses on integrative health, interprofessional collaboration, and the role of CAM in public health.

Another area of focus in curriculum reform is the development of cultural competence. Given that many CAM practices have roots in specific cultural or ethnic traditions, it is essential for practitioners to be culturally sensitive and aware of the diverse health beliefs and practices of patients

DOI: 10.1201/9781003327202-3

from different backgrounds. This includes understanding how cultural factors can influence health behaviors, treatment acceptance, and health outcomes. With the rise of technology in healthcare, CAM programs are also beginning to incorporate digital health tools into their curriculum. This can include training in telehealth, electronic health records, and the use of health apps and wearables. As technology plays an increasingly significant role in healthcare delivery, CAM practitioners must be adept at using these tools to enhance patient care. Finally, curriculum reform in CAM programs often involves fostering a commitment to lifelong learning. The field of CAM is constantly evolving, with new research, therapies, and technologies emerging. Graduates are encouraged to engage in continuous professional development through workshops, conferences, and advanced certifications, ensuring they remain at the forefront of their field. Curriculum and curriculum reform in CAM programs are driven by the need to provide a comprehensive, evidence-based, and culturally sensitive education that prepares practitioners for the unique challenges of CAM practice. By integrating traditional knowledge with modern science, emphasizing clinical experience, and adapting to the changing landscape of healthcare, these programs strive to cultivate skilled, ethical, and adaptable practitioners who can contribute positively to the health and well-being of their patients and the wider community.

The curriculum for CAM education varies widely, reflecting the diversity of practices and philosophies in the field. These models aim to provide comprehensive training that equips students with the knowledge, skills, and competencies needed to practice CAM safely and effectively. Here are some of the key components and approaches found in CAM curriculum models:

1. **Foundational Science Education**: Similar to conventional medical training, CAM curricula often start with foundational courses in basic sciences such as biology, chemistry, anatomy, and physiology. This foundation is crucial for understanding the human body and the mechanisms of health and disease.
2. **Core CAM Theory and Practice**: Students learn the specific theories, philosophies, and practices of their chosen CAM modality. This could include the principles of traditional Chinese medicine (TCM), Ayurvedic medicine, naturopathy, homeopathy, or others. These courses go in-depth into the historical, philosophical, and practical aspects of the specific CAM approach.
3. **Clinical Skills and Applications**: Practical skills are a cornerstone of CAM education. This includes training in specific techniques (such as acupuncture, herbal medicine preparation, massage techniques, or chiropractic adjustments) and clinical decision-making. Students often engage in supervised clinical experiences to apply their knowledge in real-world settings.
4. **Integrative Health Concepts**: Many CAM programs include training on how to integrate CAM practices with conventional medicine. This includes understanding when and how to refer patients to other healthcare professionals and how to work within a multidisciplinary healthcare team.
5. **Research and Evidence-Based Practice**: Given the growing emphasis on evidence-based practice in healthcare, CAM curricula increasingly incorporate coursework in research methods and critical appraisal of scientific literature. This equips students to understand and contribute to research in their field and to apply evidence-based principles in their practice.
6. **Ethical, Legal, and Professional Development**: Courses covering ethics, legal issues, and professional development are integral. These classes address the ethical responsibilities of CAM practitioners, legal regulations affecting practice, and skills for managing a healthcare practice.
7. **Public Health and Wellness Education**: CAM education often includes components on public health, wellness promotion, and disease prevention, emphasizing the role of CAM in maintaining health and preventing illness.

8. **Cultural Competence and Global Perspectives**: Recognizing the diverse cultural roots and global applications of many CAM modalities, curricula often cover cultural competence and the global context of health and wellness.
9. **Specialized Electives and Advanced Study**: Many programs offer electives or advanced study areas allowing students to specialize further in areas like pediatric CAM, sports medicine, women's health, or geriatrics within the CAM framework.
10. **Lifelong Learning and Continuing Education**: The field of CAM, like all areas of healthcare, is constantly evolving. Curricula often emphasize the importance of ongoing learning and professional development.
11. **Interdisciplinary Collaboration**: Some CAM curricula include opportunities for interdisciplinary learning, where CAM students learn alongside or in collaboration with students in conventional medical fields. This fosters mutual understanding and respect between different healthcare disciplines.

The design of CAM curriculum models is influenced by factors such as the specific CAM modality, the regulatory environment of the region, and the educational standards set by accrediting bodies. As the interest in and acceptance of CAM grows, these curricula continue to evolve, incorporating new scientific findings and pedagogical approaches to meet the needs of students and the populations they will serve.

HISTORICAL PERSPECTIVE ON COMPLEMENTARY AND ALTERNATIVE MEDICINE (CAM) CURRICULUM

The historical perspective on the curriculum of CAM is as diverse and rich as the field itself. The evolution of CAM education reflects a confluence of cultural, scientific, and philosophical developments over time. Here's an overview of this historical journey:

ANCIENT AND TRADITIONAL ROOTS

1. **Traditional Origins**: Historically, many CAM practices have roots in ancient civilizations. For example, TCM and Ayurveda emerged several thousand years ago in China and India, respectively. These systems developed comprehensive theories and practices, including herbal medicine, acupuncture (in the case of TCM), and yoga and dietary practices (in Ayurveda).
2. **Oral and Apprenticeship Models**: Initially, the knowledge of these traditional systems was passed down orally or through apprenticeships. The education was deeply entrenched in the local culture and spirituality, often intertwined with religious practices.

FORMALIZATION AND EXPANSION

3. **Establishment of Schools**: As these practices gained popularity, the need for more formalized education emerged. By the late 19th and early 20th centuries, schools dedicated to specific CAM practices, such as naturopathy and chiropractic, began to appear, particularly in Western countries.
4. **Integration of Western Science**: During the 20th century, there was a gradual incorporation of Western scientific principles into CAM curricula. This was partly in response to the increasing demand for evidence-based practices. Schools began to include basic sciences like anatomy, physiology, and biochemistry, alongside traditional CAM teachings.

REGULATORY AND ACADEMIC RECOGNITION

5. **Regulation and Standardization**: The latter part of the 20th century saw efforts to regulate and standardize CAM education. This included the establishment of accrediting bodies and the development of formal educational standards and licensing requirements in many countries.

6. **Research and Academia**: The late 20th and early 21st centuries also witnessed a growing interest in CAM research. Universities and research institutions began to conduct studies on the efficacy and mechanisms of CAM practices, leading to a more evidence-based approach to the curriculum.

CONTEMPORARY DEVELOPMENTS

7. **Integration with Conventional Medicine**: More recently, there has been a move toward integrating CAM education with conventional medical education. This includes offering CAM courses in medical schools and developing integrative medicine programs that combine both approaches.

8. **Globalization and Cross-Cultural Exchange**: Globalization has facilitated a cross-cultural exchange of medical knowledge. CAM curricula increasingly incorporate a global perspective, recognizing and respecting the diverse origins and applications of different practices.

9. **Digital and Distance Learning**: The rise of digital technology has introduced online and distance learning options in CAM education, making it more accessible to a broader audience.

10. **Public Health and Community Wellness**: Modern CAM curricula often include public health education, emphasizing the role of CAM in promoting community wellness and preventive health.

FUTURE TRENDS

11. **Evidence-Based and Patient-Centric Approach**: The future of CAM education seems to be moving toward a more evidence-based, patient-centric approach. This involves a continuous update of the curriculum based on the latest research findings and a focus on holistic patient care.

12. **Interdisciplinary Collaboration**: There is an increasing emphasis on interdisciplinary education, where CAM practitioners are trained to collaborate effectively with conventional healthcare providers.

In summary, the curriculum of CAM has evolved from being deeply rooted in traditional and cultural practices to incorporating elements of modern science and evidence-based medicine. This evolution reflects the dynamic nature of the field, adapting to changing societal needs, scientific advancements, and global influences. As CAM continues to grow in popularity and acceptance, its educational models are likely to further evolve, integrating new discoveries and adapting to the changing landscape of healthcare.

TYPES OF CURRICULUM MODELS IN COMPLEMENTARY AND ALTERNATIVE MEDICINE (CAM) PROGRAMS

CAM programs exhibit a variety of curriculum models, each designed to cater to the unique aspects of CAM practices and the educational needs of their students. These models vary based on factors like the specific CAM modality being taught, the educational standards of the region, and the integration with conventional healthcare education. Here are some of the prevalent types of curriculum models in CAM programs:

1. **Stand-Alone CAM Programs**
   - *Specialized Focus*: These programs are dedicated entirely to a specific CAM modality, such as acupuncture, naturopathy, or chiropractic care.
   - *Deep Dive into CAM Theories*: Students receive in-depth training in the theories, philosophies, and techniques specific to that modality.
   - *Clinical Practice Emphasis*: Significant emphasis is placed on developing practical skills through clinical rotations or internships.
2. **Integrative CAM Programs**
   - *Combination Approach*: These programs combine CAM education with conventional medical teachings, offering a more integrative approach to healthcare.
   - *Interdisciplinary Learning*: Students learn both CAM and conventional medicine practices, understanding how to integrate these approaches into patient care.
   - *Focus on Collaboration*: Emphasis is placed on developing skills for collaboration within a diverse healthcare team.
3. **Comprehensive Holistic Programs**
   - *Holistic Health Focus*: These programs cover a wide range of CAM practices, offering a broad understanding of holistic health.
   - *Multimodality Training*: Students are trained in various CAM modalities, learning to tailor a combination of treatments to individual patient needs.
   - *Wellness and Prevention*: There's often a strong emphasis on wellness, prevention, and lifestyle management.
4. **Research-Oriented CAM Programs**
   - *Evidence-Based Focus*: These programs emphasize research in CAM, teaching students to critically evaluate and conduct scientific studies.
   - *Integration of Modern Science*: There's a strong component of modern biomedical sciences, bridging CAM practices with contemporary scientific understanding.
   - *Preparation for Academic Careers*: Graduates are often prepared for careers in academic research or as educators in the field of CAM.
5. **Online and Distance Learning CAM Programs**
   - *Flexible Learning Options*: Designed for remote learning, these programs offer flexibility for students who cannot attend traditional classes.
   - *Blended Practical Experience*: Often combined with in-person workshops or clinical experiences to ensure practical skill development.
   - *Technology-Enabled Learning*: Utilize digital platforms for interactive learning, virtual simulations, and online community engagement.
6. **Continuing Education and Professional Development in CAM**
   - *Lifelong Learning*: Aimed at existing healthcare professionals, these programs offer additional training in specific CAM modalities.
   - *Skill Enhancement*: Focus on enhancing specific skills or knowledge areas, keeping professionals updated with the latest developments in CAM.
   - *Certification and Specialization*: Often lead to certifications or specialized credentials in areas like herbal medicine, homeopathy, or mind-body therapies.
7. **Community and Public Health-Oriented CAM Programs**
   - *Public Health Integration*: These programs integrate CAM practices with public health principles.
   - *Community Health Focus*: Students learn how to apply CAM in community settings, focusing on health promotion and disease prevention.
   - *Cultural Competence*: Emphasis on understanding and respecting diverse health beliefs and practices in various communities.

Each of these curriculum models caters to different educational goals and career paths in the field of CAM. As the interest and acceptance of CAM grow in the healthcare sector, these educational models continue to evolve, incorporating new scientific findings and pedagogical approaches to meet the needs of students and the communities they will serve.

RATIONALITY AND IMPORTANCE OF COMPLEMENTARY AND ALTERNATIVE MEDICINE (CAM) PROGRAMS CURRICULUM REFORM

The rationality and importance of curriculum reform in CAM programs are grounded in several key factors that reflect the evolving landscape of healthcare, the growing body of scientific research in CAM, and the changing needs of patients and healthcare systems. Here are some of the main reasons driving these reforms:

1. **Increasing Demand and Popularity**
   - *Public Interest*: There's a growing interest and usage of CAM therapies among the general population. This increased demand necessitates a well-trained workforce of CAM practitioners.
   - *Patient-Centered Care*: As patients increasingly seek out CAM options, healthcare providers must be knowledgeable about these therapies to offer comprehensive, patient-centered care.
2. **Integration with Conventional Medicine**
   - *Holistic Approach*: There's a growing recognition of the benefits of a holistic approach to health, integrating both conventional and CAM therapies.
   - *Collaborative Care*: The need for CAM practitioners to collaborate effectively with conventional healthcare professionals calls for a curriculum that fosters understanding and respect for both modalities.
3. **Evidence-Based Practice**
   - *Scientific Validation*: With ongoing research, many CAM practices are being scientifically validated, requiring the incorporation of this new evidence into educational programs.
   - *Critical Thinking*: Teaching CAM practitioners to critically evaluate research and apply evidence-based practices enhances the quality and safety of the care they provide.
4. **Standardization and Quality Assurance**
   - *Regulatory Requirements*: As more countries establish regulatory frameworks for CAM practices, there is a need to standardize education to meet these requirements.
   - *Professional Credibility*: Standardizing and updating the curriculum helps in maintaining and enhancing the professional credibility of CAM practices.
5. **Global Health Perspectives**
   - *Cultural Sensitivity*: As CAM includes practices from various cultures, educating practitioners about the cultural origins and significance of these practices is crucial for culturally sensitive care.
   - *Global Exchange of Knowledge*: The curriculum must evolve to include global perspectives on health, recognizing the contributions and practices of different cultures.
6. **Technological Advancements**
   - *Digital Health*: Incorporating the latest digital health tools and technologies in CAM education ensures practitioners are adept at using modern healthcare resources.
   - *Distance Learning*: The expansion of online and distance learning platforms necessitates curriculum reforms to adapt to these new modes of education.

7. **Public Health and Prevention**
   - *Preventive Healthcare*: CAM often emphasizes preventive health measures and life-style interventions, aligning with broader public health goals.
   - *Community Health Needs*: CAM programs need to address the specific health needs of communities, requiring a curriculum that is responsive to these needs.
8. **Continuous Professional Development**
   - *Lifelong Learning*: The field of CAM, like all areas of healthcare, is continually evolving. A modern curriculum should foster an ethos of lifelong learning and adaptation.
9. **Ethical and Legal Considerations**
   - *Ethical Practice*: As CAM practices vary widely, ethical considerations, particularly in areas such as informed consent and treatment efficacy, are critical.
   - *Legal Knowledge*: Practitioners need to be aware of the legal aspects of practicing CAM, which vary by region and are subject to change.

Curriculum reform in CAM programs is essential to ensure that the education and training of CAM practitioners are relevant, evidence-based, and aligned with the current healthcare landscape. Such reforms not only enhance the quality of CAM education but also ensure that CAM practitioners are well-equipped to meet the health needs of the population effectively and safely, in a manner that complements and integrates with conventional healthcare systems.

COMPLEMENTARY AND ALTERNATIVE MEDICINE (CAM) PROGRAMS CURRICULUM REFORM: CHALLENGES

Curriculum reform in CAM programs faces several challenges. These obstacles stem from the unique nature of CAM practices, the diversity of educational standards, and the integration of CAM into the broader healthcare system. Understanding these challenges is crucial for effective curriculum development and reform. Here are some of the primary challenges:

1. **Balancing Traditional Practices and Modern Science**
   - *Integrating Evidence-Based Medicine*: One of the biggest challenges is integrating the traditional knowledge and practices of CAM with the principles of evidence-based medicine. This requires a careful balance that respects traditional wisdom while embracing scientific rigor.
   - *Updating Content with Emerging Research*: Continuously updating the curriculum to reflect the latest research and evidence in CAM can be demanding, especially given the vast and varied nature of CAM practices.
2. **Standardization across Diverse Modalities**
   - *Diverse Practices*: CAM encompasses a wide range of practices, from acupuncture to herbal medicine, each with its own theoretical foundations and techniques. Developing a standardized curriculum that accommodates this diversity is challenging.
   - *Varied Educational Standards*: There's a lack of uniformity in educational standards across different countries and regions, making international standardization difficult.
3. **Regulatory and Accreditation Issues**
   - *Navigating Regulation*: The regulatory environment for CAM varies greatly by region and is often in flux, complicating curriculum development.
   - *Accreditation and Quality Assurance*: Ensuring that CAM programs meet the accreditation standards of relevant bodies, which may have differing criteria, is a complex task.
4. **Clinical Training and Experience**
   - *Practical Training Opportunities*: Providing adequate clinical training opportunities for students in CAM disciplines can be challenging, especially in regions where CAM is less integrated into the healthcare system.

- *Quality of Clinical Supervision*: Ensuring high-quality clinical supervision and finding experienced practitioners to mentor students is another significant challenge.

5. **Interprofessional Education and Collaboration**
   - *Integrating with Conventional Medicine Curricula*: Developing curricula that promote collaboration and understanding between CAM and conventional medical practitioners is challenging yet essential for integrative healthcare.
   - *Overcoming Skepticism*: There's often skepticism and a lack of understanding about CAM practices among conventional healthcare providers, which can impede collaborative efforts.

6. **Resources and Funding**
   - *Financial Constraints*: CAM programs often face financial constraints, limiting resources available for curriculum development, faculty recruitment, and research activities.
   - *Access to Educational Materials*: There is sometimes a lack of comprehensive, high-quality educational materials and textbooks specific to CAM disciplines.

7. **Cultural and Ethical Considerations**
   - *Cultural Sensitivity*: Ensuring that the curriculum is culturally sensitive and appropriate, especially when dealing with CAM practices rooted in specific cultural traditions, is challenging.
   - *Ethical Training*: Imparting ethical considerations specific to CAM, such as the use of endangered species in certain traditional medicines or unproven therapies, is critical and complex.

8. **Technology Integration**
   - *Adopting New Technologies*: Incorporating modern technologies, such as telemedicine and digital health tools, into CAM education requires ongoing adaptation and technical support.
   - *Online and Distance Learning*: Developing effective online and distance learning programs for hands-on CAM practices is a challenging task.

9. **Research and Development in CAM**
   - *Encouraging Research Literacy*: Instilling a strong foundation in research methodology within CAM curricula is challenging due to the historically practical and experiential nature of many CAM practices.
   - *Facilitating CAM-Specific Research*: There is a need for more dedicated research to validate and understand the mechanisms of CAM therapies. However, securing funding and institutional support for such research can be difficult.

10. **Aligning with Public Health Goals**
    - *Public Health Integration*: Effectively integrating CAM education with public health priorities, such as disease prevention and health promotion, requires a curriculum that goes beyond individual patient care.
    - *Addressing Global Health Issues*: Adapting CAM education to address global health challenges and disparities in healthcare access and delivery presents unique difficulties.

11. **Patient Safety and Risk Management**
    - *Safety Training*: Ensuring that CAM practitioners are thoroughly trained in patient safety, particularly for practices that involve physical manipulation or herbal remedies, is crucial.
    - *Risk-Benefit Analysis*: Teaching students to perform a comprehensive risk-benefit analysis for CAM therapies, especially in cases where scientific evidence is limited, is a complex aspect of curriculum development.

12. **Interdisciplinary and Experiential Learning**
    - *Designing Interdisciplinary Courses*: Creating courses that effectively combine CAM with other disciplines, such as psychology, nutrition, and physical therapy, requires innovation and interprofessional collaboration.

- *Providing Experiential Learning*: Developing opportunities for students to experience real-world applications of CAM, including internships and community engagement programs, can be logistically and financially demanding.

13. **Reflecting Societal and Environmental Changes**
    - *Adapting to Societal Needs*: As societal attitudes and needs change, CAM curricula must adapt to remain relevant and responsive to these shifts.
    - *Environmental Sustainability*: Incorporating principles of sustainability, especially in practices like herbal medicine that rely on natural resources, is an emerging area of focus in CAM education.

14. **Developing Critical Thinking and Decision-Making Skills**
    - *Critical Analysis*: Teaching students to critically analyze both traditional practices and contemporary research findings in CAM is essential but can be challenging, given the often qualitative and subjective nature of traditional knowledge.
    - *Clinical Decision-Making*: Developing curricular components that enhance decision-making skills in complex clinical scenarios involving CAM therapies is crucial for producing competent practitioners.

15. **Global and Local Relevance**
    - *Balancing Global with Local*: Ensuring that CAM education is globally relevant while still being locally applicable and respectful of indigenous knowledge and practices requires a nuanced approach.
    - *Cross-Cultural Competence*: Developing curricula that foster cross-cultural competence and an understanding of health beliefs and practices across different cultures is vital in a globally connected world.

16. **Blending Modern Technology with Traditional Practices**
    - *Incorporating Modern Tools*: Integrating modern technological tools, such as advanced diagnostic equipment or digital health records, into traditional CAM practices can be challenging, especially when preserving the authenticity of these practices.
    - *Teaching Technological Competence*: Preparing students to be competent in using technology while maintaining the hands-on, personalized nature of CAM treatments requires a careful balance in curriculum design.

17. **Adaptation to Healthcare Policy and Legislation**
    - *Policy Changes*: Rapid changes in healthcare policies and regulations can necessitate frequent updates to the curriculum, ensuring that it remains compliant and relevant.
    - *Legal Implications*: Educating students about the legal implications and responsibilities associated with CAM practice, which can vary significantly from conventional medical practice, is a complex but necessary component.

18. **Addressing Skepticism and Bias**
    - *Overcoming Misconceptions*: There is often a need to address and overcome skepticism and bias against CAM, both among students coming from a conventional medicine background and the general public.
    - *Building Credibility*: Establishing and maintaining the credibility of CAM programs in the eyes of the broader medical and academic communities is a persistent challenge.

19. **Financial and Economic Considerations**
    - *Cost-Effectiveness of CAM Education*: Ensuring that CAM education is cost-effective, both for institutions and students, is a significant challenge, especially when resources are limited.
    - *Economic Viability*: Making the case for the economic viability and potential return on investment of CAM practices to stakeholders and potential students is important for the sustained growth of these programs.

20. **Addressing Diverse Learning Needs**
    - *Accommodating Different Learning Styles*: CAM education must cater to a diverse range of learning styles, from hands-on experiential learning to theoretical study, which requires a flexible and adaptable curriculum.
    - *Language and Accessibility Barriers*: Overcoming language barriers and ensuring that CAM education is accessible to students from various backgrounds and with different abilities is an important consideration.

21. **Ensuring Quality and Consistency in Teaching**
    - *Faculty Training and Development*: Ensuring that faculty members are well-trained and up-to-date with both traditional CAM practices and modern medical knowledge can be challenging.
    - *Consistency across Institutions*: Achieving consistency in the quality of education across different institutions offering CAM programs is essential for the credibility of these qualifications.

22. **Responding to Changing Health Needs**
    - *Emerging Health Trends*: The curriculum must be agile enough to respond to emerging health trends and crises, such as global pandemics, where CAM might play a role.
    - *Chronic Disease Management*: With the rise of chronic diseases, integrating CAM practices into long-term disease management and wellness programs is becoming increasingly important.

23. **Promoting Ethical and Sustainable Practices**
    - *Sustainability in CAM Resources*: Ensuring that the sourcing and use of natural resources in CAM practices, such as herbs and animal products are sustainable and ethical.
    - *Ethical Patient Care*: Instilling a strong sense of ethics regarding patient care, particularly in areas where CAM practices are not yet fully supported by scientific evidence, is crucial.

24. **Addressing the Scope and Depth of CAM Modalities**
    - *Comprehensive Coverage*: Ensuring the curriculum covers the breadth and depth of various CAM modalities can be challenging due to their vast and diverse nature.
    - *Depth vs. Breadth Dilemma*: Striking the right balance between providing a broad overview of multiple CAM practices and delving deeply into specific modalities is a significant pedagogical challenge.

25. **Cultivating Critical and Reflective Practitioners**
    - *Fostering Critical Reflection*: Encouraging students to engage in critical reflection and self-assessment regarding their practice and beliefs about health and healing is vital.
    - *Developing Reflective Practitioners*: Teaching students to be reflective practitioners who continually evaluate and improve their practice based on new evidence, patient feedback, and personal growth.

26. **Ensuring Cultural Relevance and Responsiveness**
    - *Respecting Traditional Origins*: CAM education must respect and preserve the cultural and traditional origins of various practices, which can sometimes be challenging in a modern, globalized context.
    - *Cultural Adaptation*: Adapting CAM practices to different cultural contexts while maintaining their integrity and effectiveness is an ongoing challenge.

27. **Enhancing Interpersonal and Communication Skills**
    - *Patient-Practitioner Relationship*: Developing strong interpersonal and communication skills is crucial for CAM practitioners, as many CAM therapies involve a significant amount of direct patient interaction.
    - *Health Literacy and Education*: Training practitioners to effectively educate and communicate with patients about CAM therapies, including their benefits and limitations.

28. **Adapting to Patient Demographics and Needs**
    - *Demographic Changes*: Adapting the curriculum to reflect changes in patient demographics, such as an aging population or increasing diversity in health beliefs and practices.
    - *Personalized Medicine*: Moving toward a more personalized medicine approach that considers individual differences in lifestyle, environment, and genetics in CAM practice.
29. **Incorporating Interdisciplinary and Collaborative Research**
    - *Multidisciplinary Research*: Encouraging and incorporating interdisciplinary research that spans CAM and other healthcare disciplines into the curriculum.
    - *Collaborative Learning*: Promoting collaborative learning experiences that involve students and faculty from different health disciplines to foster a comprehensive understanding of health and wellness.
30. **Addressing Environmental Health and Sustainability**
    - *Environmental Health*: Integrating concepts of environmental health and its impact on overall wellness and the effectiveness of CAM therapies.
    - *Sustainable Practices*: Educating about sustainable practices in the procurement and use of natural resources for CAM therapies, particularly in herbal medicine and naturopathy.
31. **Keeping Pace with Rapid Technological Advancements**
    - *Emerging Technologies*: Continuously updating the curriculum to include emerging technologies relevant to CAM practice, such as advanced imaging techniques or health informatics.
    - *Digital Health Literacy*: Equipping students with the necessary skills to utilize digital health technologies, from telehealth platforms to health apps.
32. **Navigating the Economics of CAM Education and Practice**
    - *Economic Challenges of CAM Practice*: Addressing the economic aspects of running a CAM practice, including insurance, billing, and the financial accessibility of CAM treatments for patients.
    - *Cost of Education*: Managing the rising costs of CAM education and ensuring that it remains accessible to a diverse student body.
33. **Aligning with Global Health Initiatives**
    - *Global Health Challenges*: Aligning CAM education with global health initiatives and challenges, such as addressing global health disparities and contributing to international health goals.
    - *International Collaborations*: Facilitating international collaborations and exchanges to enrich CAM education and practice through a global lens.

Overcoming these challenges requires collaborative efforts among educators, practitioners, regulatory bodies, and students. It also calls for innovative solutions, continuous evaluation, and adaptation of the curriculum to ensure that CAM education remains relevant, effective, and integrated with the broader healthcare landscape. By addressing these challenges, CAM programs can better prepare practitioners to meet the evolving health needs of the population in a safe, ethical, and effective manner.

COMPLEMENTARY AND ALTERNATIVE MEDICINE (CAM) PROGRAMS CURRICULUM REFORM: RECOMMENDATIONS

Curriculum reform in CAM programs is essential to align these educational offerings with current healthcare needs, scientific advancements, and societal expectations. Here are some recommendations for effective curriculum reform in CAM programs (Al-Worafi, 2020a, b; 2023; 2024a, b; Bodeker, 2005; Hasan et al., 2019):

1. **Strengthen Integration with Conventional Medicine**
   - *Collaborative Learning*: Encourage joint learning opportunities for CAM and conventional medicine students to foster mutual understanding and respect.
   - *Integrative Health Models*: Include courses that specifically address how CAM can be integrated into conventional healthcare settings.
2. **Emphasize Evidence-Based Practice**
   - *Research Literacy*: Incorporate training in research methods and critical evaluation of evidence, enabling students to discern and apply scientifically valid information.
   - *Update Curriculum Regularly*: Ensure the curriculum reflects the latest scientific research and clinical guidelines in CAM therapies.
3. **Enhance Clinical Training**
   - *Hands-On Experience*: Expand opportunities for practical, supervised clinical experience in diverse settings, including hospitals and community clinics.
   - *Interprofessional Clinical Rotations*: Facilitate clinical rotations that involve interprofessional teams to simulate real-world healthcare environments.
4. **Focus on Holistic Patient Care**
   - *Patient-Centered Approach*: Emphasize the importance of holistic patient care, considering physical, emotional, mental, and spiritual aspects of health.
   - *Communication Skills*: Teach effective communication techniques for interacting with patients, including active listening and empathy.
5. **Incorporate Global and Cultural Perspectives**
   - *Cultural Competence*: Include coursework on cultural sensitivity and the diverse origins of various CAM practices.
   - *Global Health*: Introduce global health concepts, highlighting how CAM can address health disparities and contribute to global wellness.
6. **Foster Ethical and Professional Standards**
   - *Ethics Education*: Provide comprehensive education on the ethical aspects of CAM practice, including informed consent and professional conduct.
   - *Legal and Regulatory Knowledge*: Ensure students are aware of the legal and regulatory frameworks governing CAM practices in their region.
7. **Adopt Technological Advancements**
   - *Digital Health Technologies*: Integrate training on the use of health technology, such as telemedicine, health informatics, and electronic health records.
   - *Online Learning Platforms*: Utilize digital platforms for theoretical components, enabling flexible learning opportunities.
8. **Encourage Lifelong Learning and Continuing Education**
   - *Continuing Education*: Stress the importance of ongoing education and professional development in the rapidly evolving field of CAM.
   - *Adaptive Curriculum Design*: Design curricula that are flexible and adaptable to accommodate future changes in the field.
9. **Develop Interdisciplinary and Collaborative Research Skills**
   - *Research Opportunities*: Encourage participation in interdisciplinary research projects that combine CAM and conventional medicine.
   - *Critical Analysis*: Teach students how to critically analyze and conduct research, fostering a culture of evidence-based practice.
10. **Promote Public Health and Prevention**
    - *Community Health*: Include public health principles in the curriculum, emphasizing the role of CAM in disease prevention and health promotion.

- *Wellness and Lifestyle Education*: Teach students about the importance of lifestyle factors in health and disease, consistent with many CAM philosophies.

11. **Address Sustainability and Environmental Health**
    - *Sustainable Practices*: Educate on sustainable and ethical sourcing and use of natural resources, particularly in herbal medicine and other nature-based therapies.
    - *Environmental Impact*: Discuss the environmental impact of health practices and the role of CAM in promoting ecological health.

12. **Address Business and Practice Management Skills**
    - *Practice Management*: Offer courses on the business aspects of running a CAM practice, including marketing, financial management, and entrepreneurship.
    - *Professional Development*: Provide guidance on career paths, professional networking, and navigating the healthcare marketplace.

13. **Strengthen Patient Safety and Risk Management Education**
    - *Safety Protocols*: Emphasize training in safety protocols specific to various CAM practices, such as proper dosing in herbal medicine or technique in chiropractic care.
    - *Risk Management*: Educate students on identifying, assessing, and managing potential risks associated with CAM therapies.

14. **Enhance Critical Thinking and Decision-Making Skills**
    - *Problem-Solving*: Incorporate case studies and problem-solving exercises that challenge students to think critically and make informed decisions.
    - *Clinical Reasoning*: Foster clinical reasoning skills that combine traditional knowledge with modern clinical practices.

15. **Expand Access and Inclusivity in CAM Education**
    - *Diverse Student Recruitment*: Strive for a diverse student body to bring varied perspectives to CAM education and practice.
    - *Accessibility and Inclusivity*: Ensure that CAM education is accessible to all, including accommodations for students with disabilities and support for non-native language speakers.

16. **Promote Collaborative and Community-Based Learning**
    - *Community Engagement*: Develop programs that involve students in community health initiatives, where they can apply CAM practices in real-world settings.
    - *Service Learning*: Encourage service-learning projects that give students hands-on experience while benefiting the community.

17. **Foster Leadership and Advocacy Skills**
    - *Health Advocacy*: Train students to be advocates for health and wellness, including the responsible use of CAM therapies.
    - *Leadership Development*: Include leadership training to prepare students for future roles in healthcare, policy, and education.

18. **Integrate Environmental and Planetary Health Concepts**
    - *Planetary Health*: Discuss the interconnections between human health and the health of the planet, emphasizing sustainable healthcare practices.
    - *Eco-friendly Practices*: Teach the importance of eco-friendly approaches in CAM therapies and the significance of biodiversity conservation.

19. **Address Language and Communication Barriers**
    - *Multilingual Education*: Offer resources or courses in multiple languages to cater to a linguistically diverse student population.
    - *Cultural Communication*: Teach students effective communication strategies that respect cultural differences and language barriers.

20. **Utilize Modern Teaching Methodologies**
    - *Blended Learning*: Combine traditional face-to-face teaching with online and digital learning methods for a more flexible learning experience.

- *Interactive Learning*: Employ interactive teaching methods, such as simulations, workshops, and group discussions, to enhance engagement and understanding.

21. **Enhance Alumni and Professional Networks**
- *Alumni Engagement*: Develop strong alumni networks to provide mentorship, career opportunities, and ongoing professional support for graduates.
- *Professional Collaboration*: Create opportunities for students to interact with professionals in the field, including conferences, seminars, and networking events.

22. **Promote Ethical Sourcing and Use of CAM Materials**
- *Ethical Sourcing Education*: Educate on the ethical and sustainable sourcing of materials used in CAM practices, especially in herbal and natural therapies.
- *Responsible Use*: Teach the importance of responsible and ethical use of natural resources, considering environmental and conservation issues.

23. **Strengthen Interdisciplinary and Multimodal Education**
- *Multimodal Approaches*: Encourage learning across various CAM modalities, giving students a broad understanding of the range of available therapies and their interrelationships.
- *Interdisciplinary Collaboration*: Foster an educational environment that promotes collaboration with other healthcare disciplines, such as nursing, physiotherapy, and psychology.

24. **Addressing the Digital Divide and Technological Access**
- *Equal Access to Technology*: Ensure that all students, regardless of their background, have equal access to the digital tools and resources required for their education.
- *Training in Digital Literacy*: Provide training in digital literacy to ensure that students can effectively utilize online learning resources and digital health technologies.

25. **Enhancing Flexibility in Learning Pathways**
- *Customizable Curricula*: Offer flexible curriculum pathways that allow students to tailor their education according to their interests and career goals.
- *Modular Learning*: Implement modular learning structures that enable students to progressively build their expertise while managing other commitments.

26. **Promoting Global Exchange and Learning Programs**
- *International Exchange Programs*: Establish international exchange programs that allow students to experience CAM practices in different cultural and healthcare settings.
- *Global Health Perspectives*: Include courses that offer a global perspective on CAM, addressing international health challenges and the role of CAM in diverse healthcare systems.

27. **Focusing on Mental Health and Well-being**
- *Mental Health Education*: Integrate mental health and well-being into the curriculum, recognizing the important role CAM can play in this area.
- *Self-Care Practices*: Teach self-care practices to students, emphasizing the importance of practitioner health and wellness in the healthcare profession.

28. **Incorporating Patient and Community Voices**
- *Patient Involvement*: Involve patients and community members in curriculum development and delivery to ensure that education is aligned with patient needs and perspectives.
- *Community-Based Learning*: Encourage community-based learning experiences where students can learn directly from the communities they will serve.

29. **Addressing Chronic Diseases and Aging Populations**
- *Chronic Disease Management*: Include specific training on the management of chronic diseases, an area where CAM can play a significant role.
- *Geriatric CAM*: Given the aging population, incorporate geriatric CAM, focusing on the unique health needs of older adults.

30. **Promoting Entrepreneurial and Innovation Skills**
    - *Business Acumen*: Teach business and entrepreneurial skills to help students navigate the complexities of setting up and running a CAM practice.
    - *Innovation in CAM*: Encourage innovation and creativity in developing new CAM therapies and approaches, aligned with scientific principles and evidence.
31. **Promoting Environmental Health and Ecological Awareness**
    - *Eco-Conscious CAM Education*: Embed principles of ecological sustainability and environmental health within the curriculum to raise awareness about the ecological impact of health practices.
    - *Green Healthcare Practices*: Teach students about green healthcare practices and how CAM can contribute to environmentally sustainable healthcare solutions.
32. **Emphasizing the Importance of Personal Development**
    - *Self-Reflection and Growth*: Encourage self-reflection and personal growth as part of the curriculum, recognizing their importance in developing empathetic and effective healthcare practitioners.
    - *Mindfulness and Stress Management*: Include training in mindfulness, stress management, and coping strategies, beneficial for both personal development and patient care.
33. **Enhancing Quality Control and Assurance in Education**
    - *Accreditation Standards*: Work toward meeting and exceeding accreditation standards to ensure the highest quality of CAM education.
    - *Continuous Quality Improvement*: Implement continuous quality improvement processes to regularly assess and enhance the curriculum.
34. **Addressing the Unique Challenges of Online CAM Education**
    - *Effective Online Pedagogy*: Develop effective online teaching strategies and materials tailored to the unique aspects of CAM education.
    - *Hybrid Learning Models*: Explore hybrid models that combine online theoretical learning with in-person practical training.
35. **Fostering a Culture of Respect and Inclusivity**
    - *Inclusive Environment*: Cultivate an inclusive educational environment that respects and values the diversity of students, faculty, and CAM practices.
    - *Respect for Different Health Beliefs*: Teach students to respect and understand a variety of health beliefs and practices, fostering an inclusive approach to patient care.
36. **Strengthening Community Health and Outreach Programs**
    - *Community Partnerships*: Build partnerships with community organizations to provide students with real-world experiences in delivering CAM services.
    - *Health Promotion Activities*: Encourage students to participate in health promotion and disease prevention activities within the community.
37. **Prioritizing Safety in CAM Practices**
    - *Safety Training*: Focus on rigorous safety training, especially for practices involving physical interventions or herbal and natural supplements.
    - *Adverse Effects Education*: Educate about potential adverse effects and interactions of CAM therapies, emphasizing the importance of reporting and managing these issues.
38. **Expanding Career and Postgraduate Opportunities**
    - *Career Guidance*: Provide robust career guidance to help students navigate various career paths in CAM.
    - *Postgraduate Studies*: Offer options for postgraduate study and specialization, allowing students to deepen their expertise in specific areas of CAM.
39. **Encouraging Ethical Marketing and Public Representation**
    - *Ethical Marketing Skills*: Teach students ethical marketing skills to accurately represent CAM practices and their benefits without overstating claims.

- *Public Education and Advocacy*: Encourage students to participate in public education and advocacy, promoting accurate and responsible information about CAM.
40. **Building Resilience and Adaptability in Healthcare**
    - *Adaptability Skills*: Equip students with skills to adapt to changing healthcare landscapes, including new health challenges and technological advancements.
    - *Resilience Training*: Incorporate resilience training to prepare students for the demands and challenges of a career in healthcare.

Reforming the CAM curriculum requires a multifaceted approach that addresses current healthcare trends, technological advancements, and the holistic nature of CAM practices. By incorporating these recommendations, CAM education can produce competent, ethical, and well-rounded practitioners capable of contributing positively to the healthcare system and the well-being of patients.

CONCLUSION

In conclusion, the reform of CAM program curricula is a multifaceted and dynamic undertaking that must continuously adapt to the evolving landscape of healthcare, education, and societal needs. The recommended reforms aim to create a holistic, evidence-based, and patient-centered educational framework that integrates traditional CAM wisdom with modern medical knowledge and practices. Key aspects of these reforms include the integration of CAM with conventional medicine, emphasizing evidence-based practice, enhancing clinical training, and focusing on holistic patient care. Additionally, incorporating global and cultural perspectives, fostering ethical and professional standards, and adopting technological advancements are crucial. It's also essential to encourage lifelong learning and continuing education, as well as to promote public health and prevention. The challenges of curriculum reform are significant, ranging from balancing traditional practices with scientific rigor, ensuring standardization and quality assurance, to navigating regulatory and accreditation landscapes. However, by addressing these challenges through collaborative efforts and innovative solutions, CAM education can evolve to produce competent, ethical, and adaptable practitioners. These reforms are vital for CAM to maintain relevance and efficacy in the modern healthcare arena. By producing well-rounded practitioners who are skilled in their respective modalities and adept at navigating complex healthcare environments, CAM can continue to play a valuable and complementary role in promoting health and wellness globally. The ultimate goal is to enrich the healthcare system with diverse, effective, and culturally sensitive approaches to health and healing, contributing positively to the well-being of individuals and communities worldwide.

REFERENCES

Al-Worafi, Y.M. (Ed.). (2020a). *Drug Safety in Developing Countries: Achievements and Challenges.* Academic Press.

Al-Worafi, Y.M. (2020b). Herbal medicines safety issues. In Al-Worafi, YM. (ed), *Drug Safety in Developing Countries* (pp. 163–178). Academic Press.

Al-Worafi, Y.M. (2023). *Patient Safety in Developing Countries: Education, Research, Case Studies.* CRC Press.

Al-Worafi, Y.M. (Ed.). (2024a). *Handbook of Medical and Health Sciences in Developing Countries.* Springer, Cham.

Al-Worafi, Y.M. (2024b). Complementary and alternative medicine (CAM) in developing countries. In: Al-Worafi, Y.M. (eds), *Handbook of Medical and Health Sciences in Developing Countries.* Springer, Cham. https://doi.org/10.1007/978-3-030-74786-2_301-1

Bodeker, G. (2005). *WHO Global Atlas of Traditional, Complementary and Alternative Medicine* (Vol. 1). World Health Organization.

Hasan, S., Al-Omar, M.J., AlZubaidy, H., and Al-Worafi, Y.M. (2019). Use of medications in Arab countries. In Laher, I. (ed), *Handbook of Healthcare in the Arab World* (p. 42). Springer, Cham.

4 Complementary and Alternative Medicine (CAM) Education
Admission Criteria & Requirements

INTRODUCTION

Gaining entry into complementary and alternative medicine (CAM) education programs involves a demanding and competitive selection process that emphasizes not only academic prowess but also relevant experience and distinct personal attributes. Prospective students must undertake a series of steps and fulfill certain criteria to earn their place in these esteemed programs. A key requirement for admission into CAM programs is a robust academic foundation, particularly in subjects like biology, chemistry, physics, and mathematics. Most institutions have a minimum grade point average (GPA) threshold, often with an emphasis on science and math courses. Additionally, standardized tests, akin to the Medical College Admission Test in the U.S., are frequently a prerequisite. These tests evaluate an applicant's knowledge and capabilities in areas pertinent to CAM. Academic performance and test scores are initial measures of an applicant's suitability and preparedness for the challenging curriculum that lies ahead. Yet, academic accomplishments alone are insufficient for securing admission. CAM programs place a high value on practical experience and exposure in healthcare environments. Participation in volunteer activities, internships, or research projects in relevant fields is encouraged, allowing applicants to gain a more profound understanding of the healthcare landscape. These experiences are crucial for demonstrating commitment and dedication to the field of CAM. Admissions committees look for candidates who extend their learning beyond the classroom to actively contribute to their communities and exhibit a true passion for healthcare. The personal statement or essay is another vital component of the application process. This narrative provides a platform for candidates to express their unique qualities, motivations, and ambitions It is an opportunity to share their personal journey and the experiences that have driven them toward a career in CAM. Admissions committees seek candidates who can articulate their goals, showcase empathy and compassion, and display their capacity to handle the rigors and responsibilities of healthcare professions. The personal statement is an avenue for applicants to distinguish themselves and leave a lasting impression on the selection panel.

Letters of recommendation are equally significant, usually coming from professors, mentors, or professionals in the healthcare field who have closely interacted with the applicant. These letters offer insights into the applicant's character, work ethic, and potential for success in CAM. Strong recommendations can significantly boost an applicant's chances by providing external validation of their skills and qualities.

Moreover, many CAM education programs include interviews as part of their selection process. These interviews allow committees to assess an applicant's communication skills, professionalism, and interpersonal attributes. Candidates must demonstrate effective communication, critical thinking, and ethical decision-making skills. Interviews also give applicants the chance to ask questions and determine their fit with the program, as aligning the student's and institution's values is crucial.

It's important to note that selection criteria and admission processes can vary between institutions and countries. Some programs may have specific prerequisites or additional requirements, such as

DOI: 10.1201/9781003327202-4

completing certain courses or demonstrating proficiency in a language. Prospective applicants must thoroughly research their desired programs to ensure they meet all the necessary criteria.

The competition for places in CAM education programs is intense, with many highly qualified candidates vying for limited spots. To enhance their chances of success, aspirants should prepare early, pursuing academic and extracurricular excellence and engaging in activities that underscore their dedication to healthcare.

In summary, the admission process for CAM education is comprehensive and stringent. It demands a blend of academic excellence, relevant experience, personal qualities, and a compelling personal narrative. Prospective students must fulfill academic prerequisites, excel in standardized tests, gain practical experience, obtain strong letters of recommendation, and perform impressively in interviews. While competition is fierce, those who exhibit extraordinary attributes, commitment, and a sincere passion for healthcare have a greater likelihood of gaining admission to these distinguished programs.

ADMISSION TO COMPLEMENTARY AND ALTERNATIVE MEDICINE (CAM) UNDERGRADUATE PROGRAMS: REQUIREMENTS

Admission to undergraduate programs in CAM typically involves a set of requirements tailored to assess both the academic preparedness and the personal commitment of applicants to the field. Here are the key requirements for undergraduate CAM programs:

1. **High School Diploma or Equivalent**: Applicants must have completed high school or an equivalent education. This forms the basic educational foundation for further study in CAM.
2. **Strong Academic Record**: A strong performance in high school, especially in science subjects such as biology, chemistry, and physics, is often required. This demonstrates the applicant's ability to handle the scientific aspects of CAM education.
3. **Minimum GPA**: Many programs specify a minimum GPA from high school to ensure applicants have the academic ability to succeed in an undergraduate program.
4. **Standardized Test Scores**: Some institutions may require standardized test scores such as the Scholastic Assessment Test. These scores can be used to assess general academic abilities and readiness for college-level education.
5. **Letters of Recommendation**: Letters from high school teachers or counselors that speak of the applicant's academic abilities, character, and interest in health and wellness can strengthen an application.
6. **Personal Statement or Essay**: A personal statement allows applicants to articulate their interest in CAM, their career aspirations, and why they are drawn to a particular program. This is a chance to showcase passion, personal experiences, and understanding of CAM principles.
7. **Relevant Extracurricular Activities**: Involvement in extracurricular activities related to health, wellness, or community service can be advantageous. This might include clubs, sports, volunteer work, or any activities that demonstrate a commitment to the holistic and integrative approach of CAM.
8. **Interviews**: Some programs may conduct interviews to get a better sense of the applicant's motivations, communication skills, and suitability for a career in CAM.
9. **Introductory Courses or Workshops**: Prior completion of introductory courses or workshops in CAM-related subjects, either in high school or through other programs, can be beneficial.
10. **Healthcare Exposure**: Experience in healthcare settings, even at a basic level, can be valuable. This might include shadowing a healthcare professional, volunteering in healthcare facilities, or participating in health-related community service.

11. **Cultural Competence**: Demonstrating an understanding and appreciation of cultural diversity, especially as it pertains to health beliefs and practices, can be important in CAM programs, which often encompass a wide range of cultural perspectives.
12. **Language Proficiency**: For programs in non-English speaking countries or those with a multicultural focus, proficiency in additional languages may be beneficial.
13. **Ethical Awareness**: An understanding of ethical considerations in healthcare, particularly those relevant to CAM practices, can be a part of the evaluation process.
14. **Technical Skills**: Basic computer literacy and technical skills may be required, as many programs incorporate digital learning tools and online coursework.
15. **Financial Planning**: Understanding the costs associated with the program and having a plan for financial management, including awareness of potential scholarships or financial aid, is important for applicants.

Each CAM program may have its unique set of requirements and emphases, so it's crucial for prospective students to research individual programs thoroughly. This ensures they not only meet the necessary academic and technical standards but also align with the program's specific focus and philosophy in the field of CAM.

ADMISSION TO COMPLEMENTARY AND ALTERNATIVE MEDICINE (CAM) POSTGRADUATE PROGRAMS: REQUIREMENTS

Gaining admission into CAM programs requires meeting a set of specific criteria that blend academic, experiential, and personal elements. Here's a breakdown of the typical requirements:

1. **Academic Excellence**: Applicants must have a solid academic foundation, often with a focus on sciences such as biology, chemistry, physics, and mathematics. A minimum GPA is usually set by institutions to ensure candidates possess the requisite academic prowess.
2. **Standardized Testing**: Similar to traditional medical programs, some CAM programs may require standardized tests. These tests assess knowledge and aptitude in relevant areas; ensuring applicants have a foundational understanding of the sciences and other pertinent subjects.
3. **Practical Experience**: Given the hands-on nature of CAM practices, experience in healthcare settings is highly valued. This can include internships, volunteer work, or research projects in relevant fields, providing insight into healthcare and demonstrating commitment to the CAM domain.
4. **Personal Statement or Essay**: An essential component, the personal statement allows candidates to express their motivations, experiences, and aspirations in pursuing a career in CAM. It's an opportunity to convey individuality and passion beyond academic achievements.
5. **Letters of Recommendation**: Strong endorsements from educators, healthcare professionals, or mentors who can attest to the applicant's character, work ethic, and potential in the CAM field are crucial. These letters provide a third-party perspective on the applicant's capabilities.
6. **Interviews**: Many programs conduct interviews to evaluate an applicant's communication skills, professionalism, and suitability for a career in CAM. This face-to-face interaction helps assess the candidate's interpersonal skills and ethical reasoning.
7. **Specific Prerequisites**: Depending on the program, there may be additional prerequisites such as specific coursework, certifications, or demonstrated proficiency in certain areas relevant to CAM.

8. **Cultural and Ethical Sensitivity**: Given the diverse nature of CAM practices, understanding and respecting cultural differences and ethical considerations in healthcare can be an implicit requirement.

9. **Language Proficiency**: For programs in non-English speaking countries or those that cater to a diverse population, proficiency in a second language might be beneficial.

10. **Personal Qualities**: Traits like empathy, compassion, and a genuine interest in holistic and alternative approaches to health are often sought after in candidates.

11. **Relevant Extracurricular Activities**: Engagement in extracurricular activities that demonstrate a commitment to health and wellness can be advantageous. This includes participation in clubs, organizations, or community services related to health, nutrition, fitness, mindfulness, or holistic care.

12. **Work Experience in a Related Field**: Having work experience in healthcare or a field related to CAM can strengthen an application. This experience demonstrates a practical understanding of health and wellness and shows commitment to the field.

13. **Research Experience**: Involvement in research, especially in topics relevant to CAM, can be a significant advantage. This experience not only shows a candidate's ability to engage with scientific methods but also their interest in advancing the understanding of CAM practices.

14. **Certifications or Training**: Certain programs may value or require certifications in related areas such as yoga, massage therapy, acupuncture, herbal medicine, or other holistic practices. These certifications can demonstrate both skill and dedication to specific CAM modalities.

15. **Understanding of CAM Philosophies**: A demonstrated understanding of and alignment with the philosophies underlying various CAM practices can be crucial. This might include an awareness of the holistic approach to health, integrative medicine principles, or the historical and cultural contexts of certain CAM practices.

16. **Statement of Purpose**: Some programs might require a more detailed statement of purpose, where applicants need to articulate their reasons for choosing a CAM career path, their understanding of the field, and how they foresee their role in the future of healthcare.

17. **Portfolio of Practices**: For programs focusing on specific CAM therapies (like naturopathy, herbalism, or mind-body practices), a portfolio showcasing personal and professional experiences with these therapies can be beneficial.

18. **Global Health Exposure**: Exposure to health practices and challenges in various cultural and global contexts can be an asset, particularly in CAM programs that emphasize a global perspective on health and wellness.

19. **Ethical Reasoning and Decision Making**: Demonstrating the ability to navigate complex ethical scenarios, particularly those unique to CAM practices, can be crucial in the admission process.

20. **Adaptability and Open-Mindedness**: Qualities like adaptability, open-mindedness, and a willingness to learn and embrace diverse health practices are important in the evolving field of CAM.

21. **Financial Planning and Scholarships**: Understanding the financial aspects of CAM education, including tuition, potential scholarships, and financial aid, is important. Some programs may require applicants to submit a financial plan or apply for scholarships as part of the admission process.

22. **Networking and Professional Connections**: Building a network within the CAM community, attending relevant conferences, or engaging with professionals in the field can enhance an applicant's understanding of the industry and potentially support their application.

It's important for prospective students to research specific CAM programs thoroughly, as requirements can vary significantly between institutions and countries. The field of CAM encompasses a
wide range of practices and philosophies, making it crucial for applicants to align their personal and
professional goals with the focus and ethos of the program they are applying to.

ADMISSION TO COMPLEMENTARY AND ALTERNATIVE
MEDICINE (CAM) PROGRAMS: EQUITABLE ACCESS

Ensuring equitable access to CAM programs involves addressing various barriers and implementing inclusive admission practices. This approach aims to create a diverse and representative student
body, acknowledging that CAM benefits from a range of perspectives and backgrounds. Here are
key strategies and considerations for equitable access in CAM program admissions:

1. **Holistic Admission Processes**: Adopting a holistic review process allows admissions
   committees to consider a wide range of factors beyond academic metrics. This includes
   evaluating an applicant's personal experiences, community service, leadership qualities,
   and overcoming personal or systemic challenges.
2. **Outreach and Awareness Programs**: Institutions can create outreach programs targeting underrepresented communities to raise awareness about CAM fields and educational
   opportunities. These programs can provide information, mentorship, and resources to
   encourage applications from diverse backgrounds.
3. **Scholarships and Financial Aid**: Providing scholarships and financial aid specifically
   aimed at underrepresented or economically disadvantaged students can help alleviate the
   financial barriers to higher education in CAM.
4. **Cultural Competency Training**: Incorporating cultural competency as a part of the curriculum and admissions process ensures that faculty and staff are equipped to understand
   and value diversity, which is reflected in their admission decisions.
5. **Support Services**: Offering robust support services, including academic advising, tutoring, and mental health services, can help ensure that all students, regardless of their background, have the support they need to succeed.
6. **Flexible Admission Criteria**: Recognizing that traditional metrics like standardized test
   scores may not fully capture an applicant's potential, especially for students from varied educational backgrounds or life experiences, can lead to a more inclusive admission
   process.
7. **Community Partnerships**: Establishing partnerships with community organizations,
   schools, and healthcare providers in diverse communities can aid in identifying and supporting potential applicants.
8. **Representation in Recruitment**: Ensuring that recruitment teams and materials reflect
   diversity and inclusivity can make CAM programs more appealing to a broader range of
   applicants.
9. **Language and Accessibility Considerations**: Providing admission materials and assistance in multiple languages and ensuring websites and application processes are accessible
   to people with disabilities are crucial for inclusivity.
10. **Feedback and Continuous Improvement**: Institutions should regularly review their
    admission processes and outcomes to identify any biases or barriers and make necessary
    adjustments to promote equity.
11. **Mentorship Programs**: Developing mentorship programs where current students or
    alumni from underrepresented groups mentor prospective or incoming students can foster
    a more welcoming and inclusive environment.

12. **Alumni Engagement**: Engaging diverse alumni in recruitment and outreach efforts can provide relatable role models for prospective students and help demonstrate the program's commitment to diversity.

13. **Broad Marketing Strategies**: Marketing CAM programs through a variety of channels and platforms can reach a more diverse audience. This includes using social media, community events, and collaboration with organizations that work with underrepresented groups.

14. **Anti-Bias Training**: Providing anti-bias training for admissions staff and faculty involved in the selection process can help reduce unconscious biases that might affect admission decisions.

15. **Student-Led Initiatives**: Encouraging and supporting student-led diversity and inclusion initiatives can create a more welcoming environment for underrepresented students and provide valuable insights for the administration on improving equity.

16. **Accessibility of Application Materials**: Ensuring that application materials are easily accessible and understandable to people from various educational backgrounds and abilities is key. This might involve simplifying application procedures and offering guidance on how to complete them.

17. **Community Engagement and Feedback**: Engaging with communities to understand their needs and receiving feedback on admission practices can help tailor approaches to be more inclusive and responsive to the needs of diverse populations.

18. **Diverse Admissions Committee**: Having a diverse group of individuals on admissions committees can bring different perspectives to the evaluation process, helping to reduce biases and promote a more inclusive selection.

19. **Needs-Based Assessment**: Implementing a needs-based assessment process can help identify candidates who may require additional support, whether it is academic, financial, or personal, and provide the necessary resources to help them succeed.

20. **Pipeline Programs**: Establishing pipeline programs for high school students from underrepresented groups can foster interest and preparedness for CAM education. These programs can include summer camps, workshops, and internships that introduce students to CAM fields.

21. **Targeted Recruitment Efforts**: Focusing recruitment efforts on geographic areas, schools, and communities that are typically underrepresented in CAM fields can help diversify the applicant pool.

22. **Flexible Learning Options**: Offering flexible learning options, such as part-time study, online courses, or evening classes, can make CAM education more accessible to those who may have work or family commitments.

23. **Reducing Language Barriers**: Offering language support for non-native English speakers, such as translation services or English as a Second Language resources, can help reduce barriers for these applicants.

24. **Addressing Systemic Barriers**: Recognizing and addressing systemic barriers within the institution and the broader educational system that may hinder equitable access for certain groups.

25. **Partnerships with Minority-Serving Institutions**: Collaborating with minority-serving institutions, such as Historically Black Colleges and Universities or Hispanic-Serving Institutions can help attract a diverse range of applicants.

26. **Inclusive Campus Environment**: Creating an inclusive and welcoming campus environment that respects and celebrates diversity can encourage underrepresented students to apply and feel valued in the program.

27. **Regular Review of Admission Trends**: Regularly reviewing admission trends to monitor the effectiveness of equitable access initiatives and making data-driven decisions to further improve inclusivity.

28. **Student Organizations and Clubs**: Supporting and promoting student organizations and clubs that focus on diversity and inclusion can provide a supportive community for under-represented students.

29. **Disability Accommodations**: Ensuring that the program is accessible to students with disabilities, including providing necessary accommodations during the application process and throughout their education.

30. **Socioeconomic Considerations**: Being mindful of socioeconomic factors that might impact an applicant's ability to access resources, prepare for entrance exams, or obtain certain experiences that are valued in the admissions process.

By implementing these strategies, CAM programs can work toward creating a more equitable and inclusive environment, ensuring that students from all backgrounds have the opportunity to pursue education and careers in CAM.

ADMISSION TO COMPLEMENTARY AND ALTERNATIVE MEDICINE (CAM) PROGRAMS: CHALLENGES

Admission to CAM programs can present several challenges for both applicants and institutions. These challenges often stem from the unique nature of CAM education, its position in the broader healthcare education landscape, and the diversity of applicants. Understanding these challenges is crucial for both prospective students and educational institutions to navigate the admission process effectively. Here are some key challenges:

1. **Limited Recognition and Misunderstanding of CAM**: Despite growing popularity, CAM practices may still face limited recognition in mainstream healthcare. This can lead to misunderstandings about the rigor and validity of CAM programs, affecting both applicants' decisions and institutions' ability to attract students.

2. **Varied Accreditation Standards**: The standards and processes for accrediting CAM programs can vary significantly, leading to confusion about the quality and recognition of different programs. Applicants may find it challenging to discern which programs are reputable and will meet their career needs.

3. **Balancing Traditional and Alternative Approaches**: CAM programs often need to balance traditional scientific and medical education with alternative health practices. This can be challenging in terms of curriculum development and ensuring that students receive a well-rounded education.

4. **High Competition in Certain Areas**: In some regions, the demand for CAM education may exceed the available spots, leading to highly competitive admission processes. This can be daunting for applicants, especially those without a strong conventional academic background.

5. **Financial Barriers**: The cost of education in CAM programs can be a significant barrier for many applicants. Unlike more traditional medical programs, CAM students may have fewer opportunities for scholarships, grants, or financial aid.

6. **Diverse Applicant Backgrounds**: CAM programs often attract applicants from a wide range of backgrounds and experiences. Catering to this diversity in terms of educational preparedness, cultural perspectives, and career goals can be challenging for admissions committees.

7. **Standardized Testing Biases**: Reliance on standardized tests in the admission process can disadvantage applicants from varied educational backgrounds or those who may not perform well in standardized testing environments.

8. **Ensuring Adequate Practical Training**: CAM education heavily relies on practical, hands-on training. Ensuring that all students have adequate access to such training, especially in programs with limited resources or those located in areas with fewer healthcare facilities can be challenging.

9. **Cultural and Ethical Sensitivities**: Given the diverse and often culturally rooted practices in CAM, programs must navigate various cultural and ethical sensitivities both in the admission process and in curriculum content.

10. **Regulatory and Legal Constraints**: Depending on the location, there might be regulatory or legal constraints regarding the practice of certain CAM modalities. This can impact the scope of what CAM programs can offer and what students can expect to practice post-graduation.

11. **Lack of Standardized Curriculum**: The absence of a standardized curriculum in CAM education means that programs can vary widely in terms of content, depth, and focus, making it difficult for applicants to compare programs and choose the best fit for their interests and career goals.

12. **Integrating Research and Evidence-Based Practices**: Incorporating research and evidence-based practices into CAM education, and demonstrating this integration to applicants, can be challenging given the varying levels of scientific evidence supporting different CAM modalities.

13. **Public Perception and Professional Acceptance**: The perception of CAM in the public and professional domains can influence both the appeal of CAM programs to potential applicants and the acceptance of CAM graduates in the healthcare workforce.

14. **Adapting to Technological Advances**: Keeping pace with technological advances and integrating them into CAM education, such as telemedicine and digital health tools, can be challenging for some institutions.

15. **Global and Cultural Relevance**: Ensuring that the curriculum and training are globally relevant and respectful of the diverse cultural origins of many CAM practices can be a complex task for educational institutions.

16. **Quantifying Experience and Skills**: Evaluating the non-academic experiences and skills of applicants, which are often crucial in CAM practices, can be challenging. This includes assessing hands-on healing skills, empathy, intuition, and other less tangible qualities that are important in CAM professions.

17. **Interdisciplinary Integration**: CAM often involves an interdisciplinary approach, integrating various health disciplines. Creating a curriculum that effectively blends these different areas while maintaining a focus on CAM's unique principles can be challenging for institutions.

18. **Meeting Diverse Learning Needs**: CAM programs attract students with a wide range of learning styles and educational backgrounds. Developing teaching methods that cater to this diversity while ensuring a consistent level of education can be a significant challenge.

19. **Navigating Online Education**: With the increasing availability of online education, CAM programs face the challenge of offering comprehensive online or hybrid learning experiences that are as effective as traditional, in-person training, especially for hands-on practices.

20. **Balancing Evidence-Based Science with Traditional Knowledge**: Striking a balance between modern, evidence-based science and traditional, sometimes anecdotal, knowledge within CAM practices can be complex, both in terms of curriculum development and in setting admission criteria that value both aspects.

21. **Ensuring Ethical Practice**: Teaching and maintaining ethical standards, especially in a field with diverse and sometimes unconventional practices, is crucial. CAM programs must ensure that students understand and adhere to ethical practices, which can be challenging to instill through admission processes and education.

22. **Global Standards and Mobility**: With the global interest in CAM, ensuring that qualifications are recognized internationally can be challenging. This affects students who wish to practice in different countries or regions post-graduation.

23. **Clinical Placement Challenges**: Securing sufficient and quality clinical placements for CAM students can be difficult, especially in areas where CAM is less integrated into mainstream healthcare systems.

24. **Research Opportunities**: Providing adequate opportunities for research in CAM fields, which is essential for the ongoing development and validation of CAM practices, can be limited due to funding challenges or skepticism in the broader scientific community.

25. **Addressing Misinformation**: There's a need to address and correct misinformation about CAM that applicants might have encountered. Ensuring that students enter programs with accurate and realistic expectations is essential for their future success.

26. **Coping with Rapid Changes in Healthcare**: The healthcare sector is rapidly evolving, and CAM programs must continuously adapt their curriculum and admission strategies to stay relevant and effective.

27. **Cultivating Critical Thinking**: Encouraging critical thinking and discernment in students, essential for navigating the sometimes conflicting information within CAM fields, is a key educational challenge.

28. **Building Resilience and Stress Management**: CAM practitioners often deal with patients in chronic pain or with long-term illnesses. Training students to manage this emotionally demanding aspect of their future profession starts from the admission process and continues throughout their education.

29. **Language and Communication Skills**: For CAM practices that originate from different cultures, language can be a barrier. Programs need to ensure students have or develop the necessary communication skills to understand and practice these modalities effectively.

30. **Sustainability and Environmental Awareness**: With many CAM practices emphasizing natural resources (like herbs and minerals), teaching and practicing sustainability is a growing concern, and incorporating this into both the curriculum and admissions criteria is increasingly important.

31. **Alignment with Professional Standards**: Ensuring that CAM programs align with professional standards and preparing students for certification and licensure can be challenging, particularly as these standards may vary significantly between regions and types of CAM practices.

32. **Cultural Appropriation Concerns**: CAM often includes practices rooted in specific cultural or indigenous traditions. Programs must navigate the delicate balance of teaching these practices respectfully and ethically, avoiding cultural appropriation.

33. **Managing Expectations**: Applicants may have varied expectations about what CAM education entails. Programs need to manage these expectations realistically, especially regarding career prospects and the scientific basis of different CAM modalities.

34. **Integration of Technological Advances**: Incorporating the latest technological advances into CAM education, such as digital health records or telemedicine, while maintaining the hands-on, personal nature of CAM practice can be challenging.

35. **Adaptation to Healthcare Reforms**: As healthcare systems and policies evolve, CAM programs must adapt to ensure that their education remains relevant and compliant with new healthcare reforms and regulations.

36. **Fostering Interprofessional Collaboration**: Encouraging collaboration between CAM and conventional healthcare students and professionals can be challenging but is essential for the future of integrated healthcare.

37. **Demonstrating Clinical Efficacy**: There is an ongoing challenge in demonstrating the clinical efficacy of various CAM modalities through scientific research and evidence-based practices, which is often a point of scrutiny in admissions and curriculum development.
38. **Addressing Scope of Practice Limitations**: Educating students about the legal limitations of their scope of practice in different jurisdictions is crucial, and CAM programs must incorporate this understanding into their admission and training processes.
39. **Balancing Personal Beliefs and Professional Practice**: Applicants often come to CAM with strong personal beliefs; aligning these with professional practice and ensuring they do not conflict with evidence-based care can be a delicate process.
40. **Accessibility for Non-Traditional Students**: Providing accessibility for non-traditional students, such as older students, working professionals, or those with families, in terms of scheduling, support services, and curriculum design, is an ongoing challenge.
41. **Data Privacy and Security Training**: As CAM practitioners increasingly use digital tools, training students in data privacy and security becomes essential, adding another layer of complexity to the education process.
42. **Evolving Public Health Priorities**: CAM education must evolve in response to changing public health priorities, such as the increasing focus on mental health, preventive care, and chronic disease management.
43. **Student Wellness and Burnout Prevention**: CAM programs must address student wellness and burnout, particularly as these future practitioners will be providing care in potentially high-stress environments.
44. **Developing Business and Entrepreneurial Skills**: Many CAM practitioners operate independently or in small practices. Thus, teaching business and entrepreneurial skills is important, yet this is often a gap in CAM education.
45. **Ethical Marketing of CAM Practices**: Educating students about the ethical marketing of CAM services is important, especially in a landscape where misinformation can be common.
46. **Navigating Insurance and Reimbursement Issues**: Understanding the complexities of insurance and reimbursement for CAM services is crucial for graduates, yet this can be a complex and ever-changing landscape.
47. **Cultivating a Global Perspective**: As CAM is practiced worldwide, fostering a global perspective in students is important for understanding the international context of CAM practices.
48. **Emphasizing Continuous Professional Development**: Instilling the importance of continuous professional development in students to keep up with evolving practices and theories in CAM is vital.
49. **Teaching Adaptability and Flexibility**: Given the evolving nature of CAM and healthcare in general, teaching students to be adaptable and flexible is key to ensuring they remain effective and relevant practitioners.
50. **Promoting Research Literacy**: Developing research literacy among students is essential for the ongoing evolution and validation of CAM practices, yet integrating this into a practice-focused curriculum can be challenging.

These challenges require thoughtful approaches from both educational institutions and applicants. For institutions, it involves continuously adapting and improving their programs to maintain quality, relevance, and inclusivity. For applicants, it means carefully researching and understanding the nuances of different CAM programs to find the one that best aligns with their career aspirations and educational needs.

ADMISSION TO COMPLEMENTARY AND ALTERNATIVE MEDICINE (CAM) PROGRAMS: RECOMMENDATIONS

To address the challenges in gaining admission to CAM programs and to enhance the overall process, several recommendations can be made (Al-Worafi, 2024a, b). These recommendations aim to benefit both prospective students and the institutions offering CAM programs:

For Prospective Students:

1. **Comprehensive Research by Applicants**: Prospective students should thoroughly research different CAM programs, considering factors like accreditation, curriculum, faculty expertise, clinical training opportunities, and postgraduation career support. Understanding the specific focus and strengths of each program can help in making an informed decision.
2. **Preparation for Academic Requirements**: Students should prepare to meet the academic requirements, which may include a strong foundation in sciences and possibly completion of prerequisite courses. Engaging in relevant coursework or preparatory programs can strengthen their application.
3. **Gaining Relevant Experience**: Obtaining hands-on experience in health-related fields, such as volunteering, internships, or working in healthcare settings, can provide valuable insights into the healthcare industry and enhance an application.
4. **Developing a Well-Rounded Application**: Beyond academics, applicants should demonstrate a well-rounded profile, showcasing their commitment to CAM principles through extracurricular activities, community service, and personal experiences related to health and wellness.
5. **Crafting a Compelling Personal Statement**: The personal statement is a critical component of the application. Applicants should articulate their interest in CAM, their understanding of its role in healthcare, and their personal and professional aspirations.
6. **Securing Strong Letters of Recommendation**: Obtaining recommendations from individuals who can vouch for the applicant's academic abilities, character, and suitability for a career in CAM is crucial. These could be from educators, healthcare professionals, or mentors.
7. **Preparing for Interviews**: If interviews are part of the admission process, applicants should prepare thoroughly, focusing on communicating their passion for CAM, their understanding of the field, and their interpersonal skills.
8. **Understanding Financial Commitments**: Prospective students should have a clear understanding of the financial aspects of their education, including tuition fees, living expenses, and potential financial aid options.
9. **Fostering Cultural Competence**: Applicants should demonstrate an understanding and respect for cultural diversity, which is especially important in CAM fields that encompass a wide range of cultural practices.
10. **Building a Support Network**: Engaging with current students, alumni, or professional organizations in the CAM field can provide valuable insights, guidance, and support throughout the application process.
11. **Enhancing Personal Qualities**: CAM practitioners often require qualities like empathy, patience, and strong communication skills. Applicants should focus on developing these qualities, which can be highlighted in their applications and interviews.
12. **Continuous Learning and Adaptability**: Given the evolving nature of CAM, applicants should exhibit a willingness to engage in continuous learning and adaptability to new information and practices.

For CAM Programs/Schools:

1. **Transparent Admission Criteria**: Programs should clearly communicate their admission criteria, ensuring that applicants have a clear understanding of what is required for a successful application.
2. **Holistic Admission Process**: Implementing a holistic admission process that evaluates candidates based on a combination of academic achievements, personal experiences, and qualities can help in selecting a diverse and competent cohort.
3. **Outreach and Inclusivity Efforts**: Programs should engage in outreach efforts to attract a diverse range of applicants, including those from underrepresented backgrounds, and ensure the admission process is inclusive and equitable.
4. **Providing Adequate Information and Guidance**: Offering detailed information about the program, career prospects, and the field of CAM through open days, webinars, and counseling can help applicants make informed decisions.
5. **Facilitating Financial Aid and Scholarships**: Providing information about scholarships, grants, and financial aid options can help applicants overcome financial barriers to accessing CAM education.
6. **Regular Review and Feedback of Admission Processes**: Programs should regularly review their admission processes, seeking feedback from students and adjusting practices to ensure they remain fair, relevant, and effective.

By following these recommendations, prospective CAM students can enhance their chances of gaining admission to their desired programs, and CAM institutions can ensure they attract and select candidates who are well-suited for a successful career in this field.

CONCLUSION

In conclusion, the admission process to CAM programs is multifaceted and presents unique challenges both for applicants and educational institutions. For prospective students, the key to a successful application lies in thorough preparation and a well-rounded presentation of their academic achievements, relevant experiences, personal qualities, and a deep understanding of the CAM field. Applicants should focus on demonstrating their commitment to the holistic principles of CAM, their readiness for the demands of the program, and their potential to contribute positively to the field. For CAM programs, the admission process should strive to be transparent, inclusive, and holistic, taking into account the diverse backgrounds and experiences of applicants. Institutions should ensure that their admission criteria, outreach efforts, and support systems are designed to attract and retain a wide range of capable and dedicated students. This approach not only enriches the learning environment but also contributes to the growth and evolution of the CAM field. Both parties should acknowledge the dynamic nature of CAM, understanding that the field is continually evolving with new research, practices, and technologies. This requires an ongoing commitment to learning, adaptability, and a willingness to embrace new ideas and approaches. Ultimately, the goal of the admission process in CAM programs should be to foster a generation of well-educated, compassionate, and competent CAM practitioners who are prepared to meet the healthcare needs of diverse populations and contribute to the integrated healthcare landscape of the future.

REFERENCES

Al-Worafi, Y.M. (Ed.). (2024a). *Handbook of Medical and Health Sciences in Developing Countries*. Springer, Cham.

Al-Worafi, Y.M. (2024b). Complementary and alternative medicine (CAM) in developing countries. In: Al-Worafi, Y.M. (eds), *Handbook of Medical and Health Sciences in Developing Countries*. Springer, Cham. https://doi.org/10.1007/978-3-030-74786-2_301-1

5 Complementary and Alternative Medicine (CAM) Education
Teaching Strategies

BACKGROUND

Education in complementary and alternative medicine (CAM) has undergone significant transformations in recent decades. This shift mirrors the broader trend in medical education, transitioning from a teacher-centered approach to a student-centered one (Dacre and Fox, 2000; Spencer and Jordan, 1999; Newble and Entwistle, 1986; Al-Worafi, 2022a–q). In this new paradigm, the focus has moved from the activities of the educators to the learning outcomes of the students. Facilitators of learning have taken the place of traditional lecturers, promoting interactive and small-group teaching methods over conventional didactic approaches. The goal of schools specializing in CAM is to nurture lifelong learners (Dacre and Fox, 2000; Spencer and Jordan, 1999; Newble and Entwistle, 1986; Al-Worafi, 2022a–c). The learning styles and environments significantly influence students' approach to education in this field (Dacre and Fox, 2000; Spencer and Jordan, 1999; Newble and Entwistle, 1986; Coles, 1998). Previously, education in this area followed a traditional, teacher-based model, akin to pre-university child education, where educators lectured and students passively absorbed information. However, contemporary literature suggests that a deeper learning approach, focusing on understanding the meaning of the material, is more beneficial for students in CAM. Learning styles, defined as the preferred methods individuals use in learning situations, play a crucial role in this educational context (Cassidy, 2004).

LEARNING STYLES

Examples of learning styles frameworks in CAM education are as follows (Childs-Kean et al., 2020; Al-Worafi, 2022b):

THE VARK LEARNING STYLE

The acronym VARK stands for Visual, Aural, Read/write, and Kinesthetic sensory modalities that are used for learning information (Vark Learn Limited, 2021).

KOLB LEARNING STYLE INVENTORY

Kolb's experiential learning style theory is typically represented by a four-stage learning cycle in which the learner touches all the bases (McLeod, 2017):

1. **Concrete Experience**: This is when a new experience or situation is encountered, or is a reinterpretation of an existing experience.
2. **Reflective Observation of the New Experience**: This is of particular importance if there are any inconsistencies between the experience and understanding.

DOI: 10.1201/9781003327202-5

3. **Abstract Conceptualization**: This is a reflection that gives rise to a new idea, or a modification of an existing abstract concept (the person has learned from their experience).
4. **Active Experimentation**: This is when the learner applies their idea(s) to the world around them to see what happens.

HONEY AND MUMFORD LEARNING STYLE QUESTIONNAIRE

This learning style consists of the following (Honey and Mumford, 1992):

Activist: Likes to take action
Reflector: Likes to think before they act
Theorist: Likes logic and likes to see both details and the overall picture
Pragmatist: Likes practicality and experimenting

GREGORC STYLE DELINEATOR

This learning style consists of the following (Coffield et al., 2004):

Concrete Sequential: Order and practicality
Abstract Sequential: Logic and rationales
Abstract Random: Spontaneity and emotions
Concrete Random: Originality and independence

GRASHA-REICHMANN STUDENT LEARNING STYLE SCALE

This learning style consists of the following (Novak et al., 2006):

Independent: Solo worker
Avoidant: Avoids participation
Collaborative: Works well with peers and faculty
Dependent: Works within specific guidelines
Competitive: Competition and winning
Participant: Joins all available learning activities

4MAT LEARNING STYLE

4MAT is a model for creating more dynamic and engaging learning. It is a framework for learning that helps educators deliver information in more dynamic and engaging ways. While traditional instruction may focus primarily on facts and information (what?) the 4MAT model encourages a broader array of questions to elicit much higher levels of student understanding and involvement (aboutlearning.com). This learning style consists of the following (aboutlearning.com):

WHY?

To understand the meaning and purpose, the instructor's role is to make connections between the material and the learners, to engage their attention.

WHAT?

To be satisfied with the relevance and are ready to know 'What?' At this stage, the trainer provides information and satisfies desires for facts, structure, and theory.

These first two phases represent instructor-led learning. Now the learner takes over.

HOW?

With knowledge, the question asked is 'How?' and to understand how these new insights can be applied to the real world. The focus is on the problems and how learning can be used to solve them.

WHAT IF?

Finally, to try it out, the questions asked are 'What if?', 'What else?', or 'What next?' This is when there is engagement in active experimentation, trial, and error, pushing at the boundaries—learning by doing.

TEACHING STRATEGIES IN COMPLEMENTARY AND ALTERNATIVE MEDICINE (CAM) EDUCATION

LECTURE-BASED EDUCATION

Lecture-based education is the most common teaching strategy used by medical educators to teach theory-related courses. Lecture-based learning (LBL) can be defined as that type of learning when the educator/teacher educates/teaches students by delivering presentations and PowerPoint presentations (lectures) verbally to the students and learners. However, there are advantages and disadvantages of this teaching strategy as follows (Al-Worafi, 2022p; Doing, 1997; Cashin, 1985; Bonwel, 1996):

ADVANTAGES OF LECTURE-BASED LEARNING

- Effective lecturers can communicate the intrinsic interest of a subject through their enthusiasm.
- Lectures can present material not otherwise available to students.
- Lectures can be specifically organized to meet the needs of particular audiences.
- Lectures can present large amounts of information.
- Lectures can be presented to large audiences.
- Lecturers can model how professionals work through disciplinary questions or problems.
- Lectures allow the instructor maximum control of the learning experience.
- Lectures present little risk for students.
- Lectures appeal to those who learn by listening.

DISADVANTAGES OF LECTURE-BASED LEARNING

- Lectures fail to provide instructors with feedback about the extent of student learning.
- In lectures, students are often passive because there is no mechanism to ensure that they are intellectually engaged with the material.
- Students' attention wanes quickly after fifteen to twenty-five minutes.
- Information tends to be forgotten quickly when students are passive.
- Lectures presume that all students learn at the same pace and are at the same level of understanding.

- Lectures are not suited for teaching higher orders of thinking such as application, analysis, synthesis, or evaluation; for teaching motor skills, or for influencing attitudes or values.
- Lectures are not well-suited for teaching complex, abstract material.
- Lectures require effective speakers.
- Lectures emphasize learning by listening, which is a disadvantage for students who have other learning styles

ACTIVE TEACHING STRATEGIES

There are many effective teaching strategies implemented by many medical educators around the world that can be implemented in CAM education programs/courses such as (Al-Worafi, 2022d–q, 2024a, b; McCoy et al., 2018; Al-Meman et al., 2014; Stewart et al., 2011; Wood, 2003; Barrows, 1996):

PROBLEM-BASED LEARNING (PBL), INCLUDING CASE-BASED LEARNING

Use of cases or problem sets meant to be explored in self-managed teams of students (with a facilitator); problem-based learning (PBL) sessions precede any discussion of content by the instructor. The history of PBL goes back to the 1960s when the PBL process was pioneered by Barrows and Tamblyn in the medical school program at McMaster University in Hamilton (Barrows, 1996).

PROBLEM-BASED LEARNING (PBL) STEPS

Wood (2003) reported the following steps:

Step 1: Identify and clarify unfamiliar terms presented in the scenario; a scribe lists those that remain unexplained after discussion

Step 2: Define the problem or problems to be discussed; students may have different views on the issues, but all should be considered; a scribe records a list of agreed problems

Step 3: "Brainstorming" session to discuss the problem(s), suggesting possible explanations on the basis of prior knowledge; students draw on each other's knowledge and identify areas of incomplete knowledge; a scribe records all discussions

Step 4: Review steps 2 and 3 and arrange explanations into tentative solutions; a scribe organizes the explanations and restructures if necessary

Step 5: Formulate learning objectives; the group reaches a consensus on the learning objectives; the tutor ensures learning objectives are focused, achievable, comprehensive, and appropriate

Step 6: Private study (all students gather information related to each learning objective)

Step 7: Group shares results of private study (students identify their learning resources and share their results); the tutor checks the learning and may assess the group.

Trigger material for PBL scenarios is as follows:

Paper-based clinical scenarios
Experimental or clinical laboratory data
Photographs
Video clips
Newspaper articles
All or part of an article from a scientific journal

A real or simulated patient
A family tree showing an inherited disorder
Others

TIPS TO DESIGN EFFECTIVE PROBLEM-BASED LEARNING (PBL)

Wood (2003) reported the following tips for designing effective PBL:

Learning objectives likely to be defined by the students after studying the scenario should be consistent with the faculty's learning objectives
Problems should be appropriate to the stage of the curriculum and the level of the student's understanding
Scenarios should have sufficient intrinsic interest for the students or relevance to future practice
Basic science should be presented in the context of a clinical scenario to encourage integration of knowledge
Scenarios should contain cues to stimulate discussion and encourage students to seek explanations for the issues presented
The problem should be sufficiently open so that discussion is not curtailed too early in the process
Scenarios should promote participation by the students in seeking information from various learning resources

DISADVANTAGES OF PROBLEM-BASED LEARNING (PBL)

Wood (2003) reported the following disadvantages of PBL.

TUTORS WHO CAN'T "TEACH"

Tutors who do not enjoy passing on their own knowledge and understanding may find PBL facilitation difficult and frustrating.

HUMAN RESOURCES

More staff have to take part in the tutoring process.

OTHER RESOURCES

Large numbers of students need access to the same library and computer resources simultaneously.

ROLE MODELS

Students may be deprived of access to a particular inspirational teacher who in a traditional curriculum would deliver lectures to a large group.

INFORMATION OVERLOAD

Students may be unsure how much self-directed study to do and what information is relevant and useful.

TEAM-BASED LEARNING (TBL)

Use of small student groups to facilitate discussion, case study exploration, or other aspects of content; preparation required in advance and content integrated throughout the class by the facilitator (expert). The history of team-based learning (TBL) goes back to the 1970s when Larry Michaelsen found that his class size had tripled from 40 to 120 students. He had been using a case-based Socratic teaching approach that involves facilitating problem-solving discussions. He knew that he had two major challenges; the first was to engage a large class in effective problem-solving, and the second was to give his students a reason to prepare before the class session. He developed an approach that is very close to the structure that TBL classrooms use today. He made sure that students came prepared by using an ingenious approach where students were first tested individually, and then in teams. He realized that students were actively discussing the material, which otherwise would have been covered in a lecture, and he devised the "4 S" framework for classroom activities where students worked on a Significant Problem, the Same Problem, where they had to make a Specific Choice and make a Simultaneous Report. Michaelsen found that this structured problem-solving method for in-class activities really helped to deeply engage students with the content and readily understand how to apply their learning (TBL-collaborative, 2021). In the 1990s TBL became widely recognized and exploited in business schools and many other disciplines in the USA (Michaelsen and Sweet, 2012). Several years later, the TBL strategy has been widely employed in medical, nursing, veterinary, dentistry, and health education (Parmelee et al., 2009). TBL-collaborative (2021), summarized the principles of team-based learning as detailed below (Sibley and Ostafichuk, 2015; Al-Worafi, 2022h).

ELEMENTS OF TEAM-BASED LEARNING (TBL)

The four essential elements of TBL are as follows (Sibley and Ostafichuk, 2015):

1. Teams must be properly formed and managed.
2. Students must be motivated to come to class prepared.
3. Students must learn to use course concepts to solve problems.
4. Students must be truly accountable.

PRINCIPLES OF TEAM-BASED LEARNING (TBL)

PRINCIPLE 1: GROUPS MUST BE PROPERLY FORMED AND MANAGED

Groups need to be formed in a way that enables them to do the work that they will be asked to do.
Minimizing barriers to group cohesiveness.

Distributing Member Resources: In order to function as effectively as possible, each group should have access to whatever assets exist within the whole class and not carry more than a "fair share" of the liabilities.

Learning teams should be fairly large and diverse; the teams should be comprised of 5–7 members.

Groups should be permanent; groups should be equitable based on the academic and linguistic abilities of their members.

PRINCIPLE 2: STUDENTS MUST BE MADE ACCOUNTABLE

Developing groups into cohesive learning teams requires assessing and rewarding a number of different kinds of student behavior. Students must be accountable for (a) individually preparing for group work, (b) devoting time and effort to completing group assignments, and (c) interacting with each other in productive ways. Fortunately, team learning offers opportunities to establish each of these three forms of accountability.

Accountability for individual pre-class preparation.
Accountability for contributing to their team.
Accountability for high-quality team performance.

Grading System: It is essential that we use an overall assessment system for the course that encourages the kind of student behavior that will promote learning in and from group interaction.

PRINCIPLE 3: TEAM ASSIGNMENTS MUST PROMOTE BOTH LEARNING AND TEAM DEVELOPMENT

The development of appropriate group assignments is a critical aspect of successfully implementing team learning.

PRINCIPLE 4: STUDENTS MUST RECEIVE FREQUENT AND IMMEDIATE FEEDBACK

For teams to perform effectively and to develop as a team, they must have regular and timely feedback on group performance.

Timely feedback from the Readiness Assessment Tests (RATs) is an important source of feedback that supports both learning and team development.

Timely feedback on application-focused team assignments. Providing immediate feedback on application-focused team assignments is also important for both learning and team development.

TIPS TO IMPLEMENT AN EFFECTIVE TEAM-BASED LEARNING (TBL)

1. Start with a good course design.
2. Use a "backward design" when developing TBL courses and modules.
3. Make sure to organize the module activities so that students can reach the set learning goals and the teacher (and they—the students) will know that they have done it.
4. Have application exercises that promote both deep thinking and engaged content-focused discussion.
5. Do not underestimate the importance of the readiness assurance process (RAP).
6. Orient the class on the reason for using TBL and how it is different from previous experiences they may have had with learning groups.
7. Highlight accountability as the cornerstone of TBL.
8. Providing a fair appeals process that will inspire further learning.
9. Peer evaluation is a challenge to get going, but it can enhance the accountability of the process.
10. Be clear and focused with the advanced preparation.
11. Create the team thoughtfully.
12. Several low-budget "props" facilitate the implementation of a good module.

TIPS FOR DESIGNING EFFECTIVE TEAM-BASED LEARNING (TBL) FOR ANY COURSE

Orientation is very important for the success of TBL. Orientation should be about the course objectives, learning outcomes, topics, educational resources, and the TBL process.

Secondly, providing students with course resources such as books, lecture notes, and others.

Next, make a mock TBL and get feedback from all students and peers.

Finally, design the TBL session based on the course credit hours, allocated time for the course per week, and adjust it accordingly.

A TBL session usually consists of the following:

- Individual readiness assurance test (IRAT), in this step, the students will take the exam individually and solve it in 15 minutes, for example. The time may be adjusted based on the course and need. The type of questions could be multiple-choice questions (MCQs) as this is reported or the lecturer may change it to any type such as short essays or any other form. The most important issue is to achieve the course learning outcome and to improve students' knowledge and skills. However, 5 to 15 MCQs may be an option within a time of 10–15 minutes.
- Group or team readiness assurance test (gRAT/tRAT), in this step, the same iRAT questions are re-distributed to all groups for discussion, then answering, and justifying the answers. The time allowed for gRAT is 30 minutes. The students are encouraged to vote for the best correct answer during the gRAT if there is no agreement, to enhance discussion and interaction between the group members.
- The next 20 minutes are considered for the in-place presentation of the answers. The presentation of the answer must be done simultaneously among the groups (at the same time) by voting to make it more interactive, followed by justifying and defending the answers.
- The next 20 minutes are allocated for open discussion.
- The last 10 minutes are assigned for peer evaluation and appeals. Students are encouraged to evaluate each teammate on their contributions to the team's success and their own learning. An evaluation checklist developed with both qualitative and quantitative questions would be easier and facilitate the TBL session
- If the lecture is more than two hours, there will be a break for 20 minutes, otherwise, the last phase will be conducted on another day.
- During the application activities of two hours, pre-developed cases or exercises will be distributed to the students as groups and students asked to answer them as teams.
- Grading system for the TBL session, allocated marks for each step. Example: 10 marks for iRAT, 15 marks for gRAT, 5 marks for the peer evaluation, and 20 marks for the cases, or it may be adjusted as needed.

INTERACTIVE-SPACED EDUCATION

Use of repetition of content at spaced intervals combined with testing of that content; developed and used heavily within the context of medical education.

INTERACTIVE WEB-BASED LEARNING

Use of web-based modules to deliver content and assess student understanding in an interactive format.

AUDIENCE RESPONSE SYSTEM/CLICKERS

Use of remote control devices by students to anonymously respond to multiple-choice questions posed by the instructor; can be integrated into traditional lectures, often termed "active lectures." Audience response: Individual students respond to the application of skill questions via an audience response system.

DISCUSSION-BASED LEARNING, INCLUDING DELIBERATIVE DISCUSSION

Use of communication among learners (both synchronous and asynchronous) as a teaching modality; can be used with other strategies such as case studies.

PATIENT SIMULATION

Use of human patient simulators in a laboratory environment to teach providers to respond to a variety of physiological emergencies and situations.

PROCESS-ORIENTED GUIDED INQUIRY LEARNING (POGIL) / DISCOVERY LEARNING

Use of exercises specifically designed to lead teams of students through the stages of exploring data, developing concepts based on that data, and applying the concepts.

TRADITIONAL LABORATORY EXPERIENCES

Use of traditional laboratory and benchtop experiences to provide hands-on learning experiences.

VODCAST + PAUSE ACTIVITIES

A video podcast with pause activities, appended exercises, or practice questions.

VODCAST + HYPERLINKS

A video podcast with no pause activities but includes hyperlinks to external or Web media for enrichment.

INTERACTIVE VODCAST

A vodcast that requires students to physically click through questions or interactivities (vodcasts using Flash).

INTERACTIVE MODULE

An electronic lesson, often audiovisual, requires students to complete interactivities.

CASE-BASED INSTRUCTION

The use of patient cases to stimulate discussion, questioning, problem-solving, and reasoning on issues pertaining to the basic sciences and clinical disciplines.

DEMONSTRATION

A performance or explanation of a process, illustrated by examples, realia, observable action, specimens, etc.

DISCUSSION OR DEBATE

Instructors facilitate a structured or informal discussion or debate.

GAME

An instructional method requiring the learner to participate in a competitive activity with presetrules.

INTERVIEW OR PANEL

Students interview standardized patients or experts to practice interviewing and history-taking skills.

LEARNING STATION

Students rotate through learning stations, participating in performance exercises at each station.

WORKSHEET OR PROBLEM SET

Learners work in pairs or teams to solve problems or categorize information.

CP SCHEME

An interactive exercise that encourages learners to make clinical decisions following a clinical presentation scheme (flowchart).

SIMULATION OR ROLE PLAY

A method used to replace or amplify real patient encounters with scenarios designed to replicate real healthcare situations, using lifelike mannequins, physical models, or standardized patients.

ORAL PRESENTATION

Students present on topics to their peers. Professors and peers evaluate the presentations using a specific rubric.

TEAM-BASED ACTIVITY

A collaborative learning activity that fosters team discussion, thinking, or problem-solving.

PROBLEM-BASED LEARNING

Working in peer groups, students identify what they already know, what they need to know, and how and where to access new information that may lead to the resolution of the problem.

LABORATORY OR STUDIO

Students apply knowledge in the laboratory, by engaging in a hands-on or kinesthetic activity.

FORMATIVE QUIZZES

The lesson includes a set of questions bundled together into a quiz, which allows learners to self-assess.

TECHNOLOGY-ENHANCED ACTIVE LEARNING (TEAL)

An interactive lesson integrating educational technology, such as electronic games, mobile apps, virtual simulations, electronic health records (), videoconferencing, Web exercises, or bioinstruments.

FLIPPED CLASSROOM

The traditional lecture and homework elements of a course are reversed. Short video lectures or electronic handouts are viewed by students before class. In-class time is devoted to exercises, projects, or discussions. The flipped classroom (also called reverse, inverse, or backward classroom) is a pedagogical approach in which basic concepts are provided to students for pre-class learning so that class time can apply and build upon those basic concepts. The flipped classroom can be used to prepare the students to be lifelong learners and improve their self-reading skills. Furthermore, improve their basic and clinical knowledge and skills.

INTERACTIVE LECTURE-BASED TEACHING STRATEGY

Medical educators can make the lecture interactive in many ways such as asking students questions every five to ten minutes, presenting short videos/audios, rallying, and team share groups that will attract students to the lectures. Link the theory part with life, share one's practice experience with students with mini and long cases, and give students time to think about it, and solve it. Engage all students and remember that many students may be hesitant to participate, so encourage all to participate. Remember as an educator that you are here to teach the students, and assess their needs, understanding can help also. Weekly and monthly feedback from students and colleagues can improve online teaching. Record the lectures and give them to students as well as to yourself, and colleagues. Add many active teaching strategies to the lecture such as videos, cases, and others.

BLENDED TEACHING STRATEGY

Traditionally, blended learning combines online educational materials and opportunities for interaction online with traditional place-based classroom methods.

VIDEO-BASED LEARNING

Short videos can be used as an effective teaching strategy for theory, practicals, and training.

SIMULATION

Role play and other simulation methods can be used with the help of new technologies as an effective online teaching strategy for theory.

PROJECT-BASED LEARNING

PBL is a model that can be used to prepare students for real practice. PBL gives students the opportunity to develop knowledge and skills by engaging in projects set around challenges and problems they may face in the real world.

JOURNAL CLUB

To critically evaluate recent articles in the academic literature.

CASE STUDIES DISCUSSION

It is a very important and effective teaching strategy to encourage students to read given cases individually or as teams and solve them.

SELF-DIRECTED LEARNING

This allows students to improve their skills toward self-learning.

COMMUNITY SERVICES-BASED LEARNING

Many theory courses can be used as an effective teaching strategy, which allows students to achieve the course learning outcomes while contributing to patients, the public, and society.

SEMINARS

An effective strategy that can be used online to improve students' presentation skills.

IMPORTANCE OF ACTIVE TEACHING STRATEGIES

Improve students' self-learning skills.
Improve students' understanding of the course materials.
Improve students' critical thinking skills.
Improve students' communication skills.
Improve students' attitudes toward teamwork and collaboration.
Improve students' decision-making skills.
Improve students' time management skills.
Improve students' presentation skills.
Improve students' problem-solving skills.
Improve knowledge retention.
Improve leadership skills.
Improve ability to answer the cases.

BARRIERS TO IMPLEMENTING ACTIVE TEACHING STRATEGIES

There are many barriers to implementing active teaching strategies in CAM programs as follows (Al-Worafi, 2022d–q, 2024a):

Lack of resources.
Knowledge-related barriers.
Attitude of medical educators related barriers.
Resistance of medical educators to change and implement new teaching strategies.
Resistance of medical students to change and accept new teaching strategies.
University/School culture toward new teaching strategies.
Lack of training about the new teaching strategies.
Lack of motivation to train in the new teaching strategies.
Lack of technologies to implement the new teaching strategies.

RECOMMENDATIONS FOR THE BEST PRACTICES OF TEACHING STRATEGIES IN THE COMPLEMENTARY AND ALTERNATIVE MEDICINE (CAM) EDUCATION

To optimize the teaching strategies in CAM education, it's important to consider a blend of traditional and innovative approaches that cater to the unique nature of CAM. Here are some best practice recommendations:

1. **Interactive and Experiential Learning**: CAM often involves techniques and practices that are best learned through hands-on experience. Incorporate workshops, demonstrations, and practice sessions where students can directly engage with CAM modalities.
2. **Interdisciplinary Approach**: Given the diverse nature of CAM, integrate teachings from various disciplines such as biology, psychology, nutrition, and pharmacology. This approach provides students with a comprehensive understanding of how different systems interact and affect health.
3. **Case-Based Learning**: Utilize real-life case studies to teach students how to apply CAM practices in practical scenarios. This approach enhances critical thinking and decision-making skills.
4. **Cultural Competency**: As CAM includes practices from various cultures, it's crucial to educate students about cultural sensitivities and the historical context of these practices. This approach fosters respect and understanding of the diverse origins of CAM therapies.
5. **Evidence-Based Teaching**: While respecting the traditional roots of CAM, also emphasize the importance of current research and evidence-based practice. Encourage critical evaluation of the scientific literature related to CAM.
6. **Technology Integration**: Utilize modern technology for teaching, such as virtual reality for anatomy education, online platforms for collaborative learning, and digital resources for accessing the latest research in CAM.
7. **Reflective Practice and Self-Care**: Teach students the importance of self-reflection and self-care, as CAM practitioners often emphasize holistic wellness. This can include mindfulness, meditation, and stress management techniques.
8. **Ethical and Legal Considerations**: Ensure that students are aware of the ethical and legal aspects of CAM practice, including informed consent, confidentiality, and regulatory standards.
9. **Guest Lectures and Field Experts**: Invite CAM practitioners and experts in the field to give lectures or workshops. This exposure to real-world experiences enriches the learning process.
10. **Student-Centered Learning Environment**: Foster an environment where students are encouraged to explore their interests within CAM, ask questions, and engage in discussions. This approach respects diverse viewpoints and encourages a deeper understanding of the subject matter.
11. **Continuous Professional Development**: Encourage lifelong learning and professional development, highlighting the ever-evolving nature of CAM and the importance of staying updated with new practices and research findings.
12. **Collaborative Learning**: Encourage group work and collaborative projects. This approach mirrors the interdisciplinary nature of CAM practice and fosters teamwork, communication skills, and the ability to work effectively in diverse groups.
13. **Personalized Learning Paths**: Recognize that students have different learning styles and interests. Offer elective courses or specializations within CAM to allow students to pursue areas they are particularly passionate about, such as herbal medicine, acupuncture, or mind-body therapies.

14. **Clinical Rotations and Internships**: Provide opportunities for students to gain practical experience in clinical settings. This real-world exposure is invaluable for understanding how CAM integrates into healthcare and for developing practical skills under supervision.
15. **Peer Teaching and Mentoring**: Implement peer teaching programs where more advanced students mentor beginners. This not only reinforces the knowledge of the mentors but also fosters a supportive learning community.
16. **Problem-Based Learning (PBL)**: Utilize PBL to enhance critical thinking and problem-solving skills. In this approach, students learn by actively solving complex, real-world problems, which is highly applicable in the diverse and often case-specific world of CAM.
17. **Seminars and Workshops on Emerging Trends**: Regularly schedule seminars and workshops focusing on emerging trends and innovations in CAM. This keeps the curriculum dynamic and relevant and prepares students for a rapidly evolving field.
18. **Global Health Perspectives**: Integrate global health perspectives into the curriculum to help students understand how CAM practices vary across different cultures and regions. This could include study-abroad opportunities or collaborations with international CAM institutions.
19. **Sustainability in CAM Practices**: Teach students about sustainable practices in CAM, especially in the use of natural resources like medicinal plants. This is crucial for the ethical and responsible practice of CAM.
20. **Feedback and Continuous Improvement**: Establish a robust feedback system where students can provide evaluations of their courses and instructors. Use this feedback for continuous improvement of the curriculum and teaching methods.
21. **Digital and Social Media Literacy**: Equip students with skills in digital and social media literacy, as these platforms are increasingly used for health information dissemination and professional networking in CAM.
22. **Entrepreneurial Skills**: Offer courses or workshops in business management and entrepreneurship tailored to CAM practices, as many CAM practitioners operate in private practice or small business settings.
23. **Inclusivity and Accessibility**: Ensure that the educational environment is inclusive and accessible to students with diverse backgrounds and abilities. This may include offering resources in multiple languages, ensuring physical accessibility, and providing learning materials in various formats.
24. **Research Skills Development**: Encourage students to develop strong research skills. This can include training in how to conduct clinical trials, understanding statistical methods, and evaluating the efficacy and safety of CAM therapies. This skill set is crucial for contributing to the evidence base of CAM practices.
25. **Integrating Technology in Practice**: Teach students how to integrate technology into CAM practice, such as using electronic health records, telehealth platforms for remote consultations, and apps for patient education and self-care management.
26. **Public Health Perspectives**: Incorporate public health perspectives into the curriculum to help students understand the broader impact of CAM practices on community health and wellness.
27. **Nutrition and Lifestyle Education**: Given the holistic nature of CAM, include comprehensive education on nutrition, lifestyle changes, and their role in preventive health care.
28. **Legal and Regulatory Compliance**: Apart from ethical practices, it's important to educate students on the legal and regulatory aspects specific to CAM, including licensing requirements, scope of practice, and insurance considerations.
29. **Patient Education and Communication Skills**: Emphasize the importance of effective patient education and communication. This includes training on how to explain CAM modalities to patients, addressing misconceptions, and ensuring that patients make informed decisions about their treatment options.

30. **Cross-Cultural Communication and Sensitivity Training**: As CAM draws from global traditions, provides training in cross-cultural communication and sensitivity to prepare students for working with diverse populations.
31. **Environmental and Global Health Issues**: Teach the implications of environmental factors and global health issues on CAM, including the impact of climate change on herbal medicine resources and the role of CAM in global health crises.
32. **Healthcare System Navigation**: Educate students on how CAM fits within the broader healthcare system, including how to collaborate with other healthcare professionals and navigate insurance processes.
33. **Quality Assurance in CAM Practices**: Discuss the importance of quality assurance in CAM practices, including standardization of herbal products, quality control in manufacturing, and ensuring consistency in therapeutic practices.
34. **Advanced Specialization Opportunities**: Offer advanced courses or certifications in specific CAM areas for students wishing to specialize further, such as advanced acupuncture techniques, master herbalist programs, or specialization in Ayurvedic medicine.
35. **Lifelong Learning and Continuous Education**: Instill the importance of lifelong learning and provide resources for continuous education in CAM, including conferences, online courses, and professional associations.
36. **Self-experimentation and Personal Experience**: Encourage students to engage in self-experimentation with safe CAM practices. Personal experience with these therapies can provide valuable insights and deepen their understanding.

By incorporating these strategies, CAM education can be both comprehensive and engaging, preparing students to become skilled, knowledgeable, and compassionate practitioners in the field.

CONCLUSION

In conclusion, the field of CAM education is multifaceted and dynamic, necessitating a comprehensive and adaptive approach to teaching. Key to this approach is the integration of interactive and experiential learning methods, interdisciplinary studies, and an emphasis on both traditional knowledge and evidence-based practice. By incorporating hands-on experience, case-based learning, and an understanding of global health perspectives and cultural sensitivities, CAM education can prepare practitioners who are not only skilled in various modalities but also deeply aware of the ethical, legal, and social implications of their practice. Moreover, fostering a student-centered learning environment, encouraging lifelong learning, and integrating modern technology and research skills are crucial for keeping pace with the evolving landscape of healthcare. Preparing students for the realities of CAM practice, including navigating the healthcare system, understanding public health implications, and focusing on holistic patient care, is essential. Ultimately, the goal is to produce well-rounded, knowledgeable, and compassionate CAM practitioners who are equipped to contribute positively to individual health and wellness, as well as to the broader healthcare system. This holistic approach to CAM education is vital in ensuring that CAM continues to grow as a respected and integral part of healthcare, both now and in the future.

REFERENCES

Al-Meman, A., Al-Worafi, Y.M., and Saeed, M.S. (2014). Team-based learning as a new learning strategy in pharmacy college, Saudi Arabia: Students' perceptions. *Universal Journal of Pharmacy*, 3(3), 57–65.
Al-Worafi, Y.M. (2022a). *A Guide to Online Pharmacy Education: Teaching Strategies and Assessment Methods*. CRC Press.
Al-Worafi, Y.M. (2022b). Pharmacy education: Learning styles. In: Al-Worafi, Y.M. (ed), *A Guide to Online Pharmacy Education: Teaching Strategies and Assessment Methods*. CRC Press.

Al-Worafi, Y.M. (2022c). Competencies and learning outcomes. In Al-Worafi, Y.M. (ed), *A Guide to Online Pharmacy Education: Teaching Strategies and Assessment Methods.* CRC Press.

Al-Worafi, Y.M. (2022d). Teaching the theory. In Al-Worafi, Y.M. (ed), *A Guide to Online Pharmacy Education: Teaching Strategies and Assessment Methods.* CRC Press.

Al-Worafi, Y.M. (2022e). Teaching the practice and tutorial. In Al-Worafi, Y.M. (ed), *A Guide to Online Pharmacy Education: Teaching Strategies and Assessment Methods.* CRC Press.

Al-Worafi, Y.M. (2022f). Self-learning and self-directed learning. In Al-Worafi, Y.M. (ed), *A Guide to Online Pharmacy Education: Teaching Strategies and Assessment Methods.* CRC Press.

Al-Worafi, Y.M. (2022g). Traditional and active strategies. In Al-Worafi, Y.M. (ed), *A Guide to Online Pharmacy Education: Teaching Strategies and Assessment Methods.* CRC Press.

Al-Worafi, Y.M. (2022h). Team-based learning in pharmacy education. In Al-Worafi, Y.M. (ed), *A Guide to Online Pharmacy Education: Teaching Strategies and Assessment Methods.* CRC Press.

Al-Worafi, Y.M. (2022i). Problem-based learning in pharmacy education. In Al-Worafi, Y.M. (ed), *A Guide to Online Pharmacy Education: Teaching Strategies and Assessment Methods.* CRC Press.

Al-Worafi, Y.M. (2022j). Case-based learning in pharmacy education. In Al-Worafi, Y.M. (ed), *A Guide to Online Pharmacy Education: Teaching Strategies and Assessment Methods.* CRC Press.

Al-Worafi, Y.M. (2022k). Simulation in pharmacy education. In Al-Worafi, Y.M. (ed), *A Guide to Online Pharmacy Education: Teaching Strategies and Assessment Methods.* CRC Press.

Al-Worafi, Y.M. (2022l). Project-based learning in pharmacy education. In Al-Worafi, Y.M. (ed), *A Guide to Online Pharmacy Education: Teaching Strategies and Assessment Methods.* CRC Press.

Al-Worafi, Y.M. (2022m). Flipped classes in pharmacy education. In Al-Worafi, Y.M. (ed), *A Guide to Online Pharmacy Education: Teaching Strategies and Assessment Methods.* CRC Press.

Al-Worafi, Y.M. (2022n). Educational games in pharmacy education. In Al-Worafi, Y.M. (ed), *A Guide to Online Pharmacy Education: Teaching Strategies and Assessment Methods.* CRC Press.

Al-Worafi, Y.M. (2022o). Web-based learning in pharmacy education. In Al-Worafi, Y.M. (ed), *A Guide to Online Pharmacy Education: Teaching Strategies and Assessment Methods.* CRC Press.

Al-Worafi, Y.M. (2022p). Lecture-based/interactive lecture-based learning in pharmacy education. In Al-Worafi, Y.M. (ed), *A Guide to Online Pharmacy Education: Teaching Strategies and Assessment Methods.* CRC Press.

Al-Worafi, Y.M. (2022q). Blended learning in pharmacy education. In Al-Worafi, Y.M. (ed), *A Guide to Online Pharmacy Education: Teaching Strategies and Assessment Methods.* CRC Press.

Al-Worafi, Y.M. (Ed.). (2024a). *Handbook of Medical and Health Sciences in Developing Countries.* Springer, Cham.

Al-Worafi, Y.M. (2024b). Complementary and alternative medicine (CAM) in developing countries. In: Al-Worafi, Y.M. (eds), *Handbook of Medical and Health Sciences in Developing Countries.* Springer, Cham. https://doi.org/10.1007/978-3-030-74786-2_301-1

Barrows, H.S. (1996). Problem-based learning in medicine and beyond: A brief overview. *New Directions for Teaching and Learning,* 1996(68), 3–12.

Bonwell, C.C. (1996). Enhancing the lecture: Revitalizing a traditional format. In: Sutherland, T.E., and Bonwell, C.C. (eds), *Using Active Learning in College Classes: A Range of Options for Faculty.* New Directions for Teaching and Learning No. 67. John Wiley &Sons.

Cashin, W.E. (1985). "Improving lectures" idea paper no. 14. Kansas State University, Center for Faculty Evaluation and Development, Manhattan.

Cassidy, S. (2004). Learning styles: An overview of theories, models, and measures. *Educational Psychology,* 24(4), 419–444.

Childs-Kean, L., Edwards, M., and Smith, M.D. (2020). Use of learning style frameworks in health science education. *American Journal of Pharmaceutical Education,* 84(7), ajpe7885.

Coffield, F., Moseley, D., Hall, E., Ecclestone, K., Coffield, F., Moseley, D., Hall, E., and Ecclestone, K. (2004). *Learning Styles and Pedagogy in Post-16 Learning: A Systematic and Critical Review.* Learning and Skill Research Centre, London.

Coles, C. (1998). How students learn: The process of learning. In: *Medical Education in the Millennium* (pp. 63–82). Oxford University Press.

Dacre, J.E., and Fox, R.A. (2000). How should we be teaching our undergraduates? *Annals of the Rheumatic Diseases,* 59(9), 662–667.

Doing, C.L. (1997). Advantages and disadvantages of lectures. https://wceruw.org

Honey, P., and Mumford, A. (1992). *The Manual of Learning Styles.* 3rd ed. Honey Press, Maidenhead, Berkshire, UK. https://aboutlearning.com/about-us/4mat-overview/

McCoy, L., Pettit, R.K., Kellar, C., and Morgan, C. (2018). Tracking active learning in the medical school curriculum: A learning-centered approach. *Journal of Medical Education and Curricular Development,* 5, p. 2382120518765135.

McLeod, S. (2017). Kolb's learning styles and experiential learning cycle. *Simply Psychology*, 5.

Michaelsen, L.K., and Sweet, M. (2012). Fundamental principles and practices of team-based learning. In: *Team-Based Learning for Health Professions Education: A Guide to Using Small Groups for Improving Learning* (pp. 9–34). Routledge.

Newble, D.I., and Entwistle, N.J. (1986). Learning styles and approaches: Implications for medical education. *Medical Education*, 20(3), 162–175.

Novak, S., Shah, S., Wilson, J.P., Lawson, K.A., and Salzman, R.D. (2006). Pharmacy students' learning styles before and after a problem-based learning experience. *American Journal of Pharmaceutical Education*, 70(4), 74.

Parmelee, D.X., DeStephen, D., and Borges, N.J. (2009). Medical students' attitudes about team-based learning in a pre-clinical curriculum. *Medical Education Online*, 14(1), 4503.

Sibley, J., and Ostafichuk, P. (2015). *Getting Started with Team-Based Learning*. Stylus Publishing, LLC, Sterling.

Spencer, J.A., and Jordan, R.K. (1999). Learner centred approaches in medical education. Bmj, 318(7193), 1280–1283.

Stewart, D.W., Brown, S.D., Clavier, C.W., and Wyatt, J. (2011). Active-learning processes used in US pharmacy education. *American Journal of Pharmaceutical Education*, 75(4), 68.

Team-Based Learning. Team-Based Learning Collaborative (TBL-Collaborative). (2021). https://www.team-basedlearning.org/

Vark Learn Limited, (2021). The VARK modalities. https://vark-learn.com/introduction-to-vark/the-vark-modalities/

Wood, D.F. (2003). Problem based learning. *Bmj*, 326(7384), 328–330.

6 Complementary and Alternative Medicine (CAM) Education
Assessment & Evaluation Methods

BACKGROUND

Complementary and alternative medicine (CAM) education programs have been aligning their learning outcomes and competencies with international guidelines, as well as the standards set by local and global accreditation bodies. This alignment is aimed at producing graduates who are well-equipped with the necessary knowledge and skills to excel in the healthcare sector, particularly in CAM practices. They are trained to provide patient care that is not only effective but also based on the principles of evidence-based alternative medicine. The learning outcomes and competencies in these programs are regularly updated to mirror the ongoing advancements and improvements in CAM patient care practices (Al-Worafi, 2022a, b, 2024a, and b). The evaluation methods for CAM education have shifted from traditional to competency-based assessments in many countries. Such evaluations are pivotal in ensuring the quality of CAM practitioners, confirming that they are ready and able to serve as healthcare professionals with all the requisite knowledge, skills, and competencies. The term "assessment," with its Latin origin meaning "to sit beside and judge," is used to denote the systematic gathering of information about what students should know and be able to do in the field of CAM. This often involves some form of measurement, grading, or the use of descriptors like excellent, good, average, or poor. Discussing assessment inherently involves other related concepts, particularly when determining what exactly the assessment is measuring (Allen, 2004; Al-Worafi, 2022a–g; Gibbs et al., 2006; Fulcher et al., 2012).

TERMINOLOGIES RELATED TO EVALUATION AND ASSESSMENT

ASSESSMENT

A measurement that provides evidence of learning (Lee and Ciu, 2021; AGME, 2019).

ASSESSMENT AND EVALUATION

Mavis (2014) differentiates between assessment and evaluation as follows: assessment most often refers to the measurement of individual student performance, while evaluation refers to the measurement of outcomes for courses, educational programs, or institutions. Practically speaking, students are assessed while educational programs are evaluated. However, it is often the case that aggregated student assessments serve as an important information source when evaluating educational programs.

FORMATIVE AND SUMMATIVE ASSESSMENT

Literature reported the following definitions and differentiate between formative and summative assessment (Black and Wiliam, 2010; Garrison and Ehringhaus, 2007; Dixson and Worrell, 2016; Mavis, 2014): formative assessment has been defined as "activities undertaken by teachers—and by their students in assessing themselves—that provide information to be used as feedback to modify teaching and learning activities." The purpose of formative assessment is to monitor student learning and provide ongoing feedback to staff and students. It is an assessment for learning. If designed appropriately, it helps students identify their strengths and weaknesses and can enable students to improve their self-regulatory skills so that they manage their education in a less haphazard fashion than is commonly found. It also provides information to the faculty about the areas students are struggling with so that sufficient support can be put in place. Formative assessment can be tutor-led, peer-led, or self-assessment. Formative assessments have low stakes and usually carry no grade, which in some instances may discourage the students from doing the task or fully engaging with it (Black and Wiliam, 2010; Garrison and Ehringhaus, 2007; Dixson and Worrell, 2016; Mavis, 2014).

Summative assessments are given periodically to determine at a particular point in time what students know and do not know. Many associate summative assessments only with standardized tests such as state assessments, but they are also used and are an important part of district and classroom programs. Summative assessment at the district/classroom level is an accountability measure that is generally used as part of the grading process (Black and Wiliam, 2010; Garrison and Ehringhaus, 2007; Dixson and Worrell, 2016; Mavis, 2014).

FORMATIVE ASSESSMENT

Assessments that inform feedback to guide a learner's development rather than provide a final evaluation (Lee and Ciu, 2021; AGME, 2019).

SUMMATIVE ASSESSMENT

Assessments are intended to measure the current status of competency to inform grade assignment, promotion in rank, or professional certification (Lee and Ciu, 2021; AGME, 2019).

COMPETENCY

Observable abilities that incorporate the knowledge, skills, values, and attitudes required in the care of patients and communities (Lee and Ciu, 2021; AGME, 2019).

COMPETENCY-BASED MEDICAL EDUCATION

An educational approach that uses learning outcomes called competencies to organize medical curricula and assesses the achievement of these competencies (Lee and Ciu, 2021; AGME, 2019).

COMPETENCE ASSESSMENT

Assessments in which the learner demonstrates skills in an educational setting (Lee and Ciu, 2021; AGME, 2019).

DIRECT OBSERVATION

A method of assessment in which a teacher observes some or all of a patient care encounter with the learner (Lee and Ciu, 2021; AGME, 2019).

FEEDBACK

An informed, nonevaluative, and objective appraisal of performance intended to improve clinical skills (Lee and Ciu, 2021; AGME, 2019).

VALIDITY

The evidence that an assessment measures what it is intended to measure (Lee and Ciu, 2021; AGME, 2019).

DIAGNOSTIC ASSESSMENT

Used to give information about students' prior knowledge.

Helps in designing the most suitable educational program for each student (Garrison and Ehringhaus, 2007; Alfadl, 2018).

DIRECT ASSESSMENT

Direct assessment refers to the assessment that is based on an analysis of student behaviors or products in which they demonstrate how well they have mastered learning outcomes (Allen, 2004).

INDIRECT ASSESSMENT

Indirect assessment refers to the assessment that is based on gathering information through means other than looking at actual samples of student work such as surveys, interviews, and focus groups. Indirect assessment refers to any method of collecting data that requires reflection on student learning, skills, or behaviors, rather than a demonstration of it, it is indirect evidence of student achievement and requires that faculty infer actual student abilities, knowledge, and values rather than observe direct evidence of learning or achievement.

OBJECTIVE STRUCTURED CLINICAL EXAMINATION (OSCE)

"Objective structured clinical examination (OSCE) is an approach to the assessment of clinical competence in which the components of competence are assessed in a planned or structured way with the attention being paid to the objectivity of the examination." (Harden and Gleeson, 1979). The examination consists of multiple, standard stations at which students must complete one to two specific clinical tasks, often in an interactive environment involving patient actors such as standardized patients (Harden and Gleeson, 1979; Sturpe, 2010).

PERFORMANCE ASSESSMENT

Assessments in which a learner applies competencies in the patient care setting (Lee and Ciu, 2021; AGME, 2019).

PORTFOLIO

A collection of evidence of competence that documents a learner's journey through medical training (Lee and Ciu, 2021; AGME, 2019).

RATIONALITY OF ASSESSMENT AND EVALUATION IN COMPLEMENTARY AND ALTERNATIVE MEDICINE (CAM) EDUCATION PROGRAMS

Assessment and evaluation in medical and health sciences education at the level of programs, curriculum, and courses are very important to assess the quality of CAM education, to ensure that the graduates are competent, have the required competencies, and are able to provide good patient care services.

The literature reported the following reasons for assessing students' performance (Mavis, 2014; McAleer, 2001):

- Providing feedback to students about their mastery of course content.
- Grading or ranking students for progress and promotion decisions.
- Offering encouragement and support to students (or teachers.)
- Measuring changes in knowledge, skills, or attitudes over time.
- Diagnosing weaknesses in student performance.
- Establishing performance expectations for students.
- Identifying areas for improving instruction.
- Documenting instructional outcomes for faculty promotion.
- Evaluating the extent to which educational objectives are realized.
- Encouraging the development of a new curriculum.
- Demonstrating quality standards for the public, institution, or profession.
- Articulating the values and priorities of the educational institution.
- Inform students about the location of educational resources.
- It is an integral part of the learning process in which students are informed of any weaknesses and of how to improve the quality of their performance.
- It illustrates progress and ensures a proper standard has been achieved before progressing to a higher level of training.
- It provides certification relating to a standard of performance, e.g., the award of a degree.
- It indicates to students the areas of a course that are considered important.
- It acts as a promotion technique.
- It acts as a means of selection for a career or as an entrance requirement for a course.
- It measures the effectiveness of training and identifies curriculum weaknesses.

KEY FEATURES OF STUDENT ASSESSMENT METHODS

Assessment methods should have the following features (Mavis, 2014):

1. Reliability
2. Validity
3. Feasibility
4. Acceptability
5. Educational Impact

STRENGTHS AND LIMITATIONS OF ASSESSMENT AND EVALUATION IN COMPLEMENTARY AND ALTERNATIVE MEDICINE (CAM) EDUCATION

Each assessment and evaluation method either, summative or formative has strengths/advantages and limitations/weaknesses, examples are as follows (Morningside College, 2006).

STANDARDIZED EXAMS

SELECTION OF ASSESSMENT AND EVALUATION TOOLS IN CAM EDUCATION

Literature reported the following steps for developing a good assessment instrument (Hamstra, 2014; Hamstra, 2012):

1. Determine the purpose of your assessment
   A. Formative, summative (standard setting/criteria) research

    B. Knowledge, skills, and attitudes
         (e.g., performance, teamwork, and anxiety)
2. Content validity—identify the main construct of interest and stakeholders
3. Review with content experts—focus group
    A. Representative sample: different institutions and disciplines
    B. Thematic saturation, address political issues
    C. Set preliminary standards—what does perfect/borderline performance look like?
4. Item writing/development (based on related existing tests?)
5. If necessary, train the raters (and assess inter-rater reliability)
6. Pilot test the instrument (representative sample) for validity
    A. Feasibility check—length, clarity, and cost
    B. If necessary, go back to Step 4 (modify items and pilot test again)
7. Implement modified test—measure reliability, and validity based on a larger sample
    A. Assess construct validity

DIRECT ASSESSMENT METHODS AND ITS APPLICATION IN COMPLEMENTARY AND ALTERNATIVE MEDICINE (CAM) EDUCATION

Literature reported the following examples of direct assessment methods (Allen, 2004).

PUBLISHED/STANDARDIZED TESTS

Published tests or standardized tests are instruments that have been commercially published by a test publisher. These instruments are administered and scored in a consistent or "standard" manner. The validity and reliability of the instrument are the two essential elements for defining the standard quality of the test. These tests are generally only available from the publisher and often come in the form of kits or multiple booklets. They can be very costly if purchased.

EXAMPLES OF PUBLISHED/STANDARDIZED TESTS ARE:

Biomedical books—contain questions about each topic.
CAM books—contain questions about each topic.
Others.

STEPS IN SELECTING A PUBLISHED TEST

1. Identify a possible test.
2. Consider published reviews of this test.
3. Order a specimen set from the publisher.
4. Take the test and consider the appropriateness of its format and content.
5. Consider the test's relationship to your learning outcomes.
6. Consider the depth of processing of the items (e.g., analyze items using Bloom's taxonomy).
7. Consider the publication date and currency of the items.
8. How many scores are provided? Will these scores be useful? How?
9. Look at the test manual. Were test development procedures reasonable? What is the evidence for the test's reliability and validity for the intended use?
10. If you will be using the norms, consider their relevance for your purpose.
11. Consider practicalities, e.g., timing, test proctoring, and test scoring requirements.
12. Verify that the faculty is willing to act on results.

STRENGTHS OF PUBLISHED/STANDARDIZED TESTS

- Can provide direct evidence of student mastery of learning outcomes.
- They generally are carefully developed, highly reliable, professionally scored, and nationally normed.
- They frequently provide a number of norm groups, such as norms for community colleges, liberal arts colleges, and comprehensive universities.
- Online versions of tests are increasingly available, and some provide immediate scoring.
- Some publishers allow faculty to supplement tests with their own items, so tests can be adapted to better serve local needs.

WEAKNESSES OF PUBLISHED/STANDARDIZED TESTS

- Students may not take the test seriously if test results have no impact on their lives.
- These tests are not useful as direct measures for program assessment if they do not align with local curricula and learning outcomes.
- Test scores may reflect criteria that are too broad for meaningful assessment.
- Most published tests rely heavily on multiple-choice items which often focus on specific facts, but program learning outcomes more often emphasize higher-level skills.
- If the test does not reflect the learning outcomes that faculty value and the curricula that students experience, results are likely to be discounted and inconsequential.
- Tests can be expensive.
- The marginal gain from annual testing may be low.
- Faculty may object to standardized exam scores on general principles, leading them to ignore results.

ADVANTAGES

- Convenient.
- Can be adopted and implemented quickly.
- Reduces or eliminates faculty time demands in instrument development and grading.
- Are scored objectively.
- Provide for external validity.
- Provide reference group measures.
- Can make longitudinal comparisons.
- Can test large numbers of students.

DISADVANTAGES

- Measures relatively superficial knowledge or learning.
- Unlikely to match the specific goals and objectives of a program/institution.
- Norm-referenced data may be less useful than criterion-referenced.
- May be cost-prohibitive to administer as a pre-and post-test.
- More summative than formative (may be difficult to isolate what changes are needed).
- Norm data may be user norms rather than a true national sample.
- May be difficult to receive results in a timely manner.

LOCALLY DEVELOPED TESTS

Faculty may decide to develop their own internal test that reflects the program's learning outcomes.

EXAMPLES OF LOCALLY DEVELOPED TESTS:

Multiple-choice questions (MCQs) and other questions developed by faculty in:

Biomedicals courses.
CAM-related courses.
Others.

STRENGTHS OF LOCALLY DEVELOPED TESTS

- Can provide direct evidence of student mastery of learning outcomes.
- Appropriate mixes of essay and objective questions allow faculty to address various types of learning outcomes.
- Students generally are motivated to display the extent of their learning if they are being graded on the work.
- If well-constructed, they are likely to have good validity.
- Because local faculty write the exam, they are likely to be interested in the results and willing to use them.
- Can be integrated into routine faculty workloads.
- The evaluation process should directly lead faculty into discussions of student learning, curriculum, pedagogy, and student support services.

WEAKNESSES OF LOCALLY DEVELOPED TESTS

- These exams are likely to be less reliable than published exams.
- Reliability and validity generally are unknown.
- Creating and scoring exams takes time.
- Traditional testing methods have been criticized for not being "authentic."
- Norms generally are not available.

EMBEDDED ASSIGNMENTS AND COURSE ACTIVITIES

Embedded assignments and course activities are assignments, activities, or exercises that are done as part of a class, but that are used to provide assessment data about a particular learning outcome.

EXAMPLES OF EMBEDDED ASSIGNMENTS AND COURSE ACTIVITIES ARE:

- Community-service learning (CSL) and other fieldwork activities such as awareness programs in malls, and others.
- Culminating projects, such as papers in capstone courses.
- Exams or parts of exams.
- Group projects.
- Homework assignments.
- In-class presentations.
- Student recitals and exhibitions.
- Comprehensive exams, theses, dissertations, and defense interviews.

STRENGTHS OF EMBEDDED ASSIGNMENTS AND COURSE ACTIVITIES

- Can provide direct evidence of student mastery of learning outcomes.
- Out-of-class assignments are not restricted by time constraints typical for exams.

- Students are generally motivated to demonstrate the extent of their learning if they are being graded.
- Can provide authentic assessment of learning outcomes.
- Can involve CSL or other fieldwork activities and ratings by fieldwork supervisors.
- Can provide a context for assessing communication and teamwork skills.
- Can be used for grading as well as assessment.
- Faculties who develop the procedures are likely to be interested in results and willing to use them.
- The evaluation process should directly lead faculty into discussions of student learning, curriculum, pedagogy, and student support services.
- Data collection is unobtrusive to students.

WEAKNESSES OF EMBEDDED ASSIGNMENTS AND COURSE ACTIVITIES

- Requires time to develop and coordinate.
- Requires faculty trust that the program will be assessed, not individual teachers.
- Reliability and validity generally are unknown.
- Norms generally are not available.

PERFORMANCE MEASURES

TYPES

- Essays
- Oral presentations
- Oral exams
- Exhibitions
- Demonstrations
- Performances
- Products
- Research papers
- Poster presentations
- Capstone experiences
- Practical exams
- Supervised internships & practicums

ADVANTAGES

- Can be used to assess from multiple perspectives
- Using a student-centered design can promote student motivation
- Can be used to assess transfer of skills and integration of content
- Engages students in active learning
- Encourages time on academics outside of class
- Can provide a dimension of depth not available in the classroom
- Can promote student creativity
- Can be scored holistically or analytically
- May allow probes by faculty to gain a clearer picture of student understanding or thought processes
- Can provide closing of feedback loop between students and faculty
- Can place faculty more in a mentor role than as a judge
- Can be summative or formative

- Can provide an avenue for student self-assessment and reflection
- Can be embedded within courses
- Can adapt current assignments
- Usually the most valid way of assessing skill development

DISADVANTAGES

- Usually the mostly costly approach
- Time-consuming and labor-intensive to design and execute for faculty and students
- Must be carefully designed if used to document obtainment of student learning outcomes
- Ratings can be more subjective
- Requires careful training of raters
- Inter-rater reliability must be addressed
- Production costs may be prohibitive for some students and hamper reliability
- Sample of behavior or performance may not be typical, especially if observers are present

PORTFOLIOS

Medical student portfolios are used in a variety of ways to help institutions and students assess and track learner progress. In its original definition, a portfolio is a collection of drawings or papers that represent a compilation of a person's work. With the transition to competency-based assessments and the use of frameworks such as milestones and entrustable professional activities (EPAs), portfolios have become a valuable tool in many medical schools. Student portfolios can assist institutions to meaningfully display assessment evidence, enabling longitudinal tracking, and documentation of student achievement. With the appropriate supporting processes, portfolios can foster student skills in self-assessment and ongoing professional development (Dallaghan et al., 2020). A portfolio can be generally defined as: "a purposeful collection of student work that exhibits the student's efforts, progress, and achievements in one or more areas. The collection must include student participation in selecting contents, the criteria for selection, the criteria for judging merit, and evidence of student self-reflection" (Paulson et al., 1991).

STRENGTHS OF PORTFOLIOS

- Can provide direct evidence of student mastery of learning outcomes.
- Students are encouraged to take responsibility for and pride in their learning.
- Students may become more aware of their own academic growth.
- Can be used for developmental assessment and can be integrated into the advising process to individualize student planning.
- Can help faculty identify curriculum gaps, and lack of alignment with outcomes.
- Students can use portfolios and the portfolio process to prepare for graduate school or career applications.
- The evaluation process should directly lead faculty into discussions of student learning, curriculum, pedagogy, and student support services.
- E-portfolios or CD-ROMs can be easily viewed, duplicated, and stored.

WEAKNESSES OF PORTFOLIOS

- Requires faculty time to prepare the portfolio assignment and assist students as they prepare them.
- Requires faculty analysis and, if graded, faculty time to assign grades.
- May be difficult to motivate students to take the task seriously.

- May be more difficult for transfer students to assemble the portfolio if they haven't saved relevant materials.
- Students may refrain from criticizing the program if their portfolio is graded or if their names will be associated with portfolios during the review.

POTENTIAL ADVANTAGES OF PORTFOLIOS

- Shows sophistication in student performance.
- Illustrates longitudinal trends.
- Highlight student strengths.
- Identifies student weaknesses for remediation, if timed properly.
- Can be used to view learning and development longitudinally.
- Multiple components of the curriculum can be assessed (e.g., writing, critical thinking, and technology skills).
- Samples are more likely than test results to reflect student ability when planning, input from others, and similar opportunities common to more work settings are available.
- The process of reviewing and evaluating portfolios provides an excellent opportunity for faculty exchange and development, discussion of curriculum goals and objectives, review of criteria, and program feedback.
- May be economical in terms of student time and effort if no separate assessment administration time is required.
- Greater faculty control over interpretation and use of results.
- Results are more likely to be meaningful at all levels (student, class, program, and institution) and can be used for diagnostic and prescriptive purposes as well.
- Avoids or minimizes test anxiety and other one-shot measurement problems.
- Increases power of maximum performance measures over more artificial or restrictive speed measures on test or in-class samples.
- Increases student participation (selection, revision, and evaluation) in the assessment process.
- Could match well with Morningside's mission to cultivate lifelong learning.
- Can be used to gather information about students' assignments and experiences.
- Reflective statements could be used to gather information about student satisfaction.

POTENTIAL DISADVANTAGES OF PORTFOLIOS

- The portfolio will be no better than the quality of the collected artifacts.
- Time-consuming and challenging to evaluate.
- Space and ownership challenges make evaluation difficult.
- Content may vary widely among students.
- Students may fail to remember to collect items.
- Transfer students may not be in the position to provide complete portfolios.
- Time intensive to convert to meaningful data.
- Costly in terms of evaluator time and effort.
- Management of the collection and evaluation process, including the establishment of reliable and valid grading criteria, is likely to be challenging.
- May not provide for externality.
- If samples to be included have been previously submitted for course grades, faculty may be concerned that a hidden agenda of the process is to validate their grading.
- Security concerns may arise as to whether submitted samples are the student's own work or adhere to other measurement criteria.

- Must consider whether and how graduates will be allowed continued access to their portfolios.
- Inter-rater reliability must be addressed.

BEST PRACTICES IN SELECTING THE DIRECT ASSESSMENT METHODS IN CAM EDUCATION

To select and apply valid and reliable direct assessment methods, medical educators should take into consideration the following:

Select the appropriate direct assessment methods based on the literature and experience and it should be valid and reliable as well as approved by the school curriculum committee.

If you need to implement a new assessment method, get approval from the department and curriculum committee.

Distribute the assessment methods based on the course competencies, and course learning outcomes as follows: select and apply the appropriate assessment methods for each course learning outcome, and select and apply the appropriate assessment methods for each week's lecture—laboratory and tutorial work from week one to week 16 for example.

Allocate marks for each assessment method, for each week, for each learning outcome based on the size of the learning outcome and other factors.

Revise the assessment methods on a yearly basis, and modify them if needed.

INDIRECT ASSESSMENT METHODS AND ITS APPLICATION IN CAM EDUCATION

Literature reported the following examples of indirect assessment methods (Allen, 2004).

SURVEYS

A survey is an examination of opinions, behavior, etc., made by asking people questions (Cambridge Dictionary).

Point-of-contact surveys
Online, e-mailed, registration, or grad check surveys
Keep it simple!

EXAMPLES OF SURVEYS IN CAM EDUCATION ARE

Surveys & Questionnaires to stakeholders to ask them about the graduates.
Surveys & Questionnaires to (students, alumni, employers, and the public) about any issue.

SURVEY TYPES

Checklist
Frequency
Likert scale
Linear rating scale
Others

STRENGTHS OF SURVEY

- Are flexible in format and can include questions about many issues.
- Can be administered to large groups of respondents.
- Can easily assess the views of various stakeholders.
- Usually has face validity—the questions generally have a clear relationship to the outcomes being assessed.
- Tend to be inexpensive to administer.
- Can be conducted relatively quickly.
- Responses to close-ended questions are easy to tabulate and to report in tables or graphs.
- Open-ended questions allow faculty to uncover unanticipated results.
- Can be used to track opinions across time to explore trends.
- Are amenable to different formats, such as paper-and-pencil or online formats.
- Can be used to collect opinions.

WEAKNESSES OF SURVEY

- Provide indirect evidence about student learning.
- Their validity depends on the quality of the questions and response options.
- Conclusions can be inaccurate if biased samples are obtained.
- Results might not include the full array of opinions if the sample is small.
- What people say they do or know may be inconsistent with what they actually do or know.
- Open-ended responses can be difficult and time-consuming to analyze.

INTERVIEWS

- Interviews can be conducted one-on-one, in small groups, or over the phone.
- Interviews can be structured (with specified questions) or unstructured (a more open process).
- Questions can be close-ended (e.g., multiple-choice style) or open-ended (respondents construct a response).
- Can target students, graduating seniors, alumni, employers, community members, faculty, etc.
- Can do exit interviews or pre-post interviews.
- Can focus on student experiences, concerns, or attitudes related to the program being assessed.
- Generally should be conducted by neutral parties to avoid bias and conflict of interest.

TIPS FOR EFFECTIVE INTERVIEWS

- Conduct the interview in an environment that allows the interaction to be confidential and uninterrupted.
- Demonstrate respect for the respondents as participants in the assessment process rather than as subjects. Explain the purpose of the project, how the data will be used, how the respondent's anonymity or confidentiality will be maintained, and the respondents' rights as participants. Ask if they have any questions.
- Put the respondents at ease. Do more listening than talking. Allow respondents to finish their statements without interruption.
- Match follow-up questions to the project's objectives. For example, if the objective is to obtain student feedback about student advising, don't spend time pursuing other topics.

- Do not argue with the respondent's point of view, even if you are convinced that the viewpoint is incorrect. Your role is to obtain the respondents' opinions, not to convert them to your perspective.
- Allow respondents time to process the question. They may not have thought about the issue before, and they may require time to develop a thoughtful response.
- Paraphrase to verify that you have understood the respondent's comments. Respondents will sometimes realize that what they said isn't what they meant, or you may have misunderstood them. Paraphrasing provides an opportunity to improve the accuracy of the data.
- Make sure you know how to record the data and include a backup system. You may be using a tape recorder—if so, consider supplementing the tape with written notes in case the recorder fails or the tape is faulty. Always build a system for verifying that the tape is functioning or that other data recording procedures are working. Don't forget your pencil and paper!

STRENGTHS OF INTERVIEWS

- Are flexible in format and can include questions about many issues.
- Can assess the views of various stakeholders.
- Usually has face validity—the questions generally have a clear relationship to the outcomes being assessed.
- Can provide insights into the reasons for participants' beliefs, attitudes, and experiences.
- Interviewers can prompt respondents to provide more detailed responses.
- Interviewers can respond to questions and clarify misunderstandings.
- Telephone interviews can be used to reach distant respondents.
- Can provide a sense of immediacy and personal attention for respondents.
- Open-ended questions allow faculty to uncover unanticipated results.

WEAKNESSES OF INTERVIEWS

- Generally provides indirect evidence about student learning.
- Their validity depends on the quality of the questions.
- Poor interviewer skills can generate limited or useless information.
- Can be difficult to obtain a representative sample of respondents.
- What people say they do or know may be inconsistent with what they actually do or know.
- Can be relatively time-consuming and expensive to conduct, especially if interviewers and interviewees are paid or if the no-show rate for scheduled interviews is high.
- The process can intimidate some respondents, especially if asked about sensitive information and their identity is known to the interviewer.
- Results can be difficult and time-consuming to analyze.
- Transcriptions of interviews can be time-consuming and costly.

FOCUS GROUPS

- Traditional focus groups are free-flowing discussions among small, homogeneous groups (typically from 6 to 10 participants), guided by a skilled facilitator who subtly directs the discussion in accordance with pre-determined objectives. This process leads to in-depth responses to questions, generally with full participation from all group members. The facilitator departs from the script to follow promising leads that arise during the interaction.
- Structured group interviews are less interactive than traditional focus groups and can be facilitated by people with less training in group dynamics and traditional focus group methodology. The group interview is highly structured, and the report generally provides a few core findings, rather than an in-depth analysis.

Strengths of Focus Groups

- Are flexible in format and can include questions about many issues.
- Can provide an in-depth exploration of issues.
- Usually has face validity—the questions generally have a clear relationship to the outcomes being assessed.
- Can be combined with other techniques, such as surveys.
- The process allows faculty to uncover unanticipated results.
- Can provide insights into the reasons for participants' beliefs, attitudes, and experiences.
- Can be conducted within courses.
- Participants have the opportunity to react to each other's ideas, providing an opportunity to uncover the degree of consensus on ideas that emerge during the discussion.

Weaknesses of Focus Groups

- Generally provides indirect evidence about student learning.
- Requires a skilled, unbiased facilitator.
- Their validity depends on the quality of the questions.
- Results might not include the full array of opinions if only one focus group is conducted.
- What people say they do or know may be inconsistent with what they actually do or know.
- Recruiting and scheduling the groups can be difficult.
- Time-consuming to collect and analyze data.

BEST PRACTICES IN SELECTING THE ASSESSMENT METHODS IN CAM EDUCATION

To select and apply valid and reliable assessment methods, educators should take into consideration the following:

> Select the appropriate assessment methods based on the literature and experience and it should be valid and reliable as well as approved by the school curriculum committee.
> If you need to implement a new assessment method, get approval from the department and curriculum committee.
> Distribute the assessment methods based on the course competencies, and course learning outcomes as follows: select and apply the appropriate assessment methods for each course learning outcome; and select and apply the appropriate assessment methods for each week's lectures—laboratory and tutorials from week one to week 16 as an example.
> Allocate marks for each assessment method (if applicable), for each week, and for each learning outcome based on the size of the learning outcome and other factors.
> Revise the assessment methods on a yearly basis, and modify them if needed.

FORMATIVE ASSESSMENT METHODS AND ITS APPLICATION IN CAM EDUCATION

Literature reported the following examples of formative assessment methods (DiVall et al., 2014).

Prior Knowledge Assessment

> Such as a short quiz before or at the start of a class.
> Guides lecture content informs students of weaknesses and strengths.
> Limitations: students may not be motivated to take the assessment seriously and require flexible class time to respond.

AN EXAMPLE OF PRIOR KNOWLEDGE ASSESSMENT IN CAM EDUCATION:

Quiz about the knowledge of students about CAM for asthma before the class on asthma CAM.

MINUTE PAPER

Such as a writing exercise asking students what they thought was the most important information and what they did not understand.

It will take about a minute and is usually used at the end of class and can be used at the end of any topic and in any course.

Can provide rapid feedback and require students to think and reason.

Limitations: students may expect all items to be discussed, students may use it to get a faculty member to repeat information rather than introduce new information.

EXAMPLES OF MINUTE PAPER ASSESSMENTS IN CAM EDUCATION:

After the class of asthma CAM, the educator may ask the students about:

What have you learned today?

What didn't you understand today?

Do you have any questions?

MUDDIEST POINT

Student response to a question regarding the most confusing point for a specific topic.

This quick and easy active learning activity asks students to identify the muddiest, or most confusing point in a lecture, class session, or assignment. By asking students to write this down and collecting their responses one can quickly identify the areas where the students are having difficulty. From there one can address those difficulties at the start of the next class. The effectiveness of this strategy hinges on addressing the muddiest points identified by the students. Responses may be via email, the course management system, or with a short screencast or video.

Helps students acknowledge lack of understanding and identifies problem areas for the class.

Limitations: emphasizes what students do not understand rather than what they do understand.

EXAMPLES OF MUDDIEST POINT ASSESSMENT IN CAM EDUCATION:

After the Asthma CAM class:

The medical educator may give the students cards or white papers at the end of the class and ask them: What was the muddiest point in the Asthma CAM class? What was not or least clear to you?

Collect the students' responses & cards/papers.

Review and plan responses.

Post questions and answers online on the course page.

Answer questions at the beginning of the next class.

Revise course contents if needed.

"CLICKERS" (AUDIENCE RESPONSE SYSTEM)

A question that is asked anytime during a class to gauge learning.

Providing students/faculty with immediate feedback and debriefing can improve the understanding of a concept.

Limitations: uses up classroom time, students may not be motivated to answer questions seriously.

EXAMPLES OF "CLICKERS" (AUDIENCE RESPONSE SYSTEM) ASSESSMENT IN CAM EDUCATION:

During the Asthma CAM class, the medical educator may ask students to ask questions during class time. The medical educator will give the students immediate answers and feedback.

The Elements of Effective Feedback – the SMART System are:

Specific
Measurable and Meaningful
Accurate and Actionable
Respectful
Timely

CASE STUDIES (PROBLEM RECOGNITION)

Case analysis and response to case-related questions and/or identification of a problem.

Helps develop critical thinking and problem-solving skills, and develops diagnostic skills.

Limitations: time-consuming to create, takes considerable time for students to work on them.

EXAMPLES OF CASE STUDIES (PROBLEM RECOGNITION) IN CAM EDUCATION:

Asthma CAM case studies.
Hypertension CAM case studies.
Others.

FORMATIVE PEER ASSESSMENT

Students are introduced to the assignment and criteria for assessment
Students are trained and given practice on how to assess and provide feedback
Students complete and submit a draft
Students assess the drafts of other students and give feedback
Students reflect on the feedback received and revise their work for final submission
Assignments are graded by the instructor
Instructor reflects on the activity with the class

ASSESSMENT FOR LABORATORY AND EXPERIENTIAL SETTINGS STRATEGIES

A wide variety of formative assessment tools and methodologies can be used in laboratory and experiential settings. One of the most commonly used assessment strategies in the laboratory setting is OSCEs (Miller, 1990; Epstein and Hundert, 2002; Beck et al., 1995).

ENGAGE STUDENTS IN FORMATIVE ASSESSMENT

- Explain the rationale behind formative assessment clearly—make it clear to students that through engaging with formative tasks they get to gain experience with their assessments, risk-free, and can develop far stronger skills in order to obtain better grades in the summative assessments.
- Create a link between summative and formative assessment—design formative assessments in such a way that they contribute to the summative task. This lowers the workload on the students and provides them with the necessary feedback to improve their final performance. An example of such an assessment is producing an essay plan, a structure of a literature review, and part of the essay or bibliography.

- Lower the number of summative assessments and increase the number of formative assessments—yet do not allow one single summative assessment to carry too much weight in the final grade.

OBJECTIVE STRUCTURED CLINICAL EXAMINATION (OSCE)

"OSCE is an approach to the assessment of clinical competence in which the components of competence are assessed in a planned or structured way with the attention being paid to the objectivity of the examination. (Harden and Gleeson, 1979). The examination consists of multiple, standard stations at which students must complete one to two specific clinical tasks, often in an interactive environment involving patient actors such as standardized patients (Harden and Gleeson, 1979; Sturpe, 2010).

ADVANTAGES OF OBJECTIVE STRUCTURED CLINICAL EXAMINATION (OSCE)

The main advantage of the OSCE is its ability to assess candidates' multiple dimensions of clinical competences (Sherazi, 2019):

- History taking
- Physical examination
- Medical knowledge
- Interpersonal skills
- Communication skills
- Professionalism
- Data gathering/information collection
- Understanding about disease processes
- Evidence-based decision-making
- Primary care management/clinical management skills
- Patient-centered care
- Health promotion
- Disease prevention
- Safe and effective practice of medicine

STEPS OF OBJECTIVE STRUCTURED CLINICAL EXAMINATION (OSCE)

Kachur et al. (2012) suggested the following steps for the OSCE:

Step 1. Identify Available Resources
  a. **Assemble a Team:** One or more of the following:
       Leader
       Strong motivation to develop and implement the project
       Well-connected to procure resources, including access to institutional or local clinical skills testing facilities
       Involved in medical school curriculum decision-making
       Able to communicate well and create team spirit.
       Planner
       Understands the logistics of implementing OSCEs
       Is familiar with local conditions
       Can entertain multiple options for solving problems.

Administrator

Can implement OSCE-related tasks (e.g., scheduling, SP recruitment, photocopying of station materials, and data entry)

Able to communicate well and create a team spirit

Good at troubleshooting and problem-solving

Station Developer

Has relevant clinical experience

Is familiar with performance standards

Accepts editing

Trainer

Understands simulated patient (SP) and rater roles and case requirements

Has teaching skills (e.g., provides constructive feedback) and can manage the psychosocial impact of case portrayals

Able to communicate well and create a team spirit

Simulated Patients

At least one for each station.

Interested in taking on "educational" responsibilities

Rater

At least one for each station.

Clear about OSCE goals and performance standards.

Committed to fair performance assessments.

Effective feedback provider.

Timer

At least one for each station.

Monitor

At least one for each station.

Data Manager

Data Analyst

Understands the OSCE process.

Has psychometric skills.

Understands end-users of results (e.g., learners, program).

Program Evaluator

Understands the OSCE process.

Is familiar with evaluation models (e.g., pre/post-testing).

Can develop and analyze program evaluations (e.g., surveys, focus groups).

Best Practices: Assembling a Team
- Establish a clear common goal.
- Build a team with a variety of skills.
- Schedule regular meetings to build group identity.
- Create a common repository (i.e., shared drive, secure Website) for meeting minutes, materials, and protocols.
- Look broadly for suitable sites and funding sources.
  b. Identify location
  c. Identify sources of funding and support

Step 2. Agree on Goals and Timeline

Best Practices: OSCE Planning
- Identify the date and time of OSCE.
- Make a timeline working backward from the OSCE date.
- Start early to identify potential SPs and secure training times.
- Identify the potential location of OSCE early (clinic rooms, classrooms, or simulation center).
- Secure participants' availability.

Step 3. Establish a Blueprint
Best Practices: Blueprinting
- Delineate core competencies.
- Establish performance criteria for each level of training.
- Ensure OSCE case patient age, gender, race, and prevalence of disease reflect actual clinical practice.
- Align OSCE skills and content assessed with current or new curricula.

Step 4. Develop Cases and Stations
Best Practices: Case Development
- Choose scenarios that are both common and challenging presentations for the learners.
- Ensure that cases represent the patient population in the clinical environment.
- Build specific goals and challenges in each scenario.
- Choose a post-encounter activity (i.e., feedback, supplementary exercise, or rest).
- Make sure it is possible to complete tasks in the time allotted.
- Organize a trial run with a variety of other learners to validate and fine-tune cases.

Step 5. Create Rating Forms
Best Practices: OSCE Checklists
- Develop rating items based on the blueprint and ensure that a sufficient number of items are included to reliably assess competence within the targeted domains.
- Consider using both behavior-specific items and global rating items in OSCE rating forms to achieve a balance in terms of helping raters reflect important elements of their subjective responses and to enhance their objectivity in representing what happened during the encounter and providing learners with specific and more holistic feedback.
- Develop response options for behavior-specific items that reflect observable actions and strive to match the response options to the likely variation in performance of the learner population to maximize differentiation.

Step 6. Recruit and Train Simulated Patients (SPs)
Best Practices: SP Recruitment and Training
- Search for SPs through word-of-mouth strategies (e.g., by contacting other SPs, connecting with other SP trainers, talking to clinicians and acting teachers).
- Cast the right person for each case (i.e., physical appearance, psychological profile, availability, no contraindications).
- For high-stakes programs recruit and train alternates who can step in if needed (alternates can be cross-trained to provide coverage for multiple cases).
- Put SPs into learner's positions through role-play to enhance their understanding of the case (e.g., interactive and emotional impact of SP actions) and to promote an empathic approach to learners.
- Practice all aspects of the encounter (e.g., physical exam, feedback); do not leave SP performance to chance.
- Explore the psychological and physiological impact a case has on the SP to avoid toxic side effects (e.g., getting depressed from repeatedly portraying a depressed patient, getting muscle spasms from portraying a patient who has difficulty walking).
- Train all SPs who are portraying the same case (simultaneously or consecutively) at the same time to enhance consistency in case portrayal across SPs.

Step 7. Recruit and Train Evaluators
Best Practices: Evaluator Recruitment and Training for Rating and Feedback Tasks
- Select evaluators who are willing to adopt the program values, who are consistent in their ratings, and who don't have an axe to grind.

- Bring multiple evaluators together to jointly observe a learner's performance on tape or live, compare ratings, and discuss similarities and discrepancies. Practice giving feedback (if this is expected).
- Make raters aware of potential biases and rating mistakes.
- Provide written guidelines for rating items, evaluation schemes, and station objectives/ teaching points.
- Post-OSCE, give feedback to raters about how their ratings compare with those of others (e.g., more or less lenient, lack of range).

Step 8. Implement the OSCE: Managing the Session

Step 9. Manage, Analyze, and Report Data

Step 10. Develop a Case Library and Institutionalize OSCEs

APPLICATIONS OBJECTIVE STRUCTURED CLINICAL EXAMINATION (OSCE) IN CAM EDUCATION

OSCE can be applied and implemented in many ways as follows:

Exit exam & comprehensive exam: to assess the readiness of students for the CAM practice and this will further help them in the licensing exam.

In many courses as a mini OSCE, to assess the clinical skills of students in many courses.

In internships and clerkships.

ONLINE OBJECTIVE STRUCTURED CLINICAL EXAMINATION (OSCE)

Online OSCEs can be implemented online with the help of new technologies and training. Mock online OSCEs are very important for medical educators and students.

CHALLENGES OF OBJECTIVE STRUCTURED CLINICAL EXAMINATION (OSCE)

Lack of resources.

Lack of funds.

Workforce issues.

Knowledge of medical educators about OSCE.

The attitude of medical educators about OSCE.

The attitude of medical students about OSCE.

Resistance of medical educators about OSCE.

Resistance of medical students about OSCE.

University/School culture toward the implementation of OSCE.

Lack of training about OSCE.

Lack of motivation to train about OSCE.

Lack of technologies to implement the online OSCE.

CONCLUSION

The comprehensive analysis of CAM education presented in this chapter underlines the significant strides made in aligning educational programs with international standards and accreditation criteria. This alignment ensures that CAM practitioners are well-equipped with the requisite knowledge and skills to excel in the healthcare sector. The shift toward competency-based assessments marks a pivotal advancement in evaluating the proficiency of healthcare professionals, ensuring that they meet the quality standards necessary for effective patient care. Key to this evolution is the understanding and application of various assessment and evaluation methods. The chapter elucidates the distinctions and applications of formative and summative assessments, diagnostic assessments, direct and indirect assessments, and the pivotal role of the OSCE in assessing clinical competencies. Each method brings its own strengths and limitations, and their appropriate application is crucial for the accurate measurement of student capabilities and program effectiveness. The rationality behind these assessment methods in CAM education is clear: they are integral in ensuring that graduates are not only knowledgeable but also competent in their practice and patient care. The chapter highlights the need for regular updates and revisions of these methods to keep pace with the evolving landscape of CAM practices. In conclusion, the chapter emphasizes that the future of CAM education relies heavily on the continuous improvement of assessment methods. These methods must be carefully chosen, regularly reviewed, and aligned with both educational outcomes and professional standards. This approach will ensure that CAM practitioners are well-prepared to meet the challenges of the healthcare sector, contributing effectively to patient care and the broader medical community.

REFERENCES

Accreditation Council for Graduate Medical Education (AGME). (2019) Allergy and immunology milestones. https://www.acgme.org/Specialties/Allergy-andImmunology/Milestones

Alfadl, A.A. (2018). Assessment methods and tools for pharmacy education. In: *Pharmacy Education in the Twenty First Century and Beyond* (pp. 147–168). Academic Press.

Allen, M.J. (2004). *Assessing Academic Programs in Higher Education.* Anker Publishing, Bolton, MA. Retrieved February 6, 2013. https://www.wiley.com/en-us/Assessing+Academic+Programs+in+Higher +Education-p-9781882982677

Al-Worafi, Y.M. (2022a). *A Guide to Online Pharmacy Education: Teaching Strategies and Assessment Methods.* CRC Press.

Al-Worafi, Y.M. (2022b). Competencies and learning outcomes. In Al-Worafi, Y.M. (ed), *A Guide to Online Pharmacy Education: Teaching Strategies and Assessment Methods.* CRC Press.

Al-Worafi, Y. (2022c). Assessment methods in pharmacy education: Strengths and limitations. In Al-Worafi, Y.M. (ed), *A Guide to Online Pharmacy Education: Teaching Strategies and Assessment Methods.* CRC Press.

Al-Worafi, Y.M. (2022d). Assessment methods in pharmacy education: Direct assessment. In: Al-Worafi, Y.M. (ed), *A Guide to Online Pharmacy Education: Teaching Strategies and Assessment Methods.* CRC Press.

Al-Worafi, Y.M. (2022e). Assessment methods in pharmacy education: Indirect assessment. In: Al-Worafi, Y.M. (ed), *A Guide to Online Pharmacy Education: Teaching Strategies and Assessment Methods.* CRC Press.

Al-Worafi, Y.M. (2022f). Assessment methods in pharmacy education: Formative assessment. In: Al-Worafi, Y.M. (ed), *A Guide to Online Pharmacy Education: Teaching Strategies and Assessment Methods.* CRC Press.

Al-Worafi, Y.M. (2022g). Objective structured clinical examination (OSCE) in pharmacy education. In: Al-Worafi, Y.M. (ed), *A Guide to Online Pharmacy Education: Teaching Strategies and Assessment Methods.* CRC Press.

Al-Worafi, Y.M. (Ed.). (2024a). *Handbook of Medical and Health Sciences in Developing Countries.* Springer, Cham.

Al-Worafi, Y.M. (2024b). Complementary and alternative medicine (CAM) in developing countries. In: Al-Worafi, Y.M. (ed), *Handbook of Medical and Health Sciences in Developing Countries.* Springer, Cham. https://doi.org/10.1007/978-3-030-74786-2_301-1

Beck, D.E., Boh, L.E., and O'Sullivan, P.S. (1995). Evaluating student performance in the experiential setting with confidence. *American Journal of Pharmaceutical Education*, 59, 236–236.

Black, P., and Wiliam, D. (2010). Inside the black box: Raising standards through classroom assessment. *Phi delta kappan*, 92(1), 81–90.

Dallaghan, G.L.B., Coplit, L., Cutrer, W.B., and Crow, S. (2020). Medical student portfolios: Their value and what you need for successful implementation. *Academic Medicine*, 95(9), 1457.

DiVall, M.V., Alston, G.L., Bird, E., Buring, S.M., Kelley, K.A., Murphy, N.L., Schlesselman, L.S., Stowe, C.D., and Szilagyi, J.E. (2014). A faculty toolkit for formative assessment in pharmacy education. *American Journal of Pharmaceutical Education*, 78(9), 160.

Dixson, D.D., and Worrell, F.C. (2016). Formative and summative assessment in the classroom. *Theory into Practice*, 55(2), 153–159.

Epstein, R.M., and Hundert, E.M. (2002). Defining and assessing professional competence. *Jama*, 287(2), 226–235.

Fulcher, K.H., Swain, M., and Orem, C.D., (2012). Expectations for assessment reports: A descriptive analysis. *Assessment Update* 24(1), 1.

Garrison, C., and Ehringhaus, M. (2007). Formative and summative assessments in the classroom. http://www.amle.org/Publications/WebExclusive/Assessment/tabid /1120/Default.aspx

Gibbs, T., Brigden, D., and Hellenberg, D. (2006). Assessment and evaluation in medical education. *South African Family Practice*, 48(1), 5–7.

Hamstra, S.J. (2012). Keynote address: The focus on competencies and individual learner assessment as emerging themes in medical education research. *Academic Emergency Medicine*, 19(12), 1336–1343.

Hamstra, S.J. (2014). Designing and selecting assessment instruments: Focusing on competencies. In Dath, D., and Bandeira, G. (eds), *The Royal College Program Directors Handbook: A Practical Guide for Leading an Exceptional Program*. Royal College of Physicians and Surgeons of Canada, Ottawa, ON.

Harden, R.M., and Gleeson, F.A., (1979). Assessment of clinical competence using an objective structured clinical examination (OSCE). *Medical Education*, 13(1), 39–54.

Kachur, E.K., Zabar, S., Hanley, K., Kalet, A., Bruno, J.H., and Gillespie, C.C. (2012). Organizing OSCEs (and other SP exercises) in ten steps. In: *Objective Structured Clinical Examinations* (pp. 7–34). Springer, New York, NY.

Lee, G.B., and Chiu, A.M. (2021). Assessment and feedback methods in competency-based medical education. *Annals of Allergy, Asthma & Immunology*, 128(3), 256–262.

Mavis, B. (2014). Assessing student performance. In Huggett, K. N., and Jeffries, W. B. (eds), *An Introduction to Medical Teaching* (pp. 209–241). Springer, Dordrecht.

McAleer, S. (2001). Choosing assessment instruments. In: *A Practical Guide for Medical Teachers* (p. 303313). Churchill Livingstone, Edinburgh.

Miller, G.E. (1990). The assessment of clinical skills/competence/performance. *Academic Medicine*, 65(9), S63–S67.

Morningside College. (2006). *Assessment Handbook*. Advantages and Disadvantages of Various Assessment Methods. http://www.morningside.edu/academics/research/assessment/documents/advantagesdisadvantages.pdf, March 2006

Paulson, F.L., Paulson, P.R., and Meyer, C.A. (1991). What makes a portfolio a portfolio. *Educational Leadership*, 48(5), 60–63.

Sherazi, M.H. (2019). Objective structured clinical examination introduction. In Sherazi, M.H., and Dixon, E. (eds), *The Objective Structured Clinical Examination Review* (pp. 1–12). Springer, Cham.

Sturpe, D.A., (2010). Objective structured clinical examinations in doctor of pharmacy programs in the United States. *American Journal of Pharmaceutical Education*, 74(8), 148.

7 Complementary and Alternative Medicine (CAM) Education
Online Education

BACKGROUND

Complementary and alternative medicine (CAM) encompasses a broad range of practices and therapies that are outside the realm of conventional Western medicine. CAM includes a variety of approaches like herbal medicine, acupuncture, mind-body techniques, and homeopathy, among others. The rise of interest in CAM has been significant over the past few decades, driven by a desire for holistic and preventive care, personal wellness, and a growing skepticism or dissatisfaction with conventional healthcare systems. Online education in CAM has become increasingly popular, offering flexibility and accessibility that traditional, in-person educational programs often cannot. This mode of learning is particularly appealing to those who are already healthcare professionals seeking to expand their skills, as well as individuals interested in personal health, wellness, and alternative therapies. The online format allows for a diverse range of courses and programs, ranging from introductory overviews to comprehensive, in-depth studies in specific areas of CAM. A key feature of online CAM education is its interdisciplinary approach, combining elements from various traditions and practices. This holistic approach is well-suited to the ethos of CAM, which often emphasizes treating the whole person rather than just symptoms. Coursework may include studies in nutrition, herbal medicine, acupuncture principles, energy therapies, and other subjects that are not typically covered in conventional medical education. Moreover, online CAM education often incorporates practical learning experiences. Although some aspects of CAM, like hands-on therapies, can be challenging to teach online, many programs address this by using interactive multimedia, virtual simulations, or arranging local hands-on workshops and seminars. This blend of online and in-person learning helps students gain a more comprehensive understanding and experience of CAM practices. Another important aspect of CAM online education is its focus on evidence-based practices. While CAM is often associated with traditional and anecdotal evidence, there is a growing emphasis on scientific research and evidence-based approaches in these courses. This helps in integrating CAM practices with conventional medicine and in improving the credibility and effectiveness of CAM therapies. Finally, online CAM education faces challenges such as ensuring the quality and accreditation of programs and addressing the varying legal and regulatory frameworks governing the practice of different CAM therapies around the world. However, the increasing demand and recognition of the value of CAM therapies continue to drive the development and refinement of these online educational programs.

TERMINOLOGY

There are many important terminologies related to medical and health sciences education and online education as follows (Al-Worafi, 2022a; 2024a, b):

DOI: 10.1201/9781003327202-7

COMPLEMENTARY AND ALTERNATIVE MEDICINE (CAM) EDUCATION

The term "Complementary and Alternative Medicine (CAM) education" can be used to describe teaching CAM students about CAM-related courses at the undergraduate level, postgraduate level, and continuous medical education level.

ONLINE EDUCATION

The term "online education" can be used to describe the process of teaching and learning using online facilities instead of traditional face-to-face such as the internet, learning management systems, webinar and video conferencing platforms, and other facilities (Al-Worafi, 2022a).

ONLINE TEACHING

The term "online teaching" can be used to describe the teaching process using online facilities instead of traditional face-to-face such as the internet, learning management systems, webinar and video conferencing platforms, and other facilities (Al-Worafi, 2022a).

ONLINE LEARNING

The term "online learning" can be used to describe the process of learning using online facilities instead of traditional face-to-face such as the internet, learning management systems, webinar and video conferencing platforms, and other facilities (Al-Worafi, 2022a).

DISTANCE LEARNING

The term "distance learning" can be used to describe the process of learning from outside universities, where the students can learn from their homes without the need to be on the university's campuses (Al-Worafi, 2022a).

MOBILE LEARNING

The term "mobile learning" can be used to describe the process of learning using mobile technologies (Al-Worafi, 2022a).

BLENDED LEARNING

Blended learning can be defined as the following (Graham, 2006; Garrison and Kanuka, 2004; Garrison and Vaughan, 2008; Hrastinski, 2019; Al-Worafi, 2022a): Blended learning is an approach to education that combines online educational materials and opportunities for interaction online with traditional place-based classroom methods. It requires the physical presence of both teacher and student, with some elements of student control over time, place, path, or place. Blended learning systems combine face-to-face instruction with computer-mediated instruction. Garrison and Kanuka (2004) define blended learning as "the thoughtful integration of classroom face-to-face learning experiences with online learning experiences." Thus, we can conclude that there is general agreement that the key ingredients of blended learning are face-to-face and online instruction or learning.

E-LEARNING

The term "e-learning" can be used to describe the process of learning using electronic and information technologies (Al-Worafi, 2022a).

EDUCATIONAL TECHNOLOGY

The Association for Educational Communications and Technology (AECT) has defined educational technology as "the study and ethical practice of facilitating learning and improving performance by creating, using, and managing appropriate technological processes and resources." Educational technology refers to all valid and reliable applied education sciences, such as equipment, as well as processes and procedures that are derived from scientific research and in a given context may refer to theoretical, algorithmic, or heuristic processes: it does not necessarily imply physical technology. Educational technology is the process of integrating technology into education in a positive manner that promotes a more diverse learning environment and a way for students to learn how to use technology as well as their common assignments. Educational technology (commonly abbreviated as edutech or edtech) is the combined use of computer hardware, software, and educational theory and practice to facilitate learning. When referred to by its abbreviation, edtech, it often refers to the industry of companies that create educational technology. In addition to the practical educational experience, educational technology is based on theoretical knowledge from various disciplines such as communication, education, psychology, sociology, artificial intelligence, and computer science (Richey et al., 2008; Robinson et al., 2013; Mastellos et al., 2018).

VIRTUAL LEARNING ENVIRONMENT (VLE)

A virtual learning environment (VLE) can be defined as a web-based platform for the digital aspects of courses of study, usually within educational institutions. They present resources, activities, and interactions within a course structure and provide for the different stages of assessment. VLEs also usually report on participation and have some level of integration with other institutional systems. VLEs are often referred to as learning management systems (LMS) (Weller, 2007).

LEARNING MANAGEMENT SYSTEMS (LMS)

The term learning management systems (LMS) can be defined as an effective web-based learning system of sharing study materials, making announcements, conducting evaluations and assessments, generating results, and communicating interactively in synchronous and asynchronous ways, among various other academic activities (Kant et al., 2021; Bervell and Umar, 2017; Al-Worafi, 2022a).

COURSE MANAGEMENT SYSTEMS

The term course management systems can be defined as an effective web-based learning system, where the educators can upload and share the course study materials with students, communicate with the students, evaluate the student's performance, and do other teaching and learning activities (Al-Worafi, 2022a).

WEBINAR AND VIDEO CONFERENCING PLATFORMS

Webinar and video conferencing platforms can be defined as software and tools for facilitating online communication between educators and students. Examples of webinar and video conferencing platforms are Microsoft Teams, Cisco WebEx Teams, Google Meet, and Zoom (Al-Worafi, 2022a).

WEARABLE TECHNOLOGIES

Wearable technologies can be defined as any technology that is designed to be used while worn. Common types of wearable technology include smartwatches and smartglasses. Wearable electronic devices are often close to or on the surface of the skin, where they detect, analyze, and transmit information such as vital signs, and/or ambient data which allow in some cases immediate biofeedback to the wearer (Düking et al., 2018).

WEARABLE EDUCATIONAL TECHNOLOGIES

Wearable technologies can be defined as any wearable technology used for the teaching and learning process.

ONLINE AND DIGITAL LIBRARY

An online and digital library is a collection of documents such as journal articles, books, and other educational resources organized in an electronic form and available on the Internet for students, educators, and other staff. Furthermore, access to the databases helps students, educators, and staff access the latest volumes/issues of scientific journals (Al-Worafi, 2022a).

DISTANCE EDUCATION

The term "distance education" can be used to describe the teaching of students outside the universities, and colleges when they aren't able to attend the lectures at the universities, or college campuses due to many factors such as geographical reasons, financial reasons, and others. Therefore, the students will learn/study at their home, the materials will be received by mail, or other methods of delivery, and with the help of technologies / new technologies over the last decades. Distance education history goes back to 1728, when Caleb Phillips, the teacher of the new method ofShort Hand USA sought students who wanted to learn through weekly mailed lessons. The first distance education course in the modern sense was provided by Sir Isaac Pitman in the 1840s, "who taught a system of shorthand by mailing texts transcribed into shorthand on postcards and receiving transcriptions from his students in return for correction. The element of student feedback was a crucial innovation in Pitman's system. This scheme was made possible by the introduction of uniform postage rates across England in 1840" (Wikipedia, Distance Education, 2022). Distance learning degrees were established in 1858 at the University of London as part of its external program offering. Electronic distance learning was started in the 1920s, 1930s, and forward by using films, radio, and television which allowed universities and schools to implement broadcast educational programs for public schools (Wikipedia, Distance Education, 2022).

HISTORY OF DISTANCE EDUCATION

Distance education history goes back to 1728, when Caleb Phillips, teacher of the new method of Short Hand USA sought students who wanted to learn through weekly mailed lessons. This was followed by the first distance education course in the modern sense provided by Sir Isaac Pitman in the 1840s, "who taught a system of shorthand by mailing texts transcribed into shorthand on postcards and receiving transcriptions from his students in return for correction. The element of student feedback was a crucial innovation in Pitman's system. This scheme was made possible by the introduction of uniform postage rates across England in 1840." The first correspondence courses were offered in the 1850s and were designed for people who were unable to attend traditional in-person classes. The concept of distance education has evolved over time with the advancement of technology and has become an important component of the education system. The use of technology has greatly impacted the growth of distance education. In the early 20th century, the use of radio and television allowed for the expansion of distance education. For example, the British Open University, established in 1969, used broadcasts to deliver courses to students. The development of the internet in the late 20th and early 21st centuries has had a profound impact on distance education. Online courses and degree programs have become widely available, leading to the rapid growth of distance education. Online education is more accessible, flexible, and convenient for students, making it an attractive alternative to traditional brick-and-mortar education. Today, students can choose from a wide range of online courses and degree programs offered by universities and colleges around the world. Distance education has become a valuable tool for lifelong learning and has opened up educational opportunities for people who would otherwise be unable to attend traditional in-person classes. The growth of distance education has been accompanied by some concerns and challenges, including issues related to student engagement, motivation, and academic achievement. Research has shown that online students tend to be more motivated and engaged when they have a sense of community and support. Effective distance education programs also require a high level of instructional design and effective use of technology. This includes the use of multimedia, interactive tools, and online assessments to provide an engaging and effective learning experience for students. The use of technology, such as artificial intelligence and virtual reality, has the potential to further enhance the distance education experience and improve its effectiveness and efficiency. For example, virtual reality can be used to provide students with a realistic and immersive learning experience, while artificial intelligence can be used to personalize the learning experience based on individual student needs and preferences. In conclusion, distance education has a long and rich history, and has become an important component of the education system. The use of technology has greatly impacted the growth of distance education and has made it more accessible, flexible, and convenient for students. While there are some concerns and challenges associated with distance education, the use of technology and effective instructional design has the potential to improve the effectiveness and efficiency of distance education (Anderson and Dron, 2011; Bates, 2005; Maddux, 1992; McIsaac and Gunawardena, 1996).

PHASES OF DISTANCE EDUCATION DEVELOPMENT

Phase I. Distribution of printed teaching material through the mail through the 18th and 19th centuries.

Phase II. Broadcast of educational courses and materials through films, radio, and television since the 1920s and beyond.

Phase III. Computer-assisted learning since the 1980s and beyond.

Phase IV. Internet-assisted technologies since the 1980s and beyond (Al-Worafi, 2022b).

HISTORY OF ONLINE EDUCATION

Online education, also known as e-learning, has become increasingly popular in recent years, particularly in response to the COVID-19 pandemic. The development of online education has been driven by advancements in technology and a growing demand for accessible and flexible educational opportunities. The history of online education dates back to the mid-1960s when the first computer-based distance education courses were offered. However, it wasn't until the late 1990s and early 2000s that online education began to gain widespread recognition and acceptance. With the advent of the internet, online learning became more accessible to people around the world and allowed for the creation of virtual classrooms and online educational resources. In the early days of online education, most online courses were text-based and limited in scope, but over time technology has evolved to allow for a more interactive and engaging experience. Today's online courses often use multimedia tools such as video, audio, and animation, as well as virtual labs, simulations, and other interactive features to deliver a more immersive educational experience. The timeline of the evolution of online education is as follows:

1960s: The first computer-based learning systems are developed, using mainframe computers and early forms of computer-aided instruction.

1970s–1980s: Early forms of online learning or computer-based training (CBT) emerged, delivering training and educational content via computers.

1980s: The advent of personal computers and the rise of the internet leads to the development of online learning platforms, such as PLATO and CompuServe.

1990s: The World Wide Web is invented, leading to the growth of online learning communities and the development of online course content. The Internet and the World Wide Web made it possible to deliver educational content online to a wider audience. The University of Phoenix became one of the first universities to offer online degree programs.

1999: The first massive open online course (MOOC) is launched by the University of Manitoba.

2002: The advent of the Web 2.0 era leads to the development of new learning management systems and social media platforms for education. The Western Governors University was established, offering exclusively online degree programs.

2008: Stanford University launches the first MOOC on the Coursera platform, which quickly becomes a major player in the online education industry.

2002–2010: MOOCs were introduced, allowing unlimited participation and open access via the web. The first MOOC, "Connectivism and Connective Knowledge" was offered by George Siemens and Stephen Downes in 2008.

2012: The Khan Academy becomes popular for its free, online educational videos. Stanford University offered three MOOCs in computer science, attracting 160,000 students from 190 countries. This led to the development of platforms such as Coursera, Udemy, and edX.

2014: The first fully accredited online university, Western Governors University, is established.

2015: The University of Phoenix, one of the largest online universities, is acquired by a private equity firm.

2017: Online education continues to grow rapidly, with more and more universities offering online courses and degree programs.

2018: New research shows that online education is becoming more effective and popular, with increasing numbers of students enrolling in online courses.

2020: The COVID-19 pandemic leads to a huge surge in online education, as students are forced to attend classes from home.

2020s: Online education has become an integral part of the education system, with a wide range of options available for students of all ages and levels. From free, open courses to accredited degrees, the possibilities for online education are more than at any time in history (WGU, 2020; Al-Worafi, 2022a; Bozkurt, 2019).

HISTORY OF ONLINE MEDICAL EDUCATION

Online medical education has a relatively short history compared to other programs of education, but one that has been marked by rapid growth and change. The first online medical courses were offered in the 1990s, one of the earliest examples of this was the "Virtual College of Medical Education" which was established in 1995 by Dr. David L. Newman. This college offered online courses in topics such as anatomy and pharmacology and was one of the first institutions to use the Internet to deliver medical education. The first online medical education program was established in 1996 at the University of California, San Francisco (UCSF). The program, which was designed for working professionals, provided students with access to online lectures, discussion boards, and other resources. The program was a success, attracting students from around the world and helping to establish online medical education as a viable option for medical professionals. Over the next few years, more universities and colleges began to offer online medical education programs, including the University of Kansas Medical Center, the University of Phoenix, and the University of Virginia School of Medicine. The programs were designed to provide medical professionals with the knowledge and skills they needed to succeed in their careers, and many of them focused on specific areas of medicine, such as pediatrics, internal medicine, and surgery. In the early 2000s, the growth of online medical education continued to accelerate. The number of programs and courses available online increased dramatically, and many universities and colleges began to offer full degrees in medicine and other related fields (Anderson and Dron, 2011). At the same time, the quality of online medical education programs also improved, with universities and colleges investing in advanced technology, instructional design, and other resources to ensure that their programs were effective and engaging. The development of online LMS and course content management systems (CMS) made it easier for schools to offer online medical education, and the increasing availability of high-speed internet connections made it possible for students to access these courses from anywhere in the world. One of the most significant developments in the history of online medical education was the growth of MOOCs. MOOCs are large-scale, open-enrollment online courses that are designed to provide free or low-cost education to anyone who is interested. In the early 2010s, several organizations, including Coursera and edX, began offering MOOCs in medicine and other related fields. These MOOCs provided medical professionals with access to high-quality, cutting-edge information and resources, and helped to further popularize online medical education. One of the biggest innovations in online medical education in recent years has been the use of simulation and virtual reality technology. This technology allows students to experience medical procedures and surgeries in a virtual environment, allowing them to develop their skills and knowledge in a safe and controlled setting. This has been particularly useful for students studying fields such as surgery and anesthesia, where hands-on experience is essential. Another trend in online medical education has been the growth of interprofessional education, where students from different healthcare professions learn together online. This approach has been seen as a way to encourage collaboration and interdisciplinary thinking in the healthcare field and has been widely adopted by schools and universities around the world. Today, online medical education is an established field, with thousands of programs and courses available to medical professionals. Online medical education programs are now offered by many of the world's leading universities and colleges, and they cover a wide range of topics, including anatomy, physiology, pharmacology, and epidemiology. In addition, online medical education programs are increasingly recognized by licensing bodies and other professional organizations as a valuable tool for medical professionals looking to stay up-to-date with the latest developments in their field. Since the 2000s as a result of internet technologies many medical schools worldwide launched many distance online programs for postgraduate studies and continuous medical education. Examples of online medical education:

1990s:
- Emergence of the Internet and World Wide Web (WWW)

2000:
- The first online medical courses and programs begin to be offered by universities and medical institutions.

2002:
- The Accreditation Council for Continuing Medical Education (ACCME) launches the Accreditation with Commendation program, which sets standards for online medical education.

2005:
- The Association of American Medical Colleges (AAMC) launches the Medical School Objectives Project, which outlines the competencies that medical students should achieve during their education.

2008:
- The Accreditation Council for Graduate Medical Education (ACGME) begins to accredit online medical education programs in conjunction with traditional in-person programs.

2010s:
- The widespread adoption of online learning platforms, such as Coursera and Udemy, makes online medical education more accessible to a wider range of learners.

2015:
- The American Medical Association (AMA) and the American Board of Medical Specialties (ABMS) jointly launches the Maintenance of Certification (MOC) program, which requires physicians to complete a certain amount of continuing medical education (CME) each year in order to maintain their certification.

2017:
- The Liaison Committee on Medical Education (LCME) publishes new accreditation standards that include provisions for online medical education.

2018:
- AAMC releases a report entitled "The Future of Medical Education: Innovations and Challenges," which highlights the increasing importance of online medical education and the need for more research in this area.

2020s:
- The COVID-19 pandemic leads to a rapid acceleration of online medical education, as in-person learning becomes difficult or impossible in many cases.

India: In India, online medical education has been rapidly growing in recent years. Several universities, such as the All India Institute of Medical Sciences (AIIMS) and the Jawaharlal Institute of Postgraduate Medical Education and Research (JIPMER), now offer online courses and degrees in medicine. For example, AIIMS offers an online Master of Hospital Administration (MHA) program, which provides students with the skills and knowledge they need to manage a hospital.

Brazil: In Brazil, the Universidade de São Paulo (USP) has been a leader in offering online medical courses. For instance, USP offers an online Master of Pharmacy (MPH) program, which provides students with a comprehensive understanding of pharmacy issues and how to address them.

South Africa: In South Africa, online medical education has been gaining popularity in recent years, particularly in rural areas where access to in-person medical education may be limited. For example, the University of the Witwatersrand in Johannesburg offers an online Bachelor of Medicine and Bachelor of Surgery (MBBS) program, which provides students with a comprehensive education in medicine and prepares them to practice as doctors.

Nigeria: In Nigeria, online medical education has been growing due to the increasing demand for medical professionals and the need for accessible medical education. For instance, the University of Lagos offers an online MPH program, which provides students with the skills and knowledge they need to work in pharmacy and improve the health of communities.

Ward et al. (2003) reported that Southeastern University of the Health Sciences had developed a North Miami site to service its nontraditional, post-baccalaureate, Doctor of Pharmacy (PharmD) program in 1991. Due to the students' inability to disrupt their careers and relocate for the purpose of proximity, program enrollment numbers were low. As a result of the request by the American Council on Pharmaceutical Education to maintain nontraditional PharmD programs, the university's existing infrastructure, and NSU's (Nova University and Southeastern University) mission to utilize technologically innovative delivery systems to provide convenient, high-quality education, a distance learning format was needed. Compressed videoconferencing was chosen over satellite broadcast as the delivery system for the PharmD program for 2 essential reasons: its ability to conduct high quality, live, two-way interaction at a much lower cost and NSU's previous success with this technology in the education department.

HISTORY OF ONLINE COMPLEMENTARY AND ALTERNATIVE (CAM) EDUCATION

The history of online CAM education reflects the broader evolution of both CAM practices and online education. Understanding this history involves looking at the development of CAM as a field, the growth of online learning, and how these two trajectories intersect.

1. **Early Development of CAM**: CAM has roots in ancient and traditional practices from various cultures worldwide. However, it wasn't until the late 20th century that these practices started gaining significant attention in Western countries. This growing interest was partly due to a desire for more holistic and patient-centered approaches to healthcare, as well as disillusionment with certain aspects of conventional medicine.
2. **Rise of CAM in Mainstream Healthcare**: By the 1990s and early 2000s, CAM began to gain more mainstream acceptance in Western countries. This period saw an increase in research, publications, and discussions about CAM in medical circles. Simultaneously, there was a growing consumer demand for information and education on CAM practices, not just for professionals but also for individuals seeking alternative or complementary health approaches.
3. **Advent of Online Learning**: The late 1990s and early 2000s also witnessed the rapid development of online learning technologies. The Internet revolution made it possible to disseminate information widely and quickly. Educational institutions started exploring online courses as a way to reach a larger audience and provide flexible learning options. This technological advancement was crucial for the development of online CAM education.
4. **Emergence of Online CAM Education**: As the interest in CAM grew, so did the need for structured educational programs. Initially, CAM education was primarily offered through in-person workshops, seminars, and at specialized institutions. However, with the expanding capabilities of online learning, educational providers saw an opportunity to cater to a global audience. This led to the creation of online courses and programs focusing on various CAM modalities.
5. **Expansion and Diversification**: The 2010s saw a significant expansion in online CAM education. Numerous universities, colleges, and specialized institutes began offering courses and degree programs online. These programs ranged from introductory courses in CAM to specialized training in areas like naturopathy, herbal medicine, acupuncture, and mind-body therapies. The online format made CAM education more accessible to a diverse group of learners, including healthcare professionals, therapists, and individuals seeking personal health and wellness knowledge.
6. **Integration and Accreditation Challenges**: As online CAM education grew, issues related to quality control, accreditation, and integration with conventional medical education emerged. Recognizing the importance of these programs, various accreditation bodies and professional organizations began establishing standards to ensure the quality and efficacy of online CAM education.

The COVID-19 pandemic brought unprecedented challenges and transformations to many sectors, including education and healthcare. The field of online CAM education experienced significant changes and growth during this period and going forward.

1. **Increased Demand for CAM**: The pandemic heightened interest in health, wellness, and immune system support. People worldwide became more interested in holistic and preventive health approaches, which drove a surge in demand for CAM practices. This increased public interest naturally led to a greater demand for education and knowledge in these areas.

2. **Transition to Online Platforms**: With social distancing measures and lockdowns in place, educational institutions, including those specializing in CAM, were forced to pivot rapidly to online learning. For many CAM educational programs, this accelerated a trend that was already underway—the transition from in-person to digital learning environments. Institutions that previously offered limited online options expanded their digital offerings, while those already in the online space refined and improved their courses.

3. **Innovation in Online Learning**: The pandemic era was marked by rapid innovation in online learning technologies. CAM education platforms began incorporating more interactive and immersive technologies, like live webinars, virtual reality (VR), and augmented reality (AR) to enhance the learning experience. These technologies helped in delivering practical aspects of CAM education, which traditionally relied heavily on in-person, hands-on experiences.

4. **Focus on Mental Health and Well-being**: The pandemic caused a significant increase in mental health issues. This led to a greater emphasis within CAM education on practices that support mental and emotional well-being, such as mindfulness, meditation, yoga, and stress management techniques. Online CAM courses started incorporating these elements more prominently, reflecting the broader societal focus on mental health.

5. **Global Reach and Accessibility**: The shift to online education made CAM courses more accessible to a global audience. People from different parts of the world could now access a variety of courses that were previously limited by geographic location. This democratization of education helped in spreading CAM knowledge and practices to a wider audience.

6. **Challenges and Adaptations**: Despite the benefits, the shift to online learning also presented challenges. Ensuring the quality of education, providing practical hands-on experience, and maintaining student engagement in a virtual environment were significant hurdles. CAM educational institutions adapted by developing new pedagogical strategies, incorporating hybrid models (combining online and in-person elements), and enhancing student support systems.

7. **Post-Pandemic Trends**: As the world transitions into the post-pandemic phase, online CAM education is likely to retain its prominence. The convenience, flexibility, and broad reach of online learning are too significant to be overshadowed by the return to in-person options. Hybrid models of education that combine the best of online and in-person learning are expected to become more prevalent. Additionally, there will likely be a continued focus on integrating CAM education with conventional medical training, reflecting a more holistic approach to health and wellness.

In conclusion, the COVID-19 pandemic served as a catalyst for significant growth and transformation in online CAM education, trends that are expected to continue and evolve in the post-pandemic world.

ADVANTAGES AND DISADVANTAGES OF ONLINE COMPLEMENTARY AND ALTERNATIVE (CAM) EDUCATION

The importance of distance and online education has increased during the last decades, especially after 2019. Online education and learning had/has great impact on the continuity of teaching and learning cycles worldwide; saved the medical and health sciences schools from closing for at least one year because of the lockdown after the COVID-19 pandemic; nowadays, technologies such as online education platforms, social media, mobile technologies can help to achieve the medical and health sciences programs learning objectives/outcomes and provide efficient and convenient ways to achieve learning goals for programs. Online medical education postgraduate programs are very important for working healthcare professionals; they can save money and time and achieve their degrees while they are working as it will be flexible, easy, and convenient for them. Online medical education courses can help registered healthcare professionals obtain the required continuous medical hours (CMEs), without being absent from their work as they can take it during their weekends, and off-hours at less cost. Online medical and health sciences courses and programs can help medical and health sciences schools worldwide to increase their income with less cost. One of the key benefits of online education is accessibility. Online courses can be taken from anywhere with an internet connection, making education more accessible to people in remote or rural areas. Online education also provides more flexibility in terms of scheduling, allowing students to complete coursework at their own pace and on their own schedule. This makes it easier for working individuals or those with family commitments to pursue their education. Another advantage of online education is cost. Online courses are often less expensive than traditional, on-campus courses, and many online programs offer scholarships or financial aid to help students pay for their education. Additionally, students can save money on transportation, housing, and other expenses associated with traditional on-campus programs. Despite its many benefits, online education has faced criticism and skepticism. Some argue that online courses lack the interaction and community found in traditional classroom environments and that online students miss out on the benefits of face-to-face interaction with instructors and peers. Additionally, some are concerned about the quality and rigor of online courses, particularly when it comes to degree programs offered by for-profit institutions. In response to these concerns, many institutions and organizations have worked to improve the quality of online education and make it a more effective and engaging learning experience. For example, some universities have developed hybrid programs that combine online coursework with in-person classes or hands-on experiences, offering students the best of both worlds (Kusmaryono et al., 2021). The future of online education is promising, and advancements in technology are likely to continue to shape the landscape of education. For example, the use of artificial intelligence and machine learning is expected to play a larger role in online education, providing students with personalized and adaptive learning experiences. Additionally, VR and AR are becoming increasingly popular in education and are expected to offer students new and innovative ways to learn. In conclusion, the development of online education has been driven by advancements in technology and a growing demand for flexible and accessible learning experiences. Despite some criticism, online education has become a popular and effective way to pursue a degree or advance one's career, providing students with access to a high-quality education regardless of their location or schedule. As technology continues to evolve, it is likely that online education will become even more accessible, flexible, and engaging, offering students new and exciting opportunities to learn and grow.

ADVANTAGES OF ONLINE COMPLEMENTARY AND ALTERNATIVE (CAM) EDUCATION

There are many advantages of online CAM education as follows (Al-Worafi, 2022a,c, 2024a):

SUPPORTING THE CONTINUITY OF TEACHING AND LEARNING

Online education provides innovative and resilient solutions in times of crisis to medical education worldwide. The majority of medical and health sciences schools/departments worldwide have adapted and implemented online education successfully with the help of technologies and new technologies, which saved the teaching and learning process during the COVID-19 lockdown.

Otherwise, medical education would have had to stop for one year or more because of a pandemic. However, many low-income countries couldn't adapt, or implement online education due to many reasons.

COST SAVING

Medical and health sciences schools can save money with online teaching as well as the students with online learning. Savings on indirect costs such as cost of transportation, accommodation, actual laboratory sessions, and other costs can help medical and health sciences schools budgets as well as students.

CONVENIENT

Online medical education from home can be convenient for many educators and students.

IMPROVE TECHNOLOGY KNOWLEDGE

Online medical education is a good opportunity for medical educators and students to improve their knowledge toward technologies, new technologies.

IMPROVE TECHNOLOGY SKILLS

Online medical education is a good opportunity for medical educators and students to improve their skills toward technologies, new technologies.

SELF-LEARNING

Online medical education is a good opportunity for medical students to improve their self-learning related skills to be good healthcare professionals with lifelong learning skills.

ACCESSIBLE AND AFFORDABLE

Online medical education is a good opportunity for those working or who can't travel to obtain their postgraduate studies, a good example of this is the online master's degrees.

SCHEDULE

Online medical education helps medical schools arrange the schedule with more flexibility with time, especially for laboratories and tutorials.

SAFETY

Online medical education is safer than face-to-face education, medical educators and students teach and learn from their homes, which helps in reducing contact and the management and control of COVID-19 for example. There are no laboratories which means there is no risk for any potential risk due to exposure to chemicals and others.

INFRASTRUCTURE

Online medical education requires less infrastructure than traditional education (Al-Worafi, 2022d).

TEACHING STRATEGIES

Online medical education is a good opportunity to apply and practice more active teaching strategies.

ASSESSMENT & EVALUATION METHODS

Online medical education is a good opportunity to apply and practice online assessment and evaluation methods, explore the barriers, and make solutions.

COMMUNITY SERVICES

Online medical education is a good opportunity to apply and practice online community services for the public and patients.

ENVIRONMENT

Online medical education is beneficial to the environment.

WORKFORCE ISSUES

Medical and health sciences schools can hire many international faculty members as part-time, visiting and adjunct professors, and lecturers easier.

COMMUNICATION AND COLLABORATION

Online medical education is a good opportunity to practice online/virtual communication and collaboration, which improves the skills of communication and collaboration.

TIME MANAGEMENT

Online medical education benefits educators and students related to time management.

DISADVANTAGES & PROBLEMS OF ONLINE COMPLEMENTARY AND ALTERNATIVE (CAM) EDUCATION

Online CAM education has many disadvantages and problems as follows (Al-Worafi, 2022a,c):

STUDENTS' ENGAGEMENT

A number of students faced the problem of engagement due to many causes such as home size and others.

LACK OF INFRASTRUCTURE AND RESOURCES

Many medical and health sciences schools around the world, especially in low-income and middle-income countries faced problems of lack of infrastructure and resources such as:

INTERNET

The Internet plays a very important and vital role in online medical education. Internet facilitates the teaching and learning process which makes distance/online medical education more effective and easier than at any time in history. Medical educators and students need the Internet to make communication easy; deliver classes; upload/download lecture notes and other educational materials/resources; for exams; assignments; presentations; search for medical-related information; training; patient care services; and health promotion, awareness, and services. Without access to the internet, online teaching and learning will stop. Many medical educators and students, especially in low-income and middle-income countries face this problem.

COMPUTERS AND LAPTOPS

Using Computers and laptops is very important and essential for online teaching and learning. It provides flexible and effective access to online teaching and learning for medical educators and students. Many medical educators and students, especially in low-income and middle-income countries face this problem.

SMARTPHONES, TABLETS, AND NETBOOKS

Using smartphones, tablets, and netbooks is very important and essential for online teaching and learning. It provides flexible and effective access to online teaching and learning for medical educators and students. It helps students to download many educational resources and others. Furthermore, it allows access at any time throughout the day, inside or outside the home. It also helps educators to prepare, revise, and access lecture notes, literature, educational resources, and others at any time and any place. Many medical educators and students, especially in low-income and middle-income countries face this problem.

ONLINE & DIGITAL LIBRARY

An online and digital library is a collection of documents such as journal articles, books, and other educational resources organized in an electronic form and available on the Internet for students, educators, and other staff. Furthermore, it gives access to databases which helps students, educators, and staff to access the latest volumes/issues of scientific journals. Many medical educators and students, especially in low-income and middle-income countries face this problem.

LACK OF/POOR COMMUNICATION SKILLS

Many medical educators as well as students have this problem.

LACK OF/POOR TECHNOLOGY USE SKILLS

Many medical educators as well as students have this problem.

ASSESSMENT AND EVALUATION

There are many problems related to assessment and evaluation such as plagiarism, cheating, and others.

WORKLOAD

Many medical educator's workloads have been increased due to many reasons such as more activities, more meetings, and others.

LACK OF/ INSUFFICIENT TRAINING

Many medical educators as well as students faced this problem.

LACK OF/ INSUFFICIENT TECHNICAL SUPPORT

Many medical educators as well as students faced this problem.

QUALITY OF ONLINE EDUCATION

This is a problem in many countries around the world, especially in low-income and middle-income countries.

SELF-MOTIVATION

Many medical educators as well as students faced this problem.

DISTRACTION

Many medical educators as well as students faced this problem.

ATTITUDE

Many medical educators as well as students have a negative attitude toward online learning.

ONLINE COMPLEMENTARY AND ALTERNATIVE (CAM) EDUCATION: TEACHING STRATEGIES

There are many effective teaching strategies that can be used in online CAM education such as (Al-Worafi, 2022e–u):

INTERACTIVE LECTURE-BASED TEACHING STRATEGY

Despite the revolution in the teaching strategies and many active teaching strategies having been developed, and implemented successfully in medical education around the world, lecture-based teaching strategy remains the backbone of medical education and the preferred teaching strategy among medical educators in many countries. However, medical educators can make the online lecture interactive in many ways such as asking students questions every five to ten minutes, presenting short videos/audio, rallying, and team-sharing groups that will make students attracted to the lectures. By linking the theory part with life, sharing one's practice experiences with students with mini and long cases, and giving students time to think about it, and solve it. Engage all students while remembering that many students may be hesitant to participate and encouraging them all to participate. Remember as a medical educator that you are teaching students, assessing their needs, and understanding them can help also. Weekly and monthly feedback from students and colleagues can improve online teaching. Recording the lectures and giving them to students as well as to oneself and colleagues. Feedback is the key to success in teaching theory online.

ONLINE BLENDED TEACHING STRATEGY

Traditionally, blended learning combines online educational materials and opportunities for interaction online with traditional place-based classroom methods. It requires the physical presence of both teacher and student, with some elements of student control over time, place, path, or pace. However, as a result of the lockdown during the COVID-19 pandemic, the face-to-face part required to be modified to be online with the help of new technologies such as mobile technologies, telecommunication, and online meetings that help to teach theory online.

TEAM-BASED LEARNING (TBL) TEACHING STRATEGY

Team-based learning (TBL) has been implemented successfully in medical education teaching in many developed countries and developing countries. Therefore, implementing and redesigning team-based learning with the help of new technologies can be an effective online teaching strategy for theory.

PROBLEM-BASED LEARNING (PBL) TEACHING STRATEGY

Problem-based learning (PBL) has been implemented successfully in medical education teaching in many developed countries and developing countries. Therefore, redesigning problem-based learning with the help of new technologies such as online peer-to-peer platforms can be an effective online teaching strategy for theory.

VIDEO-BASED LEARNING

Short videos can be used as an effective online teaching strategy for theory.

SIMULATION

Role play and other simulation methods can be used with the help of new technologies as an effective online teaching strategy for theory.

PROJECT-BASED LEARNING

Project-based learning is a model in which students use online methods for assignments and other projects and this can be used as an effective online teaching strategy for theory.

JOURNAL CLUB

To critically evaluate recent articles in the academic literature can be used as an effective online teaching strategy for theory.

CASE STUDIES DISCUSSION

It is a very important and effective online teaching strategy, to encourage students to read the given cases individually or as teams and solve them.

SELF-DIRECTED LEARNING

This allows students to improve their skills toward self-learning.

FLIPPED TEACHING

This is an effective strategy and can be used online, for example, to ask students to watch videos related to the course and to summarize it in a report.

COMMUNITY SERVICES-BASED LEARNING

Many theory courses can be used with this effective teaching strategy, which allows students to achieve the course learning outcomes while contributing to the patients, the public, and society.

SEMINARS

An effective strategy and can be used online to improve the students' presentation skills.

ONLINE COMPLEMENTARY AND ALTERNATIVE (CAM) EDUCATION: ASSESSMENT AND EVALUATION METHODS

There are many direct and indirect assessment methods that can be used in online CAM education programs and courses (Al-Worafi, 2022a,v,w).

DIRECT ASSESSMENT METHODS AND ITS APPLICATION IN THE ONLINE COMPLEMENTARY AND ALTERNATIVE (CAM) EDUCATION

There are many direct assessment methods that can be used in CAM education programs/courses such as follows:

PUBLISHED/STANDARDIZED TESTS

Published tests are published or standardized tests are instruments that have been commercially published by a test publisher. These instruments are administered and scored in a consistent, or "standard" manner. The validity and reliability of the instrument are two essential elements for defining the standard quality of the test. These tests are generally only available from the publisher and often come in the form of kits or multiple booklets. They can be very costly if purchased.

LOCALLY DEVELOPED TESTS

Faculty may decide to develop their own internal tests that reflect the program's learning outcomes such as the multiple-choice Questions (MCQs) and other questions developed by faculty for CAM courses.

EMBEDDED ASSIGNMENTS AND COURSE ACTIVITIES

Embedded assignments and course activities are assignments, activities, or exercises that are done as part of a class, but that are used to provide assessment data about a particular learning outcome.

PORTFOLIOS

A portfolio can be generally defined as "a purposeful collection of student work that exhibits the student's efforts, progress, and achievements in one or more areas. The collection must include student participation in selecting contents, the criteria for selection, the criteria for judging merit, and evidence of student self-reflection" (Paulson et al., 1991).

INDIRECT ASSESSMENT METHODS AND ITS APPLICATION IN ONLINE COMPLEMENTARY AND ALTERNATIVE (CAM) EDUCATION

There are many indirect assessment methods that can be used in CAM education programs/courses such as:

SURVEYS

A survey is an examination of opinions, behavior, etc., made by asking people questions (Cambridge Dictionary).

> Point-of-contact surveys
> Online, e-mailed, registration, or grad check surveys
> Keep it simple!

EXAMPLES OF SURVEYS IN CAM EDUCATION ARE

Surveys and Questionnaires to stakeholders to ask them about the graduates.
Surveys and Questionnaires to (students, alumni, employers, public) about any issue.

INTERVIEWS

- Interviews can be conducted one-on-one, in small groups, or over the phone.
- Interviews can be structured (with specified questions) or unstructured (a more open process).
- Questions can be close-ended (e.g., multiple-choice style) or open-ended (respondents construct a response).
- Can target students, graduating seniors, alumni, employers, community members, faculty, etc.
- Can do exit interviews or pre-/post-interviews.
- Can focus on student experiences, concerns, or attitudes related to the program being assessed.
- Generally should be conducted by neutral parties to avoid bias and conflict of interest.

FOCUS GROUPS

- Traditional focus groups are free-flowing discussions among small, homogeneous groups (typically from six to ten participants), guided by a skilled facilitator who subtly directs the discussion in accordance with pre-determined objectives. This process leads to in-depth responses to questions, generally with full participation from all group members. The facilitator departs from the script to follow promising leads that arise during the interaction.
- Structured group interviews are less interactive than traditional focus groups and can be facilitated by people with less training in group dynamics and traditional focus group methodology. The group interview is highly structured, and the report generally provides a few core findings, rather than an in-depth analysis.

OBJECTIVE STRUCTURED CLINICAL EXAMINATION (OSCE)

The objective structured clinical examination (OSCE) is a method of assessing a student's clinical competence which is objective rather than subjective, and in which the areas tested are carefully planned by the examiners. OSCE is an approach to the assessment of clinical competence in which the components of competence are assessed in a planned or structured way with the attention being paid to the objectivity of the examination. (Harden and Gleeson, 1979). The examination consists of multiple, standard stations at which students must complete one to two specific clinical tasks, often in an interactive environment involving patient actors such as standardized patients (Harden and Gleeson, 1979).

TECHNOLOGIES AND TOOLS IN THE ONLINE COMPLEMENTARY AND ALTERNATIVE (CAM) EDUCATION

Suitable and effective technologies and tools are the keys to success in online medical education. Medical and health sciences schools/departments should use and adapt the most effective technologies and tools for teaching and learning, and prepare the educators and students with all required technologies as follows (Al-Worafi, 2022a,d).

INTERNET

The Internet plays a very important and vital role in online medical education. Internet facilitates the teaching and learning process which makes distance/online education more effective and easier than at any time in history. Medical educators and students need the internet for to make communication easy; deliver classes; upload/download lecture notes and other educational materials/resources; for exams; assignments; presentations; search for CAM-related information; training; patient care services; and health promotion, awareness, and services.

COMPUTERS AND LAPTOPS

Using Computers and laptops is very important and essential for online teaching and learning. It provides flexible and effective access to online teaching and learning for medical educators and students.

SMARTPHONES, TABLETS, AND NETBOOKS

Using smartphones, tablets, and netbooks is very important and essential for online teaching and learning. It provides flexible and effective access to online teaching and learning for medical educators and students. It helps students to download many educational resources and others. Furthermore, they can access it any time throughout the day, inside or outside the home. It also helps educators to prepare, revise, and access lecture notes, literature, educational resources, and others at any time and any place.

LEARNING MANAGEMENT SYSTEMS (LMS)

LMSs are very important in online medical education as well as higher education. They contain an effective web-based learning system of sharing study materials, making announcements, conducting evaluations and assessments, generating results, and communicating interactively in synchronous and asynchronous ways among various other academic activities (Kant et al., 2021; Bervell and Umar, 2017).

MOODLE

Moodle is an open-source LMS that provides collaborative learning environments that empower learning and teaching. It is a flexible and user-friendly platform adopted by most educational institutions and businesses of all sizes.

BLACKBOARD

Blackboard is the most popular LMS used by businesses and educational institutes worldwide, which delivers a powerful learning experience. It is easily customizable according to your organization's needs. It provides advanced features and integrates with Dropbox, Microsoft OneDrive, and school information systems. Moodle and Blackboard are two of the most famous and widely known LMSs in online education and help educators as well as students in the teaching and learning process (Momani, 2010). Momani (2010) compared the two known LMSs in terms of the pedagogical factor, learner environment, instructor tools, course and curriculum design, administrator tools, and technical specifications and reported that they have lots in common, but also have some key differences that make each one special in its own way.

WEBINAR AND VIDEO CONFERENCING PLATFORMS

Webinar & video conferencing platforms are very important in online education. Microsoft Teams, Cisco WebEx Teams, Google Meet, and Zoom are the most common Webinar and video conferencing platforms in online education (https://www.microsoft.com/en-us/microsoft-teams/group-chat-software; https://www.webex.com/; https://apps.google.com/meet/; https://www.zoom.us/).

Microsoft Teams, Cisco WebEx Teams, Google Meet, and Zoom offer multiple versions of their software based on usage requirements. This includes free versions that are great for light uses, short conference calls, and light file sharing. However, universities have paid the cost of all platforms for educators and students to facilitate teaching and learning. All platforms are used successfully in online education worldwide. Training, workshops, and writing manuals are very important for both medical educators and students in order to use these tools successfully.

YOUTUBE

YouTube can be used for streaming to the audience and to have a one-way video interaction. YouTube can be a good resource for many CAM-related educational videos. Medical educators can use YouTube for recording lectures and share it with students on YouTube where CAM students can easily access it any time, download it if needed, as well as share it with their colleagues.

FACEBOOK, TWITTER, AND INSTAGRAM

Facebook, Twitter, and Instagram can be used in online medical education in many ways as follows: allows students to interact with their mentors, access their course contents, customization and build student communities, transfer the resource materials, collaborative learning, and interaction with the colleagues as well as teachers would facilitate students to be more enthusiastic and dynamic; share health information, discuss clinical cases and listen to patient stories (Wong et al., 2019; Shafer et al., 2018; Gulati et al., 2020).

WHATSAPP

WhatsApp can be used in online education teaching and learning in many ways as follows: as a communication tool between educators and students; to share lecture notes and other educational resources; for group chats, and others (Coleman and O'Connor, 2019; Raiman et al., 2017; us Salam et al., 2021).

WEARABLE TECHNOLOGIES

Wearable technologies have a tremendous potential to improve online education, empowering students as well as instructors in their teaching and learning experiences (Borthwick et al., 2015; Subrahmanyam and Swathi, 2020). Wearable technologies can be used in online education for the following potential reasons: improved student engagement; convenience to wear as they are hands-free gadgets; effortless communication with enhanced features; facility to record videos;

teaching through AR; Teaching through VR, and wearables - learning apps (Subrahmanyam and Swathi, 2020).

ONLINE AND DIGITAL LIBRARY

An Online and digital library is a collection of documents such as journal articles, books, and other educational resources organized in an electronic form and available on the Internet for students, educators, and other staff. Furthermore, access to these databases helps students, educators, and staff access the latest volumes/issues of scientific journals.

ONLINE COMPLEMENTARY AND ALTERNATIVE (CAM) EDUCATION: ACHIEVEMENTS

The field of online CAM education has achieved several notable milestones, particularly in recent years. These achievements have contributed to the broader acceptance and integration of CAM practices in healthcare and have empowered individuals worldwide with knowledge and skills related to holistic health and wellness.

1. **Wider Accessibility and Global Reach**: One of the most significant achievements of online CAM education is its accessibility. The digital format has made it possible for people from all over the world, regardless of location, to access quality education in various CAM modalities. This global reach has not only spread the knowledge of CAM practices but has also facilitated a cross-cultural exchange of health and wellness traditions.
2. **Diverse Educational Offerings**: The range of courses and programs available online is vast and diverse, covering a wide array of CAM practices. From certificate courses in herbalism or aromatherapy to comprehensive degree programs in naturopathy, acupuncture, or integrative health, online CAM education caters to a wide spectrum of interests and professional needs.
3. **Integration with Conventional Medicine Education**: There's a growing trend of integrating CAM education with conventional medical training. This achievement is crucial for the collaborative future of healthcare. Many online programs now offer courses that combine traditional medical sciences with CAM practices, promoting a more holistic approach to health and wellness.
4. **Advancements in Online Learning Technologies**: Online CAM education has leveraged advancements in technology to enhance learning experiences. Interactive modules, virtual reality simulations for practices like acupuncture, and live-streamed interactive webinars have made online learning more engaging and effective.
5. **Research and Evidence-Based Focus**: Many online CAM programs have emphasized evidence-based practices, integrating current research into their curricula. This focus has helped legitimize CAM practices in the eyes of the broader medical community and the public, fostering a more critical and informed approach to CAM.
6. **Professional Development and Networking Opportunities**: Online CAM education platforms often provide more than just academic knowledge; they offer professional development resources, career services, and networking opportunities. This aspect is crucial for students and practitioners looking to establish themselves in the CAM field.
7. **Regulatory and Accreditation Advancements**: There has been progress in the standardization and accreditation of online CAM programs. This development is vital for ensuring the quality and credibility of CAM education and for protecting the public by ensuring practitioners are adequately trained and knowledgeable.

8. **Promotion of Personal Health and Wellness**: Beyond professional training, online CAM education has played a significant role in promoting personal health and wellness. Many individuals turn to these resources for self-care, preventive health strategies, and to complement their conventional medical care.

9. **Responding to Public Health Challenges**: Online CAM education has shown agility in responding to public health needs, such as the mental health crisis exacerbated by the COVID-19 pandemic. Programs focusing on stress reduction, mental wellness, and immune support have been particularly impactful.

In conclusion, online CAM education has made significant strides in making holistic health education accessible, integrating with conventional healthcare, leveraging technology for enhanced learning, and contributing to public health and wellness. These achievements have not only benefited those directly involved in CAM practices but have also had a broader impact on healthcare and wellness communities globally.

CHALLENGES OF ONLINE COMPLEMENTARY AND ALTERNATIVE (CAM) EDUCATION

Online CAM education, while having made significant strides, faces a range of challenges that affect its efficacy, credibility, and overall impact. These challenges are multifaceted, stemming from the nature of CAM practices, the evolving landscape of online education, and broader issues in healthcare and regulatory environments.

1. **Maintaining Practical Skill Development**: Many CAM practices, such as acupuncture, massage therapy, or hands-on energy work, require practical, hands-on experience. Translating these skill-based components effectively into an online format is challenging. While technology like virtual reality can help, it may not fully replicate the nuances of in-person practice.

2. **Quality Control and Accreditation**: Ensuring the quality and standardization of online CAM courses and programs is a significant challenge. The CAM field includes a broad spectrum of practices with varying levels of scientific evidence and acceptance. Establishing universally accepted accreditation standards is complex, and the variability in quality can lead to skepticism and credibility issues.

3. **Balancing Traditional Knowledge and Scientific Rigor**: CAM education often involves traditional knowledge passed down through generations. Balancing this with the requirements of scientific rigor and evidence-based practice in a way that respects both aspects is challenging. This balance is crucial for CAM's integration into mainstream healthcare.

4. **Technological Barriers**: While online education offers greater accessibility, technological barriers such as limited internet access, lack of digital literacy, and the cost of technology can limit access for some students, especially in low-income or rural areas.

5. **Regulatory and Legal Issues**: The legal and regulatory landscape for CAM practices varies significantly across countries and regions. This variance poses a challenge for online CAM education providers in terms of ensuring that their curriculum is relevant and compliant with the regulations of different regions.

6. **Cultural and Ethical Considerations**: CAM practices often have deep cultural roots. Online education must navigate these cultural aspects sensitively and ethically. This includes respecting intellectual property rights and cultural heritage associated with traditional medicines and practices.

7. **Interprofessional Education and Collaboration**: Encouraging collaboration between CAM and conventional healthcare professionals is vital but challenging. There's a need for

more interprofessional education that fosters mutual understanding and respect between different health disciplines.

8. **Student Engagement and Motivation**: Like all online education, keeping students engaged and motivated in a virtual learning environment can be challenging. Without the physical presence of a classroom, students may feel isolated or less accountable, which can impact their learning experience.

9. **Evolving Public Perception and Expectations**: Public perception of CAM is constantly evolving. Online CAM education providers must continuously adapt their curriculum to meet changing expectations and new health trends while maintaining scientific credibility and educational integrity.

10. **Economic and Business Models**: Developing sustainable economic models for online CAM education is challenging, especially when balancing affordability for students and the need for resources to provide high-quality education.

11. **Adapting to Rapid Technological Changes**: The field of online education is rapidly evolving, with new technologies emerging constantly. Keeping pace with these changes, such as incorporating AI, advanced simulations, or new interactive platforms, requires ongoing investment and adaptation. CAM education providers must continuously update their technological capabilities to provide an effective learning experience.

12. **Ensuring Academic Rigor and Credibility**: There is a critical need to ensure that online CAM programs are academically rigorous and credible. This involves not only incorporating scientific and evidence-based content but also employing qualified instructors, using robust teaching methodologies, and providing comprehensive assessments. Striking a balance between traditional CAM wisdom and contemporary academic standards can be challenging.

13. **Student Assessment and Evaluation**: Effectively assessing and evaluating students in an online environment, especially for practical skills, remains a challenge. Traditional testing methods may not fully capture the competencies required in CAM practices, and developing new assessment methods that are both effective and fair is an ongoing challenge.

14. **Coping with Diverse Learning Styles and Needs**: Students come with diverse learning styles, cultural backgrounds, and educational needs. Creating an online learning environment that caters effectively to this diversity is challenging, particularly when cultural nuances play a significant role in the understanding and practice of CAM therapies.

15. **Clinical Training and Internships**: For many CAM disciplines, clinical training or internships are essential components of education. Organizing these experiences in a virtual format is nearly impossible, and finding local opportunities that align with the online curriculum can be challenging for students and educators alike.

16. **Ethical and Legal Implications of Online Consultations**: As telemedicine becomes more prevalent, CAM practitioners trained online may engage in virtual consultations. Navigating the ethical and legal implications of such consultations, especially considering the non-conventional nature of many CAM therapies, is a complex challenge.

17. **Addressing Misinformation**: The internet is rife with misinformation about health and wellness, and CAM is not immune to this. Online CAM education providers have the additional challenge of combating myths and misinformation, ensuring that their content is accurate and evidence-based.

18. **Financial Sustainability**: Developing a financially sustainable model for online CAM education is challenging, especially for institutions that strive to keep tuition affordable while maintaining high-quality education and resources.

19. **Globalization vs. Localization**: While online education has a global reach, CAM practices are often deeply rooted in local traditions and ecosystems. Balancing a globalized curriculum with localized practices and knowledge is a nuanced challenge.

20. **Research and Development in CAM**: There's a need for more research and development in the field of CAM to enhance the content and credibility of online education. However, securing funding and support for research in CAM is often challenging due to its alternative status in the broader medical and scientific community.

Addressing these challenges requires innovative solutions, collaboration across disciplines, and ongoing dialogue between educators, students, practitioners, regulators, and other stakeholders in the healthcare field.

RECOMMENDATIONS FOR BEST PRACTICE IN ONLINE COMPLEMENTARY AND ALTERNATIVE (CAM) EDUCATION

To enhance the effectiveness and credibility of online CAM education, best practices should be implemented. These recommendations aim to address the inherent challenges of online learning, ensure quality education, and maintain the integrity of CAM practices:

1. **Develop Robust and Interactive Curriculum**: The curriculum should be comprehensive, integrating both theoretical knowledge and practical skills. The use of interactive tools like videos, virtual simulations, and interactive discussions can enhance the learning experience, particularly for practical skills that are harder to convey online.
2. **Incorporate Evidence-Based Practice**: Aligning CAM education with evidence-based practices is crucial. Courses should include current research, encourage critical thinking, and teach students to discern between well-supported and less credible CAM practices.
3. **Ensure Qualified Faculty**: Instructors should be well-qualified and experienced both in their CAM specialty and in online teaching methodologies. This ensures that the material is taught effectively and credibly.
4. **Promote Interprofessional Education**: Encourage collaboration and understanding between CAM and conventional healthcare practices. This can be achieved by incorporating interprofessional education in the curriculum, where students learn about the roles and expertise of different health professionals.
5. **Adopt Hybrid Models for Hands-On Training**: For disciplines that require hands-on skills, adopt a hybrid model combining online theory classes with in-person workshops or internships. This approach provides students with the necessary practical experience.
6. **Regularly Update Course Content**: CAM is a dynamic field with continuous research and developments. Regularly updating course materials ensures that students are learning the most current and relevant information.
7. **Implement Rigorous Assessment Methods**: Develop and utilize effective online assessment methods to evaluate both theoretical knowledge and practical skills. Consider using a mix of quizzes, assignments, project work, and virtual demonstrations to assess student performance.
8. **Ensure Accessibility and Inclusivity**: Make courses accessible to people with varying levels of technological access and proficiency. Additionally, consider cultural, linguistic, and physical accessibility in course design.
9. **Uphold Academic Integrity**: Implement measures to maintain academic integrity, such as using plagiarism detection software and proctoring tools for exams. It's essential to foster an environment of honesty and professionalism.
10. **Provide Strong Student Support Services**: Offer robust support services including technical assistance, academic advising, and career counseling. This helps in addressing student needs and enhances the overall learning experience.

11. **Engage in Continuous Quality Improvement**: Regularly solicit feedback from students, faculty, and industry professionals to continually improve the quality of the program. This can involve surveys, focus groups, and an ongoing review of educational outcomes.

12. **Cultivate a Community of Practice**: Create online forums, discussion boards, or social media groups where students, alumni, and faculty can interact, share experiences, and network. This builds a sense of community and provides ongoing professional support.

13. **Navigate Legal and Ethical Considerations**: Stay informed and compliant with the legal and ethical standards pertaining to online education and CAM practice. This includes respecting intellectual property, ensuring privacy in online interactions, and adhering to regulatory requirements.

14. **Promote Research and Critical Inquiry**: Encourage students to engage in research and critical inquiry as part of their education. This not only enriches their learning experience but also contributes to the broader CAM knowledge base.

15. **Adapt to Technological Advancements**: Stay abreast of technological advancements in online education and incorporate these tools to enhance the learning experience and operational efficiency.

16. **Foster Global and Cultural Competence**: Given the diverse origins of many CAM practices, it's important to educate students about the cultural, historical, and ethical contexts of these practices. This includes respecting and acknowledging traditional knowledge and practices and avoiding cultural appropriation.

17. **Collaborate with Professional Organizations**: Establish partnerships with CAM professional organizations and regulatory bodies. This collaboration can help in aligning educational content with industry standards, facilitating accreditation processes, and providing students with opportunities for professional development and certification.

18. **Utilize Blended Learning Approaches**: Combine synchronous (real-time) and asynchronous (self-paced) learning methods. This blended approach caters to different learning styles, allows for more flexibility, and can improve student engagement and learning outcomes.

19. **Promote Lifelong Learning**: Encourage students to view their education as a starting point for lifelong learning in CAM. This can involve providing resources for continuing education, workshops, and seminars, and encouraging alumni to stay connected and informed.

20. **Implement Scalable and Sustainable Models**: Design online CAM programs in a way that they are scalable and sustainable. This involves considering the long-term financial viability of the program, the scalability of teaching resources, and the environmental impact of the educational activities.

21. **Address Misinformation and Ethics**: Teach students to critically evaluate CAM information sources, particularly online, and to address misinformation responsibly. Emphasize ethical considerations in all aspects of CAM practice and education.

22. **Incorporate Mental Health and Well-being**: Given the growing focus on mental health, include modules on mental well-being, stress management, and self-care practices within CAM curricula.

23. **Customize Content for Target Audiences**: Tailor course content for different target audiences, such as healthcare professionals, CAM practitioners, or the general public. This customization ensures that the material is relevant and appropriate for the student's needs and backgrounds.

24. **Promote Research Literacy**: Develop students' skills in understanding, interpreting, and conducting research. This helps in fostering an evidence-based approach to CAM practices and contributes to the field's overall body of knowledge.

25. **Ensure Technological Reliability and Security**: Invest in reliable and secure technological infrastructure. This is crucial for protecting student information, delivering content
effectively, and ensuring that the online learning environment is stable and accessible.
26. **Practice Inclusivity in Health Perspectives**: Acknowledge and include various health
perspectives and models in the curriculum. This broadens students' understanding of
health and wellness and how CAM practices can be integrated into diverse health systems.
27. **Emphasize Practical Application and Case Studies**: Incorporate real-world scenarios
and case studies to provide practical applications of CAM principles. This helps students
understand how CAM is applied in different health contexts.
28. **Develop Strong Alumni Networks**: Create a strong alumni network to facilitate professional connections, mentorship opportunities, and continuous engagement with the field of
CAM.
29. **Regularly Review and Update Technological Tools**: Continuously evaluate and update
the technological tools and platforms used for delivering online education to ensure they
are user-friendly and incorporate the latest advancements in educational technology.

By following these recommendations, online CAM education programs can provide high-quality,
comprehensive, and credible education, preparing students to become knowledgeable and ethical
CAM practitioners.

CONCLUSION

In conclusion, online CAM education has made significant strides in providing accessible, diverse,
and comprehensive learning experiences. The achievements in this field, particularly in terms of
global reach, integration of evidence-based practices, and advancements in online learning technologies, are notable. However, these accomplishments come with their own set of challenges,
including ensuring practical skill development, maintaining quality and accreditation, and balancing traditional knowledge with scientific rigor. To address these challenges and optimize the learning experience, a set of best practices has been recommended. These include developing robust and
interactive curricula, incorporating evidence-based practices, ensuring qualified faculty, promoting interprofessional education, and adopting hybrid models for hands-on training. Additionally,
focusing on maintaining academic integrity, fostering global and cultural competence, and continually adapting to technological advancements are crucial. The future of online CAM education is
promising, yet it requires ongoing effort, innovation, and collaboration among educators, students,
professionals, and regulatory bodies. By adhering to these best practices and continuously striving for improvement, online CAM education can continue to evolve, contributing positively to the
healthcare field and empowering individuals with knowledge and skills in holistic health and wellness practices.

REFERENCES

Al-Worafi, Y.M. (2022a). *A Guide to Online Pharmacy Education: Teaching Strategies and Assessment
Methods*. CRC Press.
Al-Worafi, Y.M. (2022b). History and importance. In: Al-Worafi, Y.M. (ed), *A Guide to Online Pharmacy
Education: Teaching Strategies and Assessment Methods*. CRC Press.
Al-Worafi, Y.M. (2022c). Advantages and disadvantages. In: Al-Worafi, Y.M. (ed), *A Guide to Online Pharmacy
Education: Teaching Strategies and Assessment Methods*. CRC Press.
Al-Worafi, Y.M. (2022d). Technologies and tools. In: Al-Worafi, Y.M. (ed), *A Guide to Online Pharmacy
Education: Teaching Strategies and Assessment Methods*. CRC Press.
Al-Worafi, Y.M. (2022e). Curriculum-related issues. In: Al-Worafi, Y.M. (ed), *A Guide to Online Pharmacy
Education: Teaching Strategies and Assessment Methods*. CRC Press.

Al-Worafi, Y.M. (2022f). Pharmacy education: Learning styles. In: Al-Worafi, Y.M. (ed), *A Guide to Online Pharmacy Education: Teaching Strategies and Assessment Methods.* CRC Press.

Al-Worafi, Y.M. (2022g). Competencies and learning outcomes. In: Al-Worafi, Y.M. (ed), *A Guide to Online Pharmacy Education: Teaching Strategies and Assessment Methods.* CRC Press.

Al-Worafi, Y.M. (2022h). Teaching the theory. In: Al-Worafi, Y.M. (ed), *A Guide to Online Pharmacy Education: Teaching Strategies and Assessment Methods.* CRC Press.

Al-Worafi, Y.M. (2022i). Teaching the practice and tutorial. In: Al-Worafi, Y.M. (ed), *A Guide to Online Pharmacy Education: Teaching Strategies and Assessment Methods.* CRC Press.

Al-Worafi, Y.M. (2022j). Self-learning and self-directed learning. In: Al-Worafi, Y.M. (ed), *A Guide to Online Pharmacy Education: Teaching Strategies and Assessment Methods.* CRC Press.

Al-Worafi, Y.M. (2022k). Traditional and active strategies. In: Al-Worafi, Y.M. (ed), *A Guide to Online Pharmacy Education: Teaching Strategies and Assessment Methods.* CRC Press.

Al-Worafi, Y.M. (2022l). Team-based learning in pharmacy education. In: Al-Worafi, Y.M. (ed), *A Guide to Online Pharmacy Education: Teaching Strategies and Assessment Methods.* CRC Press.

Al-Worafi, Y.M. (2022m). Problem-based learning in pharmacy education. In: Al-Worafi, Y.M. (ed), *A Guide to Online Pharmacy Education: Teaching Strategies and Assessment Methods.* CRC Press.

Al-Worafi, Y.M. (2022n). Case-based learning in pharmacy education. In: Al-Worafi, Y.M. (ed), *A Guide to Online Pharmacy Education: Teaching Strategies and Assessment Methods.* CRC Press.

Al-Worafi, Y.M. (2022o). Simulation in pharmacy education. In: Al-Worafi, Y.M. (ed), *A Guide to Online Pharmacy Education: Teaching Strategies and Assessment Methods.* CRC Press.

Al-Worafi, Y.M. (2022p). Project-based learning in pharmacy education. In: Al-Worafi, Y.M. (ed), *A Guide to Online Pharmacy Education: Teaching Strategies and Assessment Methods.* CRC Press.

Al-Worafi, Y.M. (2022q). Flipped classes in pharmacy education. In: Al-Worafi, Y.M. (ed), *A Guide to Online Pharmacy Education: Teaching Strategies and Assessment Methods.* CRC Press.

Al-Worafi, Y.M. (2022r). Educational games in pharmacy education. In: Al-Worafi, Y.M. (ed), *A Guide to Online Pharmacy Education: Teaching Strategies and Assessment Methods.* CRC Press.

Al-Worafi, Y.M. (2022s). Web-based learning in pharmacy education. In: Al-Worafi, Y.M. (ed), *A Guide to Online Pharmacy Education: Teaching Strategies and Assessment Methods.* CRC Press.

Al-Worafi, Y.M. (2022t). Lecture-based/interactive lecture-based learning in pharmacy education. In: Al-Worafi, Y.M. (ed), *A Guide to Online Pharmacy Education: Teaching Strategies and Assessment Methods.* CRC Press.

Al-Worafi, Y.M. (2022u). Blended learning in pharmacy education. In: Al-Worafi, Y.M. (ed), *A Guide to Online Pharmacy Education: Teaching Strategies and Assessment Methods.* CRC Press.

Al-Worafi, Y.M. (2022v). Assessment methods in pharmacy education: Strengths and limitations. In: Al-Worafi, Y.M. (ed), *A Guide to Online Pharmacy Education: Teaching Strategies and Assessment Methods.* CRC Press.

Al-Worafi, Y.M. (2022w). Objective structured clinical examination (OSCE) in pharmacy education. In: Al-Worafi, Y.M. (ed), *A Guide to Online Pharmacy Education: Teaching Strategies and Assessment Methods.* CRC Press.

Al-Worafi, Y.M. (Ed.). (2024a). *Handbook of Medical and Health Sciences in Developing Countries.* Springer, Cham.

Al-Worafi, Y.M. (2024b). Complementary and alternative medicine (CAM) in developing countries. In: Al-Worafi, Y.M. (ed), *Handbook of Medical and Health Sciences in Developing Countries.* Springer, Cham. https://doi.org/10.1007/978-3-030-74786-2_301-1

Anderson, T., and Dron, J. (2011). Three generations of distance education pedagogy. *International Review of Research in Open and Distributed Learning*, 12(3), 80–97.

Bates, A.T. (2005). *Technology, E-Learning and Distance Education.* Routledge, London.

Bervell, B., and Umar, I.N. (2017). A decade of LMS acceptance and adoption research in Sub-Sahara African higher education: A systematic review of models, methodologies, milestones and main challenges. *EURASIA Journal of Mathematics, Science and Technology Education*, 13(11), 7269–7286.

Borthwick, A.C., Anderson, C.L., Finsness, E.S., and Foulger, T.S. (2015). Special article personal wearable technologies in education: Value or villain? *Journal of Digital Learning in Teacher Education*, 31(3), 85–92.

Bozkurt, A. (2019). From distance education to open and distance learning: A holistic evaluation of history, definitions, and theories. In: *Handbook of Research on Learning in the Age of Transhumanism* (pp. 252–273). IGI Global.

Coleman, E., and O'Connor, E. (2019). The role of WhatsApp(r) in medical education; a scoping review and instructional design model. *BMC Medical Education*, 19(1), 1–13.

Düking, P., Achtzehn, S., Holmberg, H.C., and Sperlich, B. (2018). Integrated framework of load monitoring by a combination of smartphone applications, wearables and point-of-care testing provides feedback that allows individual responsive adjustments to activities of daily living. *Sensors*, 18(5), 1632.

Garrison, D.R., and Kanuka, H. (2004). Blended learning: Uncovering its transformative potential in higher education. *The Internet and Higher Education*, 7(2), 95–105.

Garrison, D.R., and Vaughan, N.D. (2008). *Blended Learning in Higher Education: Framework, Principles, and Guidelines*. John Wiley & Sons, Hoboken, NJ.

Graham, C.R. (2006). Blended learning systems. In: *The Handbook of Blended Learning: Global Perspectives, Local Designs*, (Vol. 1, pp. 3–21). Wiley.

Gulati, R.R., Reid, H., and Gill, M. (2020). Instagram for peer teaching: Opportunity and challenge. *Education for Primary Care*, 31(6), 382–384.

Harden, R.M. and Gleeson, F.A. (1979). Assessment of clinical competence using an objective structured clinical examination (OSCE). *Medical Education*, 13(1), 39–54.

Hrastinski, S. (2019). What do we mean by blended learning? *TechTrends*, 63(5), 564–569. https://www.blendedlearning.org/models/

Kant, N., Prasad, K.D., and Anjali, K. (2021). Selecting an appropriate learning management system in open and distance learning: A strategic approach. *Asian Association of Open Universities Journal*, 16(1), 79–97.

Kusmaryono, I., Jupriyanto, J., and Kusumaningsih, W. (2021). A systematic literature review on the effectiveness of distance learning: Problems, opportunities, challenges, and predictions. *International Journal of Education*, 14(1), 62–69.

Maddux, C.D. (1992). *Distance Education: A Selected Bibliography* (Vol. 7). Educational Technology Publications, Englewood Cliffs.

Mastellos, N., Tran, T., Dharmayat, K., Cecil, E., Lee, H.Y., Wong, C.C.P., Mkandawire, W., Ngalande, E., Wu, J.T., Hardy, V., Chirambo, B.G., and O'Donoghue, J.M. (2018). Training community healthcare workers on the use of information and communication technologies: A randomised controlled trial of traditional versus blended learning in Malawi, Africa. *BMC Medical Education*, 18(1), 1–13.

McIsaac, M.S., and Gunawardena, C.N. (1996). Distance education. In: Jonassen, D.H. (ed.), *Handbook of Research for Educational Communications and Technology* (pp. 403–437). Simon & Schuster Macmillan, New York, NY.

Momani, A.M. (2010). Comparison between two learning management systems: Moodle and blackboard. https://papers.ssrn.com/sol3/papers.cfm?abstract_id=1608311

Paulson, F.L., Paulson, P.R., and Meyer, C.A. (1991). What makes a portfolio a portfolio. *Educational Leadership*, 48(5), 60–63.

Raiman, L., Antbring, R., and Mahmood, A. (2017). WhatsApp messenger as a tool to supplement medical education for medical students on clinical attachment. *BMC Medical Education*, 17(1), 1–9.

Richey, R.C., Silber, K.H., and Ely, D.P. (2008). Reflections on the 2008 AECT definitions of the field. *TechTrends*, 52(1), 24–25.

Robinson, R., Molenda, M., and Rezabek, L. (2013). Facilitating learning. In: Januszewski, A., and Molenda, M. (eds), *Educational Technology* (pp. 27–60). Routledge.

Shafer, S., Johnson, M.B., Thomas, R.B., Johnson, P.T., and Fishman, E.K. (2018). Instagram as a vehicle for education: What radiology educators need to know. *Academic Radiology*, 25(6), 819–822.

Subrahmanyam, V.V., and Swathi, K. (2020). Wearable technology and its role in education. *International Conference – 2020 on Distance Education and Educational Technology (ICE-CODL 2020)*, CDOL, JMI, New Delhi, 10–11 December, 2020.

us Salam, M.A., Oyekwe, G.C., Ghani, S.A., and Choudhury, R.I., (2021). How can WhatsApp(r) facilitate the future of medical education and clinical practice? *BMC Medical Education*, 21(1), 1–4.

Ward, C.T., Rey, J.A., Mobley, W.C., and Evans, C.D. (2003). Establishing a distance learning site for a traditional doctor of pharmacy program. *American Journal of Pharmaceutical Education*, 67(1/4), 153.

Weller, M. (2007). *Virtual Learning Environments: Using, Choosing and Developing your VLE*. Routledge.

Wikipedia, Distance Education. (2022).https://en.wikipedia.org/wiki/Distance_education

WGU. (2020). Past, present, and future of online education. www.wgu.edu/blog/past-present-future-online-education2001.html#close

Wong, X.L., Liu, R.C., and Sebaratnam, D.F. (2019). Evolving role of Instagram in# medicine. *Internal Medicine Journal*, 49(10), 1329–1332.

8 Complementary and Alternative Medicine (CAM) Education
Training

INTRODUCTION

The evolution of complementary and alternative medicine (CAM) education and training programs has been marked by significant growth and diversification over the past few decades. This evolution reflects broader shifts in societal attitudes toward health and wellness, an increasing demand for holistic and preventive healthcare approaches, and greater recognition of the benefits of integrating CAM with conventional medicine. Here's an overview of how CAM education and training programs have evolved:

EARLY DEVELOPMENT

- **Initial Phases**: The formal education and training in CAM began to materialize in the late 20th century, although many practices (such as herbalism, acupuncture, and Ayurveda) have been taught informally for centuries within their cultures of origin.
- **Institutional Recognition**: Early CAM programs were often offered by independent institutions specializing in specific modalities (e.g., chiropractic colleges, acupuncture schools). These programs varied greatly in length, depth, and accreditation.

EXPANSION AND INTEGRATION

- **Increased Popularity**: The late 20th and early 21st centuries saw a surge in interest in CAM, driven by patients seeking alternatives or complements to conventional medical treatments. This led to an increase in the number and variety of CAM education programs.
- **Integration into Conventional Curricula**: Recognizing the importance of providing holistic care, some medical and nursing schools began to integrate CAM topics into their curricula. This includes courses on nutrition, stress management, and natural products, aiming to produce healthcare professionals who are knowledgeable about a wide range of therapeutic options.

STANDARDIZATION AND ACCREDITATION

- **Accreditation Bodies**: With the proliferation of CAM programs, there emerged a need for standardization and quality assurance. Accreditation bodies, such as the Accreditation Commission for Acupuncture and Oriental Medicine (ACAOM) and the Council on Chiropractic Education (CCE), were established to set educational standards and accredit programs.

DOI: 10.1201/9781003327202-8

- **Certification and Licensure**: The evolution of education has been paralleled by the development of certification and licensure requirements for CAM practitioners, ensuring that they meet specific competency standards in their field.

RESEARCH AND EVIDENCE-BASED PRACTICE

- **Focus on Research**: There has been an increasing emphasis on incorporating evidence-based practice into CAM education. This includes teaching students to critically evaluate the scientific literature related to CAM therapies and to integrate research findings into their clinical practice.
- **Research Institutions**: Some CAM educational institutions have established research centers to contribute to the body of evidence supporting CAM practices and to integrate these findings into their teaching.

ONLINE AND CONTINUING EDUCATION

- **Online Learning**: The advent of online education has made CAM training more accessible to a wider audience. Many institutions now offer online courses or entire programs in CAM, catering to the needs of working professionals and students in remote areas.
- **Continuing Education**: As the field of CAM continues to evolve, there is a growing emphasis on continuing education for CAM practitioners. This allows them to stay updated with the latest research, clinical practices, and regulatory changes.

The future of CAM education and training programs is likely to be characterized by further integration into mainstream healthcare education, increased emphasis on evidence-based practice, and greater use of technology in education. As CAM continues to gain acceptance and recognition, the education and training of CAM practitioners will play a crucial role in ensuring the safety, efficacy, and quality of CAM therapies offered to the public.

RATIONALITY AND IMPORTANCE OF COMPLEMENTARY AND ALTERNATIVE MEDICINE (CAM) TRAINING DURING UNDERGRADUATE STUDIES

The integration of CAM training during undergraduate studies, particularly in health-related fields, is a growing trend that reflects the evolving landscape of healthcare. This shift toward incorporating CAM education is based on a rational understanding of its importance in providing holistic, patient-centered care, and the increasing demand for CAM therapies among the general population. Here are several key points that underline the rationality and importance of CAM training during undergraduate studies:

ENHANCING HOLISTIC CARE

- **Whole-Person Approach**: CAM training emphasizes treating the individual as a whole, considering physical, emotional, mental, and spiritual aspects. This approach aligns with the growing recognition of the importance of holistic care in promoting overall health and well-being.
- **Patient-Centered Care**: Understanding CAM modalities allows future healthcare professionals to better accommodate patients' preferences and values, fostering a more patient-centered approach to care.

MEETING PATIENT DEMAND

- **Increased Usage**: There's a significant portion of the population that uses CAM therapies alongside conventional medicine. Training in CAM enables future healthcare providers to understand and guide their patients through various treatment options responsibly.
- **Informed Decision-Making**: With proper education in CAM, graduates are better equipped to help patients make informed decisions about using CAM therapies, including understanding potential benefits and risks.

INTEGRATING CAM WITH CONVENTIONAL MEDICINE

- **Interprofessional Collaboration**: Knowledge of CAM can facilitate collaboration between conventional and CAM practitioners, leading to more comprehensive care plans for patients.
- **Evidence-Based Practice**: Training in CAM also involves critical evaluation of the evidence supporting various therapies, which is essential for integrating CAM into evidence-based clinical practice.

FILLING GAPS IN CONVENTIONAL CARE

- **Complementary Therapies**: Some CAM practices can offer solutions where conventional medicine may have limited options, especially for chronic conditions, pain management, and stress reduction.
- **Preventive Health**: Many CAM disciplines emphasize wellness and prevention, areas that are increasingly recognized as vital for reducing the burden of chronic diseases.

EDUCATIONAL INNOVATION AND CULTURAL COMPETENCE

- **Cultural Competence**: CAM training often includes the exploration of healing practices from various cultures, enhancing cultural competence and sensitivity among healthcare professionals.
- **Adaptability and Lifelong Learning**: The CAM field is dynamic, with new therapies and evidence emerging regularly. Training in CAM encourages an attitude of openness and adaptability, critical for lifelong learning in the rapidly evolving healthcare sector.

REGULATORY AND PROFESSIONAL DEVELOPMENT

- **Professional Standards**: As CAM practices become more integrated into mainstream healthcare, understanding regulatory and professional standards governing these practices is crucial for all healthcare professionals.
- **Career Opportunities**: For students interested in pursuing careers in healthcare, knowledge of CAM can open up additional pathways and opportunities in both conventional and alternative settings.

In conclusion, incorporating CAM training during undergraduate studies is a rational and important step toward preparing future healthcare professionals to meet the diverse needs of their patients, collaborate effectively across disciplines, and contribute to the ongoing evolution of healthcare practices. This approach not only enriches the educational experience but also aligns with the broader goals of improving healthcare outcomes, patient satisfaction, and the overall health of communities.

TYPES OF TRAINING FOR UNDERGRADUATE COMPLEMENTARY AND ALTERNATIVE MEDICINE (CAM) STUDENTS

Training for undergraduate students in CAM encompasses a broad spectrum of modalities and approaches. These training types are designed to equip students with the knowledge, skills, and competencies needed to practice safely and effectively, whether as standalone treatments or in conjunction with conventional medicine. Here's an overview of the types of CAM training commonly offered to undergraduate students:

1. **Traditional Systems of Medicine**
   - *Ayurveda*: Originating from India, Ayurveda training includes understanding the principles of doshas (body types), dietary practices, herbal remedies, and lifestyle adjustments for health and wellness.
   - *Traditional Chinese Medicine (TCM)*: TCM training covers acupuncture, herbal medicine, Tai Chi, Qigong, and dietary therapy, based on the concept of Qi (vital energy) and its flow through meridians in the body.
   - *Naturopathy*: Naturopathic training involves a wide range of natural therapies including nutrition, botanical medicine, hydrotherapy, and physical medicine, emphasizing the body's inherent ability to heal itself.
2. **Mind-Body Practices**
   - *Yoga*: Training in yoga not only involves the physical postures (asanas) but also breath control (pranayama), meditation, and principles of mindfulness and wellness.
   - *Meditation*: Students learn various meditation techniques, such as mindfulness meditation, transcendental meditation, and guided visualization, focusing on mental and emotional well-being.
   - *Biofeedback and Neurofeedback*: These practices involve training to control physiological processes such as heart rate, muscle tension, and brainwave patterns for health benefits.
3. **Manual Therapies**
   - *Massage Therapy*: Training covers a variety of massage techniques (e.g., Swedish, deep tissue, and Shiatsu) aimed at manipulating the body's tissues to improve health and well-being.
   - *Chiropractic*: Although often requiring postgraduate education, some undergraduate programs introduce the basics of chiropractic care, focusing on the diagnosis, treatment, and prevention of mechanical disorders of the musculoskeletal system.
   - *Osteopathy*: Undergraduate exposure might include principles of osteopathic medicine that emphasize the interrelationship of the body's nerves, muscles, bones, and organs.
4. **Energy Therapies**
   - *Reiki*: Students learn about channeling energy through the hands to promote healing, balance, and reduction of stress.
   - *Qigong*: Training includes learning about gentle movements, meditation, and breathing techniques to cultivate and balance Qi.
5. **Dietary and Herbal Approaches**
   - *Nutritional Therapy*: Training focuses on the use of diet and supplements to promote health, manage diseases, and prevent nutritional deficiencies.
   - *Herbalism*: Students learn about the medicinal properties of herbs and plants, including how to prepare and use herbal remedies for various health conditions.
6. **Integrative Health and Wellness**
   - This broader approach includes training on how to integrate various CAM practices with conventional medical treatments to provide comprehensive patient care. It emphasizes a holistic, patient-centered approach to health and wellness.

7. **Research and Evidence-Based Practice**
   - Training also involves learning how to critically evaluate the scientific evidence supporting different CAM therapies, understanding research methodologies, and applying evidence-based practices in clinical settings.
8. **Ethics and Professional Practice**
   - Students are taught professional standards, ethical considerations, and legal aspects of practicing CAM, including informed consent and confidentiality.

Undergraduate CAM training can vary widely depending on the institution, the specific CAM practices taught, and the regulatory framework of the country in which the training is provided. Some programs offer a broad overview of multiple CAM modalities, while others may focus more deeply on one particular area. The goal of CAM education at the undergraduate level is to provide a solid foundation in alternative and integrative health principles, preparing students for further professional training or to incorporate CAM knowledge into their future healthcare careers.

TYPES OF TRAINING FOR POSTGRADUATE COMPLEMENTARY AND ALTERNATIVE MEDICINE (CAM) STUDENTS

Postgraduate training in CAM offers advanced education and specialization opportunities for healthcare professionals seeking to deepen their knowledge and skills in specific CAM modalities. This level of training is designed to prepare practitioners to integrate CAM practices into their professional activities, conduct research, or become educators in the field. Here are some common types of postgraduate CAM training:

1. **Advanced Clinical Training Programs**
   - *Specialized Certifications*: Postgraduate students can pursue certifications in specific CAM practices such as advanced acupuncture, herbal medicine, or advanced massage techniques, focusing on specialized areas within these practices.
   - *Residency Programs*: Some CAM disciplines, like naturopathy and chiropractic, offer residency programs where practitioners can gain clinical experience under supervision in specialized areas such as pediatrics, oncology, or sports medicine.
2. **Master's and Doctoral Programs**
   - *Master's Degrees*: These programs often focus on specific areas of CAM, such as TCM, Ayurveda, integrative health, or wellness coaching, providing in-depth theoretical and practical knowledge.
   - *Doctoral Degrees*: Doctorate programs in CAM fields (e.g., Doctor of Naturopathic Medicine, Doctor of Acupuncture and Oriental Medicine) offer the highest level of education, preparing students for leadership roles in clinical practice, academia, or research.
3. **Integrative Medicine Fellowships**
   - *Fellowships*: Designed for licensed healthcare professionals (e.g., MDs, DOs, and nurses), these programs offer advanced training in integrating CAM practices into conventional medical care, focusing on holistic, patient-centered approaches.
4. **Research Fellowships and Degrees**
   - *Research-Oriented Training*: For those interested in CAM research, postgraduate programs offer opportunities to study the efficacy, mechanisms, and implementation of CAM therapies. Degrees like a Ph.D. in integrative health or a Master's in CAM research are available.
5. **Professional Development Workshops and Seminars**
   - *Continuing Education*: Postgraduate students and practicing professionals can participate in workshops and seminars for continuing education credits, staying updated with the latest developments in CAM practices and research.

6. **Online and Distance Learning Programs**
   - *E-Learning*: Many institutions offer online postgraduate programs in CAM, providing flexibility for professionals to advance their education while continuing to work.
7. **Interprofessional Education Programs**
   - *Collaborative Learning*: These programs are designed to foster collaboration between CAM practitioners and conventional healthcare professionals, enhancing the integration of CAM into mainstream healthcare settings.
8. **Specialty Training in Population Health**
   - *Focus on Specific Populations*: Some programs offer specialty training targeting the unique needs of specific populations, such as geriatrics, women's health, or pediatric CAM, focusing on the safe and effective application of CAM therapies for these groups.
9. **Regulatory, Ethics, and Business Management Education**
   - *Professional Practice*: Training in the legal, ethical, and business aspects of CAM practice is essential for practitioners looking to open their practices, work within healthcare organizations, or navigate the regulatory landscape of CAM.

Postgraduate CAM education and training are characterized by a focus on depth, specialization, and the integration of CAM practices into broader healthcare and research settings. These programs are crucial for the advancement of CAM, ensuring that practitioners are well-educated, skilled, and able to contribute to the evidence base and professional standards of the field.

COMPLEMENTARY AND ALTERNATIVE MEDICINE (CAM) TRAINING: ACHIEVEMENTS

CAM training has achieved significant milestones over the years, reflecting its growing acceptance and integration into mainstream healthcare. These achievements are pivotal in promoting the safe, effective, and informed use of CAM practices among healthcare providers and patients alike. Here's a look at some key accomplishments in the realm of CAM training:

1. **Institutional Recognition and Accreditation**
   - *Accredited Educational Programs*: The establishment of accreditation standards for CAM educational institutions, such as those by the CCE and the ACAOM, signifies a major step forward in ensuring the quality and consistency of CAM training.
   - *Integration into Medical Schools*: Many conventional medical schools now incorporate CAM courses into their curricula, recognizing the value of training future physicians in holistic and patient-centered care approaches.
2. **Research Advancements and Evidence-Based Practice**
   - *Research Institutions*: The creation of research centers dedicated to CAM, such as the National Center for Complementary and Integrative Health (NCCIH) in the United States, has been instrumental in advancing the scientific understanding and evidence base of CAM therapies.
   - *Increased Research Funding*: There has been a notable increase in funding for CAM research, enabling rigorous studies on the efficacy, safety, and mechanisms of CAM practices. This research is crucial for integrating CAM into evidence-based healthcare.
3. **Professional Development and Continuing Education**
   - *Continuing Education Opportunities*: The development of continuing education programs for healthcare professionals in CAM disciplines ensures that practitioners stay informed about the latest research, techniques, and best practices.
   - *Certification and Licensure*: The establishment of certification and licensure requirements for various CAM professions enhances the credibility of CAM practitioners and ensures a standard level of competency and safety for patients.

4. **Global Recognition and Collaboration**
   - *International Standards*: Efforts by international bodies, such as the World Health Organization (WHO), to develop guidelines and standards for CAM practices have contributed to their global recognition and acceptance.
   - *Cross-Cultural Exchange*: There has been an increase in the cross-cultural exchange of knowledge and practices, enriching the global CAM community and fostering a more integrated approach to health and wellness.

5. **Patient Access and Acceptance**
   - *Increased Public Interest*: The growing public interest in and acceptance of CAM has driven demand for trained CAM practitioners and informed healthcare providers who can guide patients in making safe and effective use of CAM therapies.
   - *Insurance Coverage*: In some regions, the inclusion of certain CAM practices in health insurance coverage reflects recognition of their value in health and wellness, making CAM more accessible to a wider population.

6. **Interprofessional Collaboration**
   - *Collaborative Care Models*: Training programs increasingly emphasize interprofessional education and collaborative care models, where CAM practitioners work alongside conventional healthcare professionals to provide comprehensive care.

7. **Technology and Innovation in CAM Education**
   - *Online Learning Platforms*: The expansion of online and digital resources for CAM education has made training more accessible, allowing for a broader reach and flexibility in learning.
   - *Simulation and Virtual Reality*: Innovative teaching methods, including simulation and virtual reality (VR), are being explored in CAM education to enhance hands-on learning experiences without geographical constraints.

These achievements in CAM training not only highlight the progress made in validating and standardizing CAM practices but also underscore the ongoing effort to integrate CAM into holistic, patient-centered healthcare models. As research continues to expand our understanding of CAM therapies, training programs will evolve to prepare healthcare professionals to meet the diverse needs of their patients with a broad array of evidence-based treatment options.

COMPLEMENTARY AND ALTERNATIVE MEDICINE (CAM) TRAINING: CHALLENGES

While the field of CAM has made significant strides, it also faces a number of challenges. These challenges stem from a variety of sources, including institutional, regulatory, scientific, and cultural factors. Addressing these issues is crucial for the continued integration and acceptance of CAM within the broader healthcare landscape. Here are some of the key challenges facing CAM training today:

STANDARDIZATION AND REGULATION

- **Varying Standards**: There is a wide variation in training standards and regulations for CAM practices both within and across countries. This lack of uniformity can lead to discrepancies in the quality of education and practice.
- **Accreditation**: While some CAM disciplines have established accreditation bodies, others lack formal accreditation processes, making it difficult to assess the quality of training programs.

EVIDENCE-BASED PRACTICE

- **Research Funding**: CAM research often receives less funding compared to conventional medicine, limiting the ability to conduct large-scale, high-quality studies.
- **Integrating Evidence into Practice**: Even when research is available, integrating evidence-based practices into CAM education and clinical practice can be challenging due to the diverse nature of CAM modalities and the varying levels of evidence supporting them.

CULTURAL AND PHILOSOPHICAL DIFFERENCES

- **Integrating CAM with Conventional Medicine**: Bridging the gap between the holistic, patient-centered philosophies of many CAM practices and the more disease-focused approach of conventional medicine can be challenging.
- **Respect for Traditional Knowledge**: Balancing respect for traditional knowledge and practices with the demands for scientific evidence and standardization can be difficult, especially for practices with deep cultural roots.

PROFESSIONAL RECOGNITION AND INTEGRATION

- **Healthcare System Integration**: Fully integrating CAM practitioners into mainstream healthcare systems remains a challenge, including issues related to scope of practice, referral pathways, and interprofessional collaboration.
- **Professional Acceptance**: CAM practitioners sometimes face skepticism or resistance from conventional healthcare professionals, which can hinder collaboration and integration efforts.

PUBLIC PERCEPTION AND AWARENESS

- **Misinformation**: There is a significant amount of misinformation about CAM practices, both online and in popular media, which can lead to confusion and mistrust among the public and healthcare professionals alike.
- **Consumer Education**: Educating consumers about the safe and effective use of CAM, including understanding the evidence, potential risks, and benefits, is an ongoing challenge.

ACCESS AND EQUITY

- **Access to CAM Education**: Access to high-quality CAM training can be limited, especially in regions without established educational institutions or accreditation standards.
- **Healthcare Disparities**: Ensuring equitable access to CAM services for underserved and marginalized populations poses significant challenges, including issues of affordability and cultural competence.

TECHNOLOGICAL AND METHODOLOGICAL INNOVATIONS

- **Adapting to Technology**: Keeping pace with technological advancements in education and healthcare delivery, such as telehealth and online learning, requires ongoing adaptation and investment.
- **Methodological Challenges**: Designing research studies that accommodate the individualized and holistic nature of many CAM practices, while still adhering to rigorous scientific standards, is complex.

ENHANCING RESEARCH AND EVIDENCE-BASED PRACTICE

- **Increasing Research Funding**: Advocating for more government and private sector funding specifically earmarked for CAM research can help address the evidence gap.
- **Collaborative Research Models**: Encouraging collaborations between CAM and conventional medical researchers could leverage the strengths of both fields, enhancing the quality and relevance of research findings.
- **Developing Appropriate Research Methodologies**: There's a need for innovative research methodologies that are suitable for evaluating CAM modalities, which may not always fit the conventional randomized controlled trial model. Patient-centered outcomes research (PCOR) and pragmatic trials are examples of approaches that can be more aligned with CAM practices.

STANDARDIZATION, ACCREDITATION, AND REGULATION

- **Global Standards for Education and Practice**: Developing and adopting international standards for CAM education and practice can help harmonize training and professional qualifications, making it easier to ensure quality and safety.
- **Expanding Accreditation**: Encouraging more CAM disciplines to establish and adhere to accreditation standards can improve the consistency and quality of education across the board.
- **Clearer Regulatory Frameworks**: Governments and professional bodies need to work together to create clearer regulatory frameworks that define the scope of practice, professional standards, and pathways for the integration of CAM practitioners into the healthcare system.

FOSTERING CULTURAL AND PHILOSOPHICAL INTEGRATION

- **Interprofessional Education (IPE)**: Incorporating IPE models that include both CAM and conventional healthcare students can foster mutual respect, understanding, and collaboration from an early stage in their professional development.
- **Valuing Traditional Knowledge**: Creating educational platforms that respect and integrate traditional knowledge alongside scientific evidence can enrich CAM training and practice, ensuring that these traditions are preserved and respected.

PROMOTING PROFESSIONAL RECOGNITION AND HEALTHCARE INTEGRATION

- **Building Clinical Integration Pathways**: Establishing clear pathways for the integration of CAM services into conventional healthcare settings, including hospitals and primary care practices, can enhance patient access to a broader range of treatments.
- **Advocacy and Public Policy**: Engaging in advocacy to influence public policy toward recognizing and integrating CAM professions within the healthcare system can help overcome professional barriers.

ADDRESSING PUBLIC PERCEPTION AND ACCESS ISSUES

- **Public Education Campaigns**: Launching public education campaigns to provide accurate information about CAM practices, their benefits, and limitations can help counteract misinformation and build informed consumer choices.

- **Improving Access and Equity**: Strategies to improve access to CAM services might include insurance coverage for CAM therapies, subsidies for low-income populations, and the incorporation of CAM services into community health centers.

LEVERAGING TECHNOLOGY AND INNOVATION

- **Embracing Digital Health**: CAM education and practice can benefit from embracing digital health technologies, such as telehealth platforms for remote consultations, and online learning tools for education.
- **Innovative Teaching Tools**: Utilizing VR, augmented reality (AR), and simulation-based learning can enhance the training of CAM practitioners, offering hands-on experience in a virtual environment.

Addressing these challenges requires concerted efforts from educators, practitioners, researchers, policymakers, and the public. Strategies might include increasing investment in CAM research, developing unified standards and accreditation processes, fostering interprofessional education and collaboration, and enhancing public education about CAM. Overcoming these obstacles is crucial for ensuring that CAM training produces knowledgeable, competent practitioners capable of delivering high-quality, evidence-based care to meet the diverse health needs of the population.

COMPLEMENTARY AND ALTERNATIVE MEDICINE (CAM) TRAINING: RECOMMENDATIONS

Improving CAM education and training is crucial for producing competent CAM practitioners who can contribute effectively to healthcare systems. Improving CAM education and training is essential for enhancing healthcare systems, addressing public health challenges, and promoting scientific research and innovation (Al-Worafi, 2020a,b, 2022a,b, 2023a–n, 2024a,b; Hasan et al., 2019). To enhance the effectiveness, integration, and recognition of CAM training, several recommendations can be made. These suggestions aim to address existing challenges, improve the quality of CAM education, ensure the safety and efficacy of CAM practices, and facilitate their integration into mainstream healthcare systems. Here are key recommendations for stakeholders involved in CAM training:

1. **Strengthen Accreditation and Standardization**
   - *Develop and enforce rigorous accreditation standards* for CAM educational institutions and programs to ensure consistency and quality across training programs.
   - *Promote the standardization of qualifications* and competencies for CAM practitioners to facilitate mutual recognition and mobility across regions and countries.
2. **Enhance Research and Evidence-Based Education**
   - *Increase funding and support for CAM research* to build a solid evidence base for various CAM modalities.
   - *Integrate evidence-based practice into CAM curricula*, teaching students how to critically evaluate research and apply it to clinical practice.
3. **Improve Interprofessional Education and Collaboration**
   - *Implement interprofessional education (IPE) programs* that include both CAM and conventional healthcare students to foster mutual respect, understanding, and collaboration.
   - *Create platforms for professional exchange and learning* between CAM practitioners and conventional healthcare providers to enhance patient care through integrated approaches.

4. **Expand Public and Professional Awareness**
   - *Conduct public awareness campaigns* to educate the public about the benefits, limitations, and appropriate use of CAM therapies.
   - *Offer continuing education opportunities* for healthcare professionals in both CAM and conventional medicine to stay updated on the latest developments, research, and integration strategies.

5. **Address Regulatory and Legal Frameworks**
   - *Work with regulatory bodies and professional associations* to develop clear scopes of practice, ethical guidelines, and professional standards for CAM practitioners.
   - *Advocate for policies that support the integration* of safe and effective CAM practices into healthcare systems, including insurance coverage and reimbursement models.

6. **Promote Access and Equity**
   - *Ensure equitable access to CAM education and services* by addressing barriers related to cost, geography, and cultural competence.
   - *Develop programs that respect and incorporate traditional knowledge* and practices, especially those from underserved or indigenous communities.

7. **Leverage Technology and Innovation**
   - *Incorporate digital health technologies* into CAM training and practice, such as telehealth, online learning platforms, and simulation-based education.
   - *Explore innovative teaching methods*, including VR and AR, to enhance hands-on learning experiences.

8. **Foster Global Collaboration**
   - *Establish international collaborations and partnerships* among CAM educational institutions to share knowledge, research findings, and best practices.
   - *Participate in global forums and organizations* dedicated to CAM to contribute to and benefit from global developments in CAM education and practice.

9. **Develop Specialized Training Paths**
   - Offer specialized training tracks within CAM disciplines for focused expertise, such as sports medicine, pediatrics, or oncology within acupuncture or herbal medicine.

10. **Encourage Lifelong Learning**
    - Implement mandatory continuing education requirements for CAM practitioners to encourage lifelong learning and ensure practitioners stay current with the latest advancements and research.

11. **Promote Cultural Competence**
    - Integrate cultural competence training into CAM curricula to ensure practitioners are sensitive to and knowledgeable about the cultural contexts of the therapies they practice and the diverse populations they serve.

12. **Facilitate Clinical Training Opportunities**
    - Increase the availability of clinical training placements in diverse settings, including hospitals, community health centers, and interdisciplinary clinics, to expose students to a wide range of patient populations and health conditions.

13. **Advance Clinical Research Skills**
    - Incorporate training in clinical research methodology into CAM programs to equip practitioners with the skills to contribute to the evidence base of their discipline.

14. **Strengthen Patient Communication Skills**
    - Emphasize training in patient communication, counseling, and education skills to improve patient-practitioner relationships and support informed patient decision-making.

15. **Enhance Business and Professional Skills**
    - Offer courses on business management, ethics, and professional development for CAM practitioners planning to enter private practice or navigate the healthcare system.

16. **Support Interdisciplinary Research Projects**
    - Encourage and support interdisciplinary research projects that involve collaborations between CAM and conventional medical researchers to foster innovation and integration.
17. **Utilize Patient-Centered Outcomes Research**
    - Train CAM practitioners in patient-centered outcomes research to focus on the outcomes that matter most to patients, enhancing the relevance of CAM interventions.
18. **Implement Quality Assurance Programs**
    - Develop and implement quality assurance programs within CAM practices to monitor and improve the quality of care provided to patients.
19. **Expand International Exchange Programs**
    - Create international exchange and internship programs for CAM students and practitioners to foster global learning and share diverse practices and approaches.
20. **Encourage Regulatory Participation**
    - Motivate CAM practitioners to participate in regulatory and legislative processes to influence the development of policies that affect CAM practice and integration.
21. **Develop Public Health Initiatives**
    - Engage CAM practitioners in public health initiatives, emphasizing the role of CAM in preventive care and public health promotion.
22. **Promote Environmental Sustainability**
    - Incorporate principles of environmental sustainability into CAM training, emphasizing the importance of sustainable practices in herbal medicine and other resource-dependent modalities.
23. **Advance Telehealth Competencies**
    - Train CAM practitioners in telehealth competencies to expand access to CAM services, particularly in underserved or remote areas.
24. **Foster Mentorship Programs**
    - Establish mentorship programs connecting CAM students and early-career practitioners with experienced professionals to guide their development and integration into the field.
25. **Support Student Research Initiatives**
    - Provide resources and support for student-led research projects in CAM, encouraging innovation and critical thinking from the outset of their careers.
26. **Enhance Access to Professional Resources**
    - Develop centralized resources and professional networks for CAM practitioners to access clinical guidelines, research databases, and professional development opportunities.
27. **Advocate for Inclusive Health Insurance Policies**
    - Advocate for health insurance policies that include comprehensive coverage for CAM services, improving patient access and supporting the integration of CAM into healthcare.
29. **Expand Access to Specialized CAM Libraries**
    - Create and maintain specialized libraries and online repositories for CAM research, textbooks, and multimedia resources to support students and practitioners in their continuous learning and research efforts.
30. **Strengthen Data Analytics and Informatics Skills**
    - Incorporate training in health informatics and data analytics into CAM curricula to enable practitioners to understand and utilize healthcare data effectively, enhancing patient care and research capabilities.

31. **Promote Global Health Initiatives**
    - Encourage participation in global health initiatives that leverage CAM's potential to address global health challenges, such as chronic disease management, mental health, and infectious diseases.
32. **Enhance Regulatory Science Education**
    - Offer education on regulatory science specific to CAM, including product development, quality assurance, and regulatory compliance, to foster innovation and ensure the safety and efficacy of CAM products.
33. **Foster Ethical Research Practices**
    - Emphasize ethical considerations in CAM research, including respect for traditional knowledge, informed consent, and ethical use of natural resources, to uphold high standards of research integrity.
34. **Support Social Entrepreneurship**
    - Encourage and support social entrepreneurship among CAM practitioners to address health disparities and promote health equity through innovative CAM services and products.
35. **Develop Community Outreach Programs**
    - Create community outreach and education programs that promote awareness and understanding of CAM practices, focusing on their benefits, safety, and integration with conventional care.
36. **Promote CAM in Public Health Policy**
    - Advocate for the inclusion of CAM in public health policy and programs, emphasizing its role in preventive health, wellness, and holistic care.
37. **Strengthen Leadership and Advocacy Training**
    - Offer leadership and advocacy training to CAM practitioners to empower them to become effective leaders, advocates, and change agents in healthcare.
38. **Enhance Multidisciplinary Collaboration**
    - Facilitate multidisciplinary collaboration projects that involve CAM practitioners, conventional healthcare providers, and other stakeholders to develop integrated care models and innovative healthcare solutions.
39. **Promote Patient Safety and Quality Care**
    - Implement training and guidelines focused on patient safety and quality care in CAM practice, including adverse event reporting and safety monitoring of CAM therapies.
40. **Support Language and Communication Skills**
    - Enhance language and communication skills training to ensure CAM practitioners can effectively communicate with patients from diverse cultural and linguistic backgrounds.
41. **Encourage Sustainability in CAM Practice**
    - Promote practices and principles of sustainability within CAM disciplines, particularly in the use and sourcing of natural products, to ensure environmental conservation and sustainability.
42. **Advance Health Literacy Programs**
    - Develop health literacy programs focused on CAM, helping patients and the general public make informed decisions about CAM use in conjunction with conventional medical advice.
43. **Promote Accessibility of CAM Services**
    - Work toward making CAM services more accessible to diverse populations, including those with disabilities, through adaptive services and facilities.

44. **Encourage Professionalism and Ethical Practice**
    - Strengthen the emphasis on professionalism and ethical practice in CAM training programs to ensure practitioners adhere to high ethical standards and professional conduct.

45. **Support Special Needs and Pediatric CAM Training**
    - Develop specialized training programs focused on CAM practices suitable for individuals with special needs and pediatric patients, addressing specific considerations and safety protocols.

46. **Foster Resilience and Self-Care among Practitioners**
    - Incorporate training on resilience, self-care, and wellness for CAM practitioners to support their mental health and well-being, enabling them to provide the best care to their patients.

47. **Enhance Emergency Response Training**
    - Provide CAM practitioners with training in emergency response and first aid, equipping them to respond effectively in emergency situations within their scope of practice.

48. **Promote Diversity and Inclusion**
    - Ensure CAM training programs promote diversity and inclusion, both in terms of the student body and faculty, as well as in the curriculum that respects and represents a wide array of cultural perspectives on health and healing.

Implementing these recommendations requires a coordinated effort among educators, practitioners, researchers, policymakers, and the public. By addressing the current challenges and opportunities within CAM training, the healthcare community can better harness the potential of CAM to improve health outcomes, patient satisfaction, and the overall effectiveness of healthcare systems.

CONCLUSION

The evolution and current state of CAM training reflect a dynamic and growing field that is increasingly recognized for its contribution to holistic, patient-centered care, and wellness. Despite its advancements and achievements, CAM education and practice face several challenges, including issues of standardization, evidence-based practice, regulatory and professional integration, public perception, and access to education and care. The recommendations provided aim to address these challenges through a multifaceted approach that involves enhancing educational standards, fostering research and evidence-based practice, promoting interprofessional education and collaboration, and advocating for supportive regulatory and policy frameworks. Additionally, leveraging technology, enhancing public and professional awareness, and ensuring access and equity are crucial steps toward the integration of CAM into mainstream healthcare. The implementation of these recommendations requires the concerted efforts of various stakeholders, including educators, healthcare professionals, researchers, policymakers, and the CAM community. By addressing these challenges and opportunities, we can work toward a healthcare system that fully embraces the potential of CAM to complement conventional medicine, thereby enhancing patient care, promoting health and wellness, and addressing the diverse needs of populations globally. The future of CAM training and practice is poised for further growth and integration, driven by increasing demand for holistic and preventive health approaches, advancements in CAM research, and greater acceptance within the healthcare community and society at large. As we move forward, it's essential to continue fostering an environment that supports innovation, collaboration, and evidence-based practice in CAM, ensuring that it remains a vital and respected component of comprehensive healthcare.

REFERENCES

Al-Worafi, Y.M. (Ed.). (2020a). *Drug Safety in Developing Countries: Achievements and Challenges*. Academic Press.

Al-Worafi, Y.M. (2020b). Herbal medicines safety issues. In: Al-Worafi, Y.M. (ed), *Drug Safety in Developing Countries* (pp. 163–178). Academic Press.

Al-Worafi, Y.M. (2022a). *A Guide to Online Pharmacy Education: Teaching Strategies and Assessment Methods*. CRC Press.

Al-Worafi, Y.M. (2022b). Competencies and learning outcomes. In: Al-Worafi, Y.M. (ed), *A Guide to Online Pharmacy Education: Teaching Strategies and Assessment Methods*. CRC Press.

Al-Worafi, Y.M. (2023a). *Patient Safety in Developing Countries: Education, Research, Case Studies*. CRC Press.

Al-Worafi, Y.M. (2023b). *Technology for Drug Safety: Current Status and Future Developments*. Springer Nature.

Al-Worafi, Y.M. (2023c). Patient safety-related issues: Patient care errors and related problems. In: Al-Worafi, Y.M. (ed), *Patient Safety in Developing Countries: Education, Research, Case Studies*. CRC Press.

Al-Worafi, Y.M. (2023d). Patient care errors and related problems: Preventive medicine errors & related problems. In: Al-Worafi, Y.M. (ed), *Patient Safety in Developing Countries: Education, Research, Case Studies*. CRC Press.

Al-Worafi, Y.M. (2023e). Patient care errors and related problems: Patient assessment and diagnostic errors & related problems. In: Al-Worafi, Y.M. (ed), *Patient Safety in Developing Countries: Education, Research, Case Studies*. CRC Press.

Al-Worafi, Y.M. (2023f). Patient care errors and related problems: Non-pharmacological errors & related problems. In: Al-Worafi, Y.M. (ed), *Patient Safety in Developing Countries: Education, Research, Case Studies*. CRC Press.

Al-Worafi, Y.M. (2023g). Patient care errors and related problems: Medical errors & related problems. In: Al-Worafi, Y.M. (ed), *Patient Safety in Developing Countries: Education, Research, Case Studies*. CRC Press.

Al-Worafi, Y.M. (2023h). Patient care errors and related problems: Monitoring errors & related problems. In: Al-Worafi, Y.M. (ed), *Patient Safety in Developing Countries: Education, Research, Case Studies*. CRC Press.

Al-Worafi, Y.M. (2023i). Patient care errors and related problems: Patient education and counselling errors and related problems. In: Al-Worafi, Y.M. (ed), *Patient Safety in Developing Countries: Education, Research, Case Studies*. CRC Press.

Al-Worafi, Y.M. (2023j). Patient safety-related issues: Other medication safety issues. In: Al-Worafi, Y.M. (eds), *Patient Safety in Developing Countries: Education, Research, Case Studies*. CRC Press.

Al-Worafi, Y.M. (2023k). Patient safety culture. In: Al-Worafi, Y.M. (ed), *Patient Safety in Developing Countries: Education, Research, Case Studies*. CRC Press.

Al-Worafi, Y.M. (2023l). Patient Safety education: Competencies and learning outcomes. In: Al-Worafi, Y.M. (eds), *Patient Safety in Developing Countries: Education, Research, Case Studies*. CRC Press.

Al-Worafi, Y.M. (Ed.). (2023m). *Clinical Case Studies on Medication Safety*. Academic Press.

Al-Worafi, Y.M. (Ed.). (2023n). *Comprehensive Healthcare Simulation: Pharmacy Education, Practice and Research*. Springer Nature.

Al-Worafi, Y.M. (Ed.). (2024a). *Handbook of Medical and Health Sciences in Developing Countries*. Springer, Cham.

Al-Worafi, Y.M. (2024b). Complementary and alternative medicine (CAM) in developing countries. In: Al-Worafi, Y.M. (ed), *Handbook of Medical and Health Sciences in Developing Countries*. Springer, Cham. https://doi.org/10.1007/978-3-030-74786-2_301-1

Hasan, S., Al-Omar, M.J., AlZubaidy, H., and Al-Worafi, Y.M. (2019). Use of medications in Arab countries. In: Laher, I. (ed), *Handbook of Healthcare in the Arab World* (p. 42). Springer, Cham.

9 Complementary and Alternative Medicine (CAM) Education

Quality and Accreditation

INTRODUCTION

Complementary and alternative medicine (CAM) encompasses a diverse range of medical practices and products not traditionally associated with the conventional medical curriculum. Over the years, there has been a growing interest in integrating CAM into healthcare practices, which has subsequently led to the need for structured education and training in this area. The quality of CAM education is crucial for ensuring that healthcare professionals are well-equipped with the knowledge and skills to integrate CAM practices safely and effectively into patient care. This has led to the establishment of various educational standards and accreditation processes specifically designed for CAM programs. Accreditation is a process that evaluates the quality of educational programs and institutions to ensure they meet established standards. For CAM education, accreditation serves multiple purposes: it ensures the curriculum covers essential aspects of CAM practices, promotes the safety and efficacy of these practices, and enhances the credibility of CAM professionals among their conventional medicine peers and the public. In the United States, for example, agencies such as the Accreditation Commission for Acupuncture and Oriental Medicine (ACAOM) and the Council on Chiropractic Education (CCE) are responsible for accrediting programs in their respective fields. These agencies set standards for curriculum, faculty qualifications, student outcomes, and institutional resources, among other criteria. The quality of CAM education is also enhanced through certification and continuing education requirements. Professionals in various CAM fields are often required to pass certification exams and engage in ongoing education to maintain their credentials. This continuous learning ensures that CAM practitioners stay updated with the latest research, techniques, and ethical standards in their fields. However, the integration of CAM education into mainstream health sciences education remains a challenge. This is due in part to the variability in the quality of CAM programs, differences in regulatory and licensing requirements between regions, and skepticism within the conventional medical community. Despite these challenges, the demand for well-educated CAM practitioners continues to grow, driven by consumer interest in holistic and preventive approaches to health. In summary, the quality and accreditation of CAM education are pivotal for ensuring that practitioners are competent and safe. Accreditation bodies play a vital role in maintaining educational standards, while ongoing education and certification ensure practitioners remain knowledgeable and skilled. As interest in CAM continues to rise, the development of more unified standards and greater integration with conventional medical education may further enhance the quality and acceptance of CAM practices.

QUALITY ASSURANCE, QUALITY, AND ACCREDITATION IN COMPLEMENTARY AND ALTERNATIVE MEDICINE (CAM) EDUCATION: RATIONALITY

The rationale behind emphasizing quality assurance, quality, and accreditation in CAM education stems from several key objectives, all aimed at enhancing the credibility, effectiveness, and

 DOI: 10.1201/9781003327202-9

integration of CAM practices within broader healthcare systems. These objectives include ensuring patient safety, promoting high standards of practice, facilitating professional recognition and integration, and supporting continuous improvement and innovation in CAM fields. Let's explore these aspects in more detail:

ENSURING PATIENT SAFETY

One of the primary rationales for stringent quality assurance and accreditation in CAM education is to ensure patient safety. By setting high standards for educational programs, accrediting bodies help ensure that CAM practitioners are knowledgeable, skilled, and competent to deliver safe care. This is particularly important in healthcare, where inadequate training can lead to adverse outcomes. Quality education programs teach not only CAM therapies but also the importance of understanding when to refer patients to other healthcare providers, thus ensuring comprehensive care.

PROMOTING HIGH STANDARDS OF PRACTICE

Quality assurance and accreditation mechanisms aim to elevate the standards of CAM practice. Through rigorous educational standards, these mechanisms ensure that practitioners have a strong foundation in both the theoretical underpinnings and practical applications of their modality. High standards of practice contribute to the effectiveness of CAM interventions and the professionalism of practitioners, which in turn helps to build public trust and confidence in CAM therapies.

FACILITATING PROFESSIONAL RECOGNITION AND INTEGRATION

Accreditation and quality assurance in CAM education are crucial for facilitating the professional recognition of CAM practitioners within the wider healthcare system. Recognized qualifications and adherence to established educational standards can pave the way for greater collaboration between CAM professionals and conventional medical practitioners. This integration is essential for a holistic approach to patient care, where different modalities are used complementarily to achieve the best health outcomes.

SUPPORTING CONTINUOUS IMPROVEMENT AND INNOVATION

Quality assurance processes are inherently geared toward continuous improvement. They encourage CAM educational institutions to regularly evaluate and update their curricula, teaching methods, and clinical training opportunities in response to advances in the field and feedback from students, faculty, and healthcare partners. This commitment to ongoing development and adaptation supports innovation in CAM practices, ensuring they remain relevant, evidence-based, and responsive to healthcare needs.

ENHANCING CREDIBILITY AND PUBLIC TRUST

By ensuring that CAM practitioners are well-educated and adhere to high standards of professional practice, quality assurance and accreditation mechanisms enhance the credibility of CAM professions. This is crucial for building public trust, as patients and conventional healthcare providers are more likely to respect and accept CAM modalities that are practiced by qualified, competent professionals. In conclusion, the emphasis on quality assurance, quality, and accreditation in CAM education is driven by a multifaceted rationale that underscores the importance of safety, effectiveness, professionalism, and integration in healthcare. These efforts are essential for legitimizing CAM practices, fostering interdisciplinary collaboration, and ultimately enhancing patient care and well-being.

QUALITY ASSURANCE, QUALITY, AND ACCREDITATION IN COMPLEMENTARY AND ALTERNATIVE MEDICINE (CAM) EDUCATION: IMPORTANCE AND BENEFITS TO SCHOOLS, STUDENTS, AND THE COMMUNITY

The emphasis on quality assurance, quality, and accreditation in CAM education brings significant benefits to educational institutions, students, and the wider community. These benefits are interconnected, enhancing the overall ecosystem of healthcare education and practice. Here's how each stakeholder benefits:

BENEFITS TO SCHOOLS

1. **Enhanced Reputation and Credibility**: Accreditation by recognized bodies boosts a school's reputation, making it more attractive to prospective students and faculty. It signals a commitment to high-quality education and adherence to established standards, which can also facilitate partnerships with healthcare organizations and other educational institutions.
2. **Continuous Improvement**: The process of achieving and maintaining accreditation requires schools to engage in continuous self-assessment and improvement. This encourages institutions to regularly update their curricula, teaching methodologies, and facilities to reflect the latest advancements and best practices in CAM and healthcare education.
3. **Increased Enrollment**: High-quality, accredited programs are more likely to attract students interested in pursuing careers in CAM. This can lead to increased enrollment as students seek out programs that offer the best preparation for their future professions.
4. **Funding and Resources**: Accredited institutions may have access to more resources, including funding opportunities, grants, and partnerships. These resources can support research initiatives, infrastructure development, and the expansion of educational offerings.

BENEFITS TO STUDENTS

1. **Quality Education**: Students benefit from comprehensive and up-to-date curricula that cover both the theoretical and practical aspects of CAM. This ensures they receive a well-rounded education that prepares them for the complexities of healthcare delivery.
2. **Professional Recognition**: Graduating from an accredited program often facilitates licensure, certification, and employment opportunities. Accreditation assures employers and certifying bodies that the education received meets specific professional standards.
3. **Skill Development**: Quality assurance in education emphasizes not only knowledge acquisition but also the development of critical thinking, clinical skills, and ethical professionalism. This prepares students for successful careers in CAM, equipped to provide effective and safe care.
4. **Career Opportunities**: With a solid educational foundation, students are better positioned to explore diverse career paths within the CAM and broader healthcare sectors, including clinical practice, research, education, and policy-making.

BENEFITS TO THE COMMUNITY

1. **Improved Healthcare Quality**: Well-educated CAM practitioners contribute to the overall quality of healthcare services. They bring a holistic approach to patient care, often integrating traditional and alternative therapies into conventional medical practices, which can lead to improved patient outcomes.
2. **Increased Access to CAM Services**: As the quality and credibility of CAM education improve, more practitioners enter the field, increasing the availability of CAM services.

This diversity in healthcare options can meet the growing patient demand for holistic and preventive care approaches.

3. **Public Trust and Safety**: Accreditation and quality assurance mechanisms ensure that CAM practitioners meet high safety and ethical standards, building public trust in CAM therapies. This is crucial for patients navigating their healthcare options and seeking safe, effective complementary therapies.

4. **Innovation and Research**: Quality education fosters an environment of inquiry and innovation, encouraging students and faculty to engage in research that can advance the field of CAM. This contributes to the community's health and wellness by providing evidence-based CAM therapies and practices.

In summary, the importance of quality assurance, quality, and accreditation in CAM education extends beyond the classroom, affecting the entire healthcare landscape. It ensures that educational institutions offer programs that prepare students for the challenges of healthcare delivery, support students in achieving their professional goals, and ultimately benefit the community by enhancing the quality, safety, and availability of CAM services.

KEY ELEMENTS CONTRIBUTE TO QUALITY ASSURANCE IN COMPLEMENTARY AND ALTERNATIVE MEDICINE (CAM) EDUCATION

Quality assurance in CAM education is vital for ensuring that practitioners are well-prepared to deliver safe and effective care. Several key elements contribute to this quality assurance:

1. **Accreditation:** Accreditation from recognized bodies ensures that CAM educational institutions and programs meet specific standards of quality. Accreditation processes evaluate various aspects of the program, including curriculum content, faculty qualifications, student support services, and institutional resources. This is foundational for ensuring that education in CAM fields is consistent, comprehensive, and prepares students for professional practice.

2. **Curriculum Standards:** A well-structured curriculum that covers the theoretical knowledge, practical skills, and ethical considerations relevant to CAM practices is crucial. The curriculum should be based on the latest evidence and research in the field, incorporating both traditional knowledge and contemporary scientific findings. This ensures that students receive a balanced education that respects the roots of CAM modalities while also acknowledging the importance of evidence-based practice.

3. **Qualified Faculty:** Instructors with appropriate qualifications and experience in their respective CAM fields are essential for delivering high-quality education. Faculty should possess a deep understanding of the subjects they teach, along with practical experience and the ability to convey complex concepts clearly. Their commitment to ongoing professional development and engagement with current research helps maintain the relevance and quality of the education provided.

4. **Student Assessment and Evaluation:** Robust mechanisms for assessing and evaluating student performance help ensure that graduates have achieved the required competencies. This includes both theoretical knowledge and practical skills. Effective assessment strategies might include written exams, practical demonstrations, case studies, and reflective practice journals.

5. **Regulatory Compliance:** Compliance with regulatory requirements for CAM practices in the jurisdiction where the education is provided ensures that programs prepare students for legal and professional practice. This includes understanding the scope of practice, professional ethics, and local laws governing CAM modalities.

6. **Clinical Training:** Hands-on clinical training is a critical component of CAM education, allowing students to apply theoretical knowledge in practical settings under the supervision of experienced practitioners. This exposure to real-world scenarios is essential for developing clinical skills and judgment.

7. **Continuing Education:** Encouraging or requiring ongoing professional development and continuing education for graduates promotes lifelong learning and ensures that practitioners remain current with advances in their field. This is important for maintaining the quality and relevance of CAM practices over time.

8. **Research and Evidence-Based Practice:** Integration of research findings into the curriculum and fostering a culture of inquiry among students and faculty support the development of evidence-based CAM practices. Encouraging critical thinking and the application of scientific methods to CAM enhances the field's credibility and effectiveness.

9. **Feedback and Continuous Improvement:** Systems for collecting feedback from students, faculty, and stakeholders, coupled with regular program reviews, are essential for continuous improvement. This feedback loop allows for adjustments to be made to the curriculum, teaching methods, and clinical training opportunities to enhance the quality of education over time.

Together, these elements form a comprehensive framework for ensuring quality assurance in CAM education, aiming to produce competent, ethical, and professional practitioners who can contribute positively to healthcare systems and patient care.

QUALITY ASSURANCE, QUALITY, AND ACCREDITATION IN COMPLEMENTARY AND ALTERNATIVE MEDICINE (CAM) EDUCATION: ACHIEVEMENTS AND FACILITATORS

The field of CAM has seen significant advancements in terms of quality assurance, quality, and accreditation over the years. These achievements have not only elevated the standards of CAM education but have also facilitated its integration into the broader healthcare system. Key achievements and facilitators in this process include:

ACHIEVEMENTS IN CAM EDUCATION

1. **Establishment of Accreditation Bodies**: One of the major achievements in CAM education has been the establishment of specialized accreditation bodies, such as the ACAOM and the CCE. These organizations have developed rigorous standards for CAM programs, ensuring that they meet high educational and ethical standards.

2. **Development of Standardized Curricula**: The development of standardized curricula for CAM programs has been crucial in ensuring that students receive a comprehensive and consistent education. These curricula cover a broad range of topics, from the theoretical foundations of CAM practices to clinical skills and patient management, ensuring that graduates are well-prepared for professional practice.

3. **Integration into Mainstream Healthcare Education**: CAM education has made strides in integrating with mainstream healthcare education, with some conventional medical schools offering courses or programs in CAM. This integration fosters interdisciplinary understanding and collaboration, which is essential for holistic patient care.

4. **International Recognition and Collaboration**: There has been an increase in international recognition and collaboration in CAM education, with institutions and accreditation bodies working together to share best practices and ensure that CAM education standards are globally consistent. This facilitates the mobility of CAM practitioners and ensures that CAM education is aligned with international healthcare standards.

Facilitators of Progress in CAM Education

1. **Research and Evidence-Based Practice**: The growing body of research supporting the efficacy of certain CAM practices has facilitated their acceptance and integration into healthcare. This evidence base has also informed CAM education, emphasizing the importance of an evidence-based approach in teaching and practice.
2. **Technological Advancements**: Technological advancements have facilitated the delivery of CAM education, enabling institutions to offer online and hybrid programs. This has made CAM education more accessible to a wider audience and has supported the continuous learning of CAM practitioners.
3. **Professionalization of CAM Fields**: The professionalization of CAM fields, including the development of professional standards, codes of ethics, and certification processes, has played a critical role in enhancing the quality and credibility of CAM education. This has helped to establish CAM as a legitimate and respected component of healthcare.
4. **Consumer Demand and Advocacy**: Increased consumer interest in and demand for CAM therapies have driven the expansion and improvement of CAM education. Advocacy by patients and CAM practitioners has also played a role in pushing for higher standards and greater recognition of CAM practices.
5. **Interdisciplinary Collaboration**: The promotion of interdisciplinary collaboration between CAM practitioners and conventional healthcare providers has facilitated a more integrated approach to healthcare. This collaboration has highlighted the importance of comprehensive CAM education that includes knowledge of conventional medical practices and how CAM can complement them.

In conclusion, the achievements in the quality assurance, quality, and accreditation of CAM education have been facilitated by a combination of dedicated accreditation bodies, a focus on evidence-based practice, technological advancements, the professionalization of the field, consumer demand, and interdisciplinary collaboration. These factors have collectively contributed to the recognition, acceptance, and integration of CAM within the healthcare system, ensuring that CAM practitioners are well-equipped to provide safe, effective, and integrated care to patients.

QUALITY ASSURANCE, QUALITY, AND ACCREDITATION IN COMPLEMENTARY AND ALTERNATIVE MEDICINE (CAM) EDUCATION: CHALLENGES AND BARRIERS

Despite significant advancements in the field, the path to enhancing quality assurance, quality, and accreditation in CAM education faces several challenges and barriers. Addressing these issues is crucial for further integrating CAM into the healthcare system and ensuring the highest standards of education and practice. Key challenges and barriers include:

Variability in Standards and Regulation

One of the primary challenges is the variability in educational standards and regulatory frameworks across different regions and CAM modalities. This lack of uniformity can make it difficult to ensure consistent quality in CAM education and practice. It also complicates the process of transferring credentials or practicing across different jurisdictions.

Skepticism from Conventional Medicine

There remains a degree of skepticism and resistance toward CAM within parts of the conventional medical community. This can hinder collaborative efforts, reduce opportunities for CAM education integration into mainstream healthcare curricula, and limit the acceptance of CAM practices.

LIMITED RESEARCH AND EVIDENCE BASE

Although research into CAM practices has increased, there is still a need for more high-quality, evidence-based research to support the efficacy and safety of various CAM modalities. This gap can affect the content and quality of CAM education, making it challenging to build curricula around evidence-based practices.

FINANCIAL AND RESOURCE CONSTRAINTS

Developing and maintaining high-quality CAM education programs requires significant financial and human resources. Accreditation processes, in particular, can be resource-intensive. Smaller institutions or those in developing regions may struggle to meet these demands, potentially affecting the quality of education they can provide.

CULTURAL AND PHILOSOPHICAL DIFFERENCES

The diverse nature of CAM practices, rooted in different cultural and philosophical traditions, poses challenges to standardizing education and practice. Balancing respect for traditional knowledge with the demands of modern healthcare and evidence-based practice requires careful consideration and sometimes leads to tension.

INTEGRATION INTO HEALTHCARE SYSTEMS

Integrating CAM education and practices into existing healthcare systems presents logistical and regulatory challenges. Establishing pathways for professional collaboration, referral, and shared patient care between CAM practitioners and conventional healthcare providers remains a complex issue.

ADDRESSING PUBLIC MISCONCEPTIONS

Public misconceptions about CAM can affect the demand for education and services, as well as the willingness of conventional healthcare providers to collaborate with CAM practitioners. Education and outreach are needed to address these misconceptions and promote an accurate understanding of the role and value of CAM in healthcare.

ADAPTATION TO TECHNOLOGICAL ADVANCES

Keeping pace with technological advances in healthcare and education requires continuous updates to CAM curricula and teaching methods. This can be a challenge, particularly for traditional practices with historical roots.

ENHANCED INTERDISCIPLINARY COLLABORATION

Strengthening collaborative networks between CAM and conventional healthcare education institutions can foster mutual understanding and respect. Joint educational programs, research projects, and clinical rotations can expose students and faculty to the benefits of integrative healthcare, breaking down barriers and skepticism.

INCREASED INVESTMENT IN CAM RESEARCH

Governments, non-profit organizations, and private entities could increase funding for CAM research to expand the evidence base for various modalities. High-quality, rigorous research can inform education standards, improve curriculum development, and enhance the integration of CAM into evidence-based practice.

DEVELOPMENT OF GLOBAL STANDARDS

International organizations could play a role in developing global standards for CAM education and practice. Such standards could help harmonize education quality, facilitate the mobility of practitioners, and ensure consistent patient care across borders.

LEVERAGING TECHNOLOGY FOR HYBRID EDUCATION MODELS

Innovative use of technology can address challenges in teaching practical skills through online education. Virtual reality, augmented reality, and simulation-based learning can complement traditional teaching methods, offering students immersive, interactive experiences that mimic real-life practice.

PUBLIC EDUCATION AND AWARENESS CAMPAIGNS

Educational campaigns aimed at the public and healthcare professionals can help dispel myths and misconceptions about CAM. By providing accurate information about the benefits, limitations, and potential integrations of CAM into healthcare, these campaigns can increase acceptance and demand for CAM services.

POLICY ADVOCACY FOR CAM INTEGRATION

Advocacy efforts can focus on influencing healthcare policies to better accommodate CAM practices and education. This includes working toward more inclusive insurance coverage, developing clear referral pathways between CAM practitioners and conventional healthcare providers, and ensuring that CAM is considered in health system planning and decision-making.

FOSTERING A CULTURE OF LIFELONG LEARNING

Encouraging a culture of continuous professional development among CAM practitioners can ensure that they remain at the forefront of their field. This includes not only keeping up with the latest research but also developing skills in areas such as patient communication, digital health technologies, and integrative care models.

ADDRESSING RESOURCE CONSTRAINTS

Strategies to support CAM educational institutions facing financial and resource constraints could include developing partnerships with larger universities, accessing grants and funding dedicated to CAM education, and leveraging community resources for clinical training opportunities.

STANDARDIZING ONLINE EDUCATION QUALITY

Establishing clear standards and accreditation criteria for online CAM education programs can ensure that they meet the same quality requirements as in-person programs. This might involve developing best practices for online pedagogy, student engagement, and assessment methods.

By addressing these challenges with thoughtful strategies and collaborative efforts, the field of CAM can move toward greater standardization, acceptance, and integration into the healthcare system. Such progress would not only benefit those seeking to enter CAM professions but also serve the broader goal of improving health outcomes and patient care through a more holistic and inclusive approach to medicine.

QUALITY ASSURANCE, QUALITY, AND ACCREDITATION IN COMPLEMENTARY AND ALTERNATIVE MEDICINE (CAM) EDUCATION: RECOMMENDATIONS

Improving quality assurance, quality, and accreditation in CAM education is crucial for producing competent practitioners and enhancing healthcare systems. To further enhance quality assurance, quality, and accreditation in CAM education, a series of recommendations can be proposed (Al-Worafi, 2020a,b, 2022a,b, 2023a–n, 2024a,b; Hasan et al., 2019). These suggestions aim to address existing challenges and promote the continuous improvement of CAM education and integration into the healthcare system. Here are the key recommendations:

1. **Strengthening Accreditation Standards**
   - Develop and continuously update comprehensive accreditation standards that reflect the latest best practices and research in CAM and healthcare education.
   - Encourage all CAM education programs to seek accreditation from recognized bodies to ensure consistency and quality across the field.
2. **Promoting Interprofessional Education**
   - Implement interprofessional education (IPE) programs that include CAM and conventional medicine students to foster mutual respect, understanding, and collaboration.
   - Design IPE curricula that highlight the strengths and limitations of both CAM and conventional treatments, emphasizing patient-centered care.
3. **Enhancing Research and Evidence-Based Practice**
   - Increase funding and support for research in CAM to expand the evidence base for various modalities.
   - Integrate evidence-based practice principles into CAM education, teaching students how to critically evaluate research and apply findings to clinical practice.
4. **Facilitating Regulatory Harmonization**
   - Work toward harmonizing regulations and standards for CAM practice and education across different jurisdictions to facilitate the mobility of practitioners and ensure consistent care quality.
   - Engage with regulatory bodies, professional associations, and educators to align on core competencies and standards.
5. **Expanding Access to Quality Resources**
   - Provide CAM educational institutions, especially those in resource-limited settings, with access to quality teaching materials, research databases, and technology tools.
   - Develop partnerships between well-resourced institutions and those in need of support to share knowledge, resources, and best practices.
6. **Encouraging Lifelong Learning**
   - Establish clear pathways for CAM practitioners' continuous professional development, including specialized training, advanced degrees, and cross-disciplinary learning opportunities.
   - Make lifelong learning a requirement for maintaining licensure or certification in CAM disciplines.
7. **Leveraging Technology in Education**
   - Invest in developing and implementing innovative educational technologies that can enhance the learning experience, especially for practical skills and clinical training.
   - Adopt hybrid models of education that combine online learning with hands-on clinical experiences to increase accessibility and flexibility for students.

8. **Increasing Public Awareness and Advocacy**
   - Launch public awareness campaigns to educate the public and healthcare providers about the benefits, limitations, and potential uses of CAM.
   - Advocate for policies that support the integration of CAM into healthcare systems, including insurance coverage and research funding.
9. **Building Global Collaboration**
   - Foster international collaboration among CAM educational institutions, accreditation bodies, and professional associations to share best practices, research findings, and educational resources.
   - Work toward international agreements that recognize CAM qualifications across borders, facilitating global mobility for practitioners.
10. **Ensuring Patient-Centered Care**
   - Emphasize the importance of patient-centered care in CAM education, teaching students to consider the whole person, including their physical, emotional, and social needs.
   - Train CAM practitioners to communicate effectively with patients and other healthcare providers, ensuring integrated and coordinated care.
11. **Educational Content and Pedagogy**
   - *Global Health Perspectives*: By incorporating global health perspectives, CAM education becomes more inclusive and respectful of cultural diversity, preparing practitioners to serve diverse populations effectively.
   - *Specialized Tracks*: Specialized tracks allow students to deepen their knowledge and skills in specific areas, fostering innovation and expertise that can advance the field of CAM.
   - *Ethical Practice*: A strong emphasis on ethics prepares students to navigate complex ethical issues, ensuring patient safety and trust in CAM practices.
   - *Clinical Reasoning*: Early integration of clinical reasoning enhances critical thinking and decision-making skills, crucial for effective practice in real-world settings.
   - *Environmental Health*: Highlighting environmental health underscores the interconnectedness of health and the environment, promoting practices that are sustainable and beneficial for both individual and planetary health.
12. **Faculty Development and Support**
   - *Enhanced Training*: Ongoing professional development for faculty ensures that teaching methods remain innovative, relevant, and effective.
   - *Diverse Faculty*: Diversity among faculty enriches the educational experience, offering students a wide range of perspectives and expertise.
   - *Research Support*: Supporting faculty research contributes to the CAM evidence base, enhancing the field's credibility and the quality of education.
13. **Student Support and Engagement**
   - *Supportive Learning Environment*: A supportive environment promotes student well-being and success, crucial for cultivating competent and confident CAM practitioners.
   - *Student Research*: Encouraging student research involvement fosters critical thinking and contributes to the field's body of knowledge.
   - *Peer Learning and Career Services*: Peer learning builds community and collaboration skills, while enhanced career services support students' transition into professional practice.
14. **Technological and Methodological Innovations**
   - *Blended Learning Models and Simulation Technologies*: These approaches accommodate diverse learning styles and enhance practical skill development, making education more accessible and effective.

- *Data Analytics and Digital Health Literacy*: Leveraging technology in education and practice prepares students for the digital future of healthcare, enabling data-driven decision-making and effective use of telehealth and other digital health tools.

15. **Regulatory and Professional Development**
    - *Inclusive Licensing and Professional Titles*: Working toward inclusive licensing and standardizing professional titles helps clarify the roles of CAM practitioners, facilitating integration into the healthcare system.
    - *Continuing Education and Leadership Programs*: Ensuring access to continuing education and developing leadership skills are vital for the ongoing professional growth and advocacy capacity of CAM practitioners.

16. **Community and Healthcare Integration**
    - *Community Partnerships and Public Health Initiatives*: These efforts promote the integration of CAM into broader healthcare and public health strategies, emphasizing preventive care and holistic wellness.
    - *Integrative Healthcare Models and Patient Education*: Supporting integrative models and enhancing patient education foster collaborative care approaches and informed healthcare choices.
    - *Policy Engagement and Global Health Exchanges*: Engaging in policy development and facilitating global exchanges expands the influence of CAM practices and ensures practitioners are well-rounded and globally aware.

17. **Advocacy for Research Funding**
    - *Increased Funding for CAM Research*: Advocacy for more research funding is crucial for building an evidence base that supports the efficacy, safety, and integration of CAM practices into mainstream healthcare.

Implementing these recommendations requires a collaborative effort among educators, accreditation bodies, practitioners, policymakers, and the public. By addressing these areas, the CAM community can enhance the quality and recognition of CAM education and practice, ultimately contributing to better health outcomes and a more holistic approach to healthcare.

CONCLUSION

The comprehensive set of recommendations for enhancing quality assurance, quality, and accreditation in CAM education underscores a multifaceted approach to improving CAM practices and their integration into the broader healthcare system. These recommendations span several critical areas, including educational content and pedagogy, faculty development and support, student engagement, technological and methodological innovations, regulatory frameworks, professional development, and community and healthcare integration. Together, they aim to ensure that CAM education is robust, evidence-based, culturally sensitive, and aligned with the evolving needs of the healthcare landscape. The successful implementation of these recommendations requires a collaborative and coordinated effort among various stakeholders, including educational institutions, healthcare practitioners, regulatory bodies, policymakers, and the community at large. By addressing the diverse needs and challenges within CAM education and practice, these strategies can foster a more inclusive, effective, and integrated approach to healthcare. This not only enhances the credibility and professionalism of CAM practitioners but also ensures that patients receive safe, effective, and holistic care. Moreover, these recommendations highlight the importance of embracing innovation, diversity, and interprofessional collaboration, while advocating for a healthcare model that values preventive care, patient-centered approaches, and the sustainable practice of medicine. The emphasis on continuous learning, research, and adaptation to technological advancements ensures that CAM education remains relevant and responsive to new challenges and opportunities. Ultimately, the goal is to cultivate a healthcare environment where CAM is not viewed as an alternative but as an integral component of a comprehensive care strategy that respects and utilizes the full spectrum of healing practices. Achieving this vision requires ongoing effort, open dialogue, and a commitment to quality and excellence in CAM education and practice. By moving forward with these recommendations, the CAM community can contribute to a more holistic, effective, and compassionate healthcare system that meets the diverse needs of patients and societies worldwide.

REFERENCES

Al-Worafi, Y.M. (Ed.). (2020a). *Drug Safety in Developing Countries: Achievements and Challenges.* Academic Press.

Al-Worafi, Y.M. (2020b). Herbal medicines safety issues. In: Al-Worafi, Y.M. (ed), *Drug Safety in Developing Countries* (pp. 163–178). Academic Press.

Al-Worafi, Y.M. (2022a). *A Guide to Online Pharmacy Education: Teaching Strategies and Assessment Methods.* CRC Press.

Al-Worafi, Y.M. (2022b). Competencies and learning outcomes. In: Al-Worafi, Y.M. (ed), *A Guide to Online Pharmacy Education: Teaching Strategies and Assessment Methods.* CRC Press.

Al-Worafi, Y.M. (2023a). *Patient Safety in Developing Countries: Education, Research, Case Studies.* CRC Press.

Al-Worafi, Y.M. (2023b). *Technology for Drug Safety: Current Status and Future Developments.* Springer Nature.

Al-Worafi, Y.M. (2023c). Patient safety-related issues: Patient care errors and related problems. In: Al-Worafi, Y.M. (ed), *Patient Safety in Developing Countries: Education, Research, Case Studies.* CRC Press.

Al-Worafi, Y.M. (2023d). Patient care errors and related problems: Preventive medicine errors & related problems. In: Al-Worafi, Y.M. (ed), *Patient Safety in Developing Countries: Education, Research, Case Studies.* CRC Press.

Al-Worafi, Y.M. (2023e). Patient care errors and related problems: Patient assessment and diagnostic errors & related problems. In: Al-Worafi, Y.M. (ed), *Patient Safety in Developing Countries: Education, Research, Case Studies.* CRC Press.

Al-Worafi, Y.M. (2023f). Patient care errors and related problems: Non-pharmacological errors & related problems. In: Al-Worafi, Y.M. (ed), *Patient Safety in Developing Countries: Education, Research, Case Studies.* CRC Press.

Al-Worafi, Y.M. (2023g). Patient care errors and related problems: Medical errors & related problems. In: Al-Worafi, Y.M. (ed), *Patient Safety in Developing Countries: Education, Research, Case Studies.* CRC Press.

Al-Worafi, Y.M. (2023h). Patient care errors and related problems: Monitoring errors & related problems. In: Al-Worafi, Y.M. (ed), *Patient Safety in Developing Countries: Education, Research, Case Studies.* CRC Press.

Al-Worafi, Y.M. (2023i). Patient care errors and related problems: Patient education and counselling errors and related problems. In: Al-Worafi, Y.M. (ed), *Patient Safety in Developing Countries: Education, Research, Case Studies.* CRC Press.

Al-Worafi, Y.M. (2023j). Patient safety-related issues: Other medication safety issues. In: Al-Worafi, YM. (eds) *Patient Safety in Developing Countries: Education, Research, Case Studies.* CRC Press.

Al-Worafi, Y.M. (2023k). Patient safety culture. In: Al-Worafi, Y.M. (ed), *Patient Safety in Developing Countries: Education, Research, Case Studies.* CRC Press.

Al-Worafi, Y.M. (2023l). Patient safety education: Competencies and learning outcomes. In: Al-Worafi, YM. (eds) *Patient Safety in Developing Countries: Education, Research, Case Studies.* CRC Press.

Al-Worafi, Y.M. (Ed.). (2023m). *Clinical Case Studies on Medication Safety.* Academic Press.

Al-Worafi, Y.M. (Ed.). (2023n). *Comprehensive Healthcare Simulation: Pharmacy Education, Practice and Research.* Springer Nature.

Al-Worafi, Y.M. (Ed.). (2024a). *Handbook of Medical and Health Sciences in Developing Countries.* Springer, Cham.

Al-Worafi, Y.M. (2024b). Complementary and alternative medicine (CAM) in developing countries. In: Al-Worafi, Y.M. (ed), *Handbook of Medical and Health Sciences in Developing Countries.* Springer, Cham. https://doi.org/10.1007/978-3-030-74786-2_301-1

Hasan, S., Al-Omar, M.J., AlZubaidy, H., and Al-Worafi, Y.M. (2019). Use of medications in Arab countries. In: Laher, I. (ed), *Handbook of Healthcare in the Arab World* (p. 42). Springer, Cham.

10 Complementary and Alternative Medicine (CAM) Education

Access/Equitable Access

INTRODUCTION

Complementary and alternative medicine (CAM) education represents a vital component in the broad landscape of healthcare training, focusing on non-traditional therapies and practices that exist outside the realm of conventional medicine. This education encompasses a diverse range of modalities, including herbal medicine, acupuncture, chiropractic care, naturopathy, and mind-body techniques, among others. The primary goal of CAM education is to provide healthcare professionals with a comprehensive understanding of these alternative approaches, enabling them to offer more holistic and patient-centered care. By integrating CAM education into the healthcare curriculum, future practitioners are equipped with a wider array of tools to address patient needs, fostering a more inclusive approach to health and wellness. Access to CAM education, however, presents a unique set of challenges and opportunities. On the one hand, the growing interest and acceptance of CAM practices within both the public and professional healthcare communities have led to increased demand for education and training in these areas. Institutions worldwide are gradually incorporating CAM courses and modules into their healthcare programs, reflecting this shift in perspective. On the other hand, equitable access to CAM education remains a significant hurdle. Factors such as geographic location, socioeconomic status, and educational disparities can limit the availability of and accessibility to CAM training for many aspiring healthcare professionals. This discrepancy raises concerns about the equitable distribution of knowledge and skills in the healthcare workforce, potentially impacting the quality of care provided to patients who seek or could benefit from CAM therapies.

RATIONALITY AND IMPORTANCE OF ACCESS/EQUITABLE ACCESS TO THE COMPLEMENTARY AND ALTERNATIVE MEDICINE (CAM) EDUCATION

The rationality behind promoting access and equitable access to CAM education stems from several key considerations that highlight its importance in contemporary healthcare. Firstly, the growing consumer interest and utilization of CAM therapies necessitate that healthcare providers possess a broad understanding of these practices to effectively communicate with patients, manage treatments, and provide integrated care. As patients increasingly seek CAM modalities for preventive care, chronic condition management, and overall well-being, healthcare professionals must be equipped to offer informed guidance, support informed decision-making, and safely integrate CAM approaches within conventional treatment plans when appropriate.

Equitable access to CAM education ensures that all healthcare providers, regardless of their geographical location, socioeconomic status, or educational background, have the opportunity to acquire knowledge and skills in CAM. This is crucial for several reasons:

DOI: 10.1201/9781003327202-10

1. **Enhanced Patient Care**: By understanding the nuances of CAM practices, healthcare professionals can offer more comprehensive and person-centered care. This holistic approach acknowledges the physical, emotional, spiritual, and social aspects of health, leading to improved patient outcomes and satisfaction.

2. **Informed Decision-Making**: Equipping healthcare providers with a thorough understanding of CAM allows them to better navigate the vast landscape of available treatments, discerning between evidence-based practices and those with limited or no scientific support. This informed perspective is essential for guiding patients in making safe and effective healthcare choices.

3. **Professional Collaboration**: Access to CAM education fosters greater collaboration among healthcare professionals from diverse disciplines. This interdisciplinary approach encourages the sharing of knowledge and expertise, promoting a more cohesive healthcare system that can address the multifaceted needs of patients.

4. **Health Equity**: Equitable access to CAM education addresses disparities in healthcare delivery and outcomes. By ensuring that all healthcare professionals have the opportunity to learn about CAM, the healthcare system can better serve diverse populations, including those that may rely on or prefer alternative therapies due to cultural, personal, or economic reasons.

5. **Adaptation to Healthcare Trends**: The inclusion of CAM education reflects and supports the evolving landscape of healthcare, where patient preferences and evidence-based practices guide the integration of diverse therapeutic approaches. Preparing healthcare professionals to operate within this dynamic environment is essential for maintaining relevance and effectiveness in patient care.

In summary, access and equitable access to CAM education are rational and important for developing a well-rounded, knowledgeable healthcare workforce capable of meeting the diverse needs of the population. By embracing the breadth of conventional and alternative approaches, the healthcare industry can ensure a more inclusive, effective, and person-centered care system.

ACCESS/EQUITABLE ACCESS TO COMPLEMENTARY AND ALTERNATIVE MEDICINE (CAM) EDUCATION: FACILITATORS

Facilitating access and equitable access to CAM education involves overcoming barriers and leveraging various strategies to ensure that healthcare professionals can gain the knowledge and skills needed to incorporate CAM into their practice. Several facilitators can play a crucial role in enhancing access to CAM education:

1. **Integration into Mainstream Curricula**: Incorporating CAM topics into the curricula of medical, nursing, pharmacy, and other healthcare professional schools as standard coursework can significantly increase exposure and understanding of CAM practices. This approach ensures that all students receive foundational knowledge in CAM, regardless of their specialization.

2. **Online Learning Platforms**: The development and expansion of online courses and degree programs in CAM can address geographical and logistical barriers. Online education allows for flexible learning schedules, making CAM education more accessible to a wider audience, including practicing healthcare professionals seeking to expand their skill sets.

3. **Continuing Education Opportunities**: Offering CAM education through continuing education programs allows healthcare professionals to stay updated on the latest research,

trends, and practices in CAM. Workshops, seminars, and online courses can cater to the needs of professionals at different stages of their careers.

4. **Scholarships and Financial Aid**: Providing scholarships, grants, and financial aid targeted toward CAM education can help overcome economic barriers faced by students and professionals from underrepresented or low-income backgrounds. This support can make CAM education more equitable and accessible.

5. **Partnerships and Collaborations**: Establishing partnerships between conventional medical institutions and CAM schools or organizations can facilitate knowledge exchange and resource sharing. Collaborations can also lead to the development of joint programs, research initiatives, and clinical training opportunities in CAM.

6. **Policy and Accreditation Support**: Advocacy for policies that recognize and support CAM education within healthcare training and practice can further facilitate access. Accreditation standards for CAM programs can ensure quality education, making CAM more integrated into the healthcare system.

7. **Public and Professional Awareness Campaigns**: Raising awareness about the benefits and evidence base of CAM practices among healthcare professionals and the public can increase demand for CAM education. Awareness campaigns can also highlight the importance of integrating CAM into healthcare delivery.

8. **Research and Evidence-Based Practice**: Strengthening the research foundation for CAM and disseminating findings through academic and professional channels can enhance the credibility and acceptance of CAM education. Evidence-based practices in CAM can be integrated into educational programs to ensure that healthcare professionals are equipped with scientifically validated knowledge.

By addressing these facilitators, stakeholders in healthcare education and policy can work toward creating a more inclusive, knowledgeable, and competent healthcare workforce that is well-equipped to meet the diverse needs of patients with an integrative approach to health and wellness.

ACCESS/EQUITABLE ACCESS TO COMPLEMENTARY AND ALTERNATIVE MEDICINE (CAM) EDUCATION: BARRIERS

Access and equitable access to CAM education face several barriers that can limit the availability and quality of education for healthcare professionals and students. These obstacles can hinder the integration of CAM into mainstream healthcare practices and affect the delivery of holistic patient care. Understanding these barriers is crucial for developing strategies to overcome them and enhance the accessibility of CAM education. Key barriers include:

1. **Limited Institutional Support**: Many conventional medical and healthcare training institutions may have limited curricular space or lack the institutional will to integrate CAM education. This can stem from skepticism about the efficacy of CAM practices, a focus on traditional biomedical models, or budgetary constraints.

2. **Lack of Standardized Curriculum**: The diversity of CAM practices and the varying levels of evidence supporting their efficacy can make it challenging to develop a standardized, evidence-based curriculum. This lack of standardization can result in inconsistencies in the quality and content of CAM education across different programs.

3. **Insufficient Faculty Expertise**: There may be a shortage of educators and clinicians with the necessary expertise in CAM practices and the ability to teach them effectively. This gap can limit the availability of high-quality CAM education and mentorship opportunities for students interested in these areas.

4. **Regulatory and Accreditation Challenges**: Regulatory and accreditation bodies for healthcare education often have specific requirements that may not fully accommodate CAM education. Navigating these regulations to incorporate CAM into accredited programs can be complex and time-consuming.

5. **Cultural and Perceptual Barriers**: Cultural biases and perceptions within the medical community and society at large can devalue CAM practices, leading to resistance against incorporating them into professional healthcare education and practice.

6. **Financial Constraints**: The cost of developing and implementing CAM education programs, including hiring qualified instructors, developing curriculum materials, and providing clinical training opportunities, can be prohibitive for some institutions. Additionally, students may face financial barriers to accessing specialized CAM training programs.

7. **Access to Clinical Training**: Gaining practical experience in CAM can be challenging due to a limited number of clinical training sites that integrate CAM practices. This lack of access can hinder students' ability to gain hands-on experience and confidence in applying CAM modalities.

8. **Research and Evidence Gaps**: Although research on CAM is growing, gaps in the evidence base for some CAM practices can lead to skepticism and hinder their inclusion in mainstream education. Addressing these research gaps is essential for building the credibility and acceptance of CAM within the healthcare community.

Overcoming these barriers requires a multifaceted approach that involves increasing awareness and understanding of CAM benefits, enhancing institutional support, developing evidence-based curricula, and fostering collaboration between conventional and alternative medicine educators and practitioners. Efforts to address financial and regulatory challenges, alongside initiatives to expand clinical training opportunities, are also crucial for improving access to CAM education and ensuring that healthcare professionals are well-equipped to meet the diverse needs of their patients with an integrative approach to care.

ACCESS/EQUITABLE ACCESS TO COMPLEMENTARY AND ALTERNATIVE MEDICINE (CAM) EDUCATION: ACHIEVEMENTS

Despite the barriers to access and equitable access to CAM education, significant achievements have been made in recent years that demonstrate progress in integrating CAM into healthcare training and practice. These achievements reflect the growing acceptance of CAM within the healthcare community and society at large, as well as efforts to address the demand for holistic, patient-centered care. Some notable achievements include:

1. **Inclusion in Medical Curricula**: A growing number of medical, nursing, and pharmacy schools have begun to integrate CAM topics into their curricula. This integration ranges from elective courses to required modules that cover various CAM modalities, reflecting an acknowledgment of the importance of CAM in comprehensive patient care.

2. **Accreditation of CAM Programs**: The accreditation of CAM educational programs, such as those in acupuncture, naturopathy, and chiropractic medicine, by recognized accrediting bodies has helped standardize the quality of education. This accreditation ensures that students receive a rigorous education based on established standards and competencies.

3. **Development of Online CAM Education Resources**: The expansion of online courses and degree programs in CAM has made education more accessible to a wider audience. Online platforms offer flexibility for both students and practicing healthcare professionals to learn about CAM at their own pace, regardless of their geographical location.

4. **Interprofessional Education Initiatives**: There have been efforts to promote interprofessional education that includes CAM, facilitating collaboration and understanding among healthcare professionals from different disciplines. These initiatives aim to prepare healthcare providers to work effectively in team-based care environments that value diverse therapeutic approaches.

5. **Research Funding and Publications**: Increased funding for CAM research from government bodies and private institutions has led to a growth in evidence-based studies on CAM therapies. The publication of CAM research in peer-reviewed journals enhances the legitimacy and acceptance of CAM practices within the scientific community and informs educational content.

6. **Public and Professional Organizations**: The establishment and growth of professional organizations and societies dedicated to CAM practices have played a crucial role in advocating for CAM education, research, and practice. These organizations often provide educational resources, host conferences, and offer certification programs to professionals.

7. **Global Health Initiatives**: Internationally, there have been initiatives to integrate CAM into public health strategies, recognizing the role of traditional and alternative medicine in global health. The World Health Organization (WHO), for example, has developed strategies and guidelines to support the integration of traditional and complementary medicine into national health systems.

8. **Community and Clinical Integration**: There has been an increase in hospitals and healthcare settings offering CAM services to patients, facilitated by healthcare professionals trained in these modalities. This integration allows patients to access a variety of therapeutic options within conventional care settings.

9. **Cross-Institutional Collaborations**: Partnerships between CAM institutions and traditional medical schools are fostering an environment where knowledge and practices are shared more freely. These collaborations can lead to the development of integrated curricula that offer students from both backgrounds opportunities to learn from each other and understand a broader spectrum of patient care approaches.

10. **Policy Advocacy and Support**: Advocacy efforts have led to policy changes at both the educational and governmental levels, supporting the inclusion of CAM in healthcare policies and insurance coverage. This policy support not only legitimizes CAM practices but also encourages educational institutions to invest more resources in CAM education.

11. **Expanded Clinical Trials**: The increase in clinical trials focused on CAM therapies has provided a stronger evidence base for their efficacy and safety. This growing body of research supports the integration of CAM into mainstream medical education by providing data that educators can use to inform their teachings and recommendations.

12. **Diversity and Inclusion Initiatives**: Recognizing the importance of diversity in healthcare, there have been concerted efforts to ensure that CAM education is inclusive and reflective of various cultural practices and perspectives. This includes scholarships aimed at underrepresented groups in healthcare, programs designed to explore the traditional medicine of different cultures, and efforts to diversify faculty and student bodies within CAM educational programs.

13. **Patient-Centered Care Models**: The shift toward patient-centered care models in healthcare has emphasized the importance of understanding and integrating patient preferences, including the use of CAM therapies. This shift has necessitated a corresponding change in education, where healthcare professionals are trained to discuss and incorporate CAM into personalized care plans.

14. **Professional Development and Continuing Education**: The development of continuing education programs in CAM for healthcare professionals already in practice allows for lifelong learning and the integration of new CAM practices as they emerge. These

programs ensure that the healthcare workforce remains knowledgeable about the latest CAM therapies and research findings.

15. **Increased Public Awareness and Demand**: As public interest in and demand for CAM continue to grow, healthcare systems are more inclined to incorporate CAM services and education. This demand acts as a driving force for educational institutions to prioritize CAM in their offerings and for healthcare providers to seek CAM education.

16. **Technological Innovations in Education**: Advances in educational technology, including virtual reality (VR) and augmented reality (AR), are being utilized to create immersive CAM learning experiences. These technologies can simulate real-life clinical scenarios involving CAM therapies, enhancing the depth and quality of CAM education.

17. **Global Standards for CAM Education**: Efforts to establish global standards for CAM education aim to harmonize training and practice worldwide, ensuring that all CAM practitioners meet a baseline level of competency and knowledge. This global approach facilitates the mobility of CAM practitioners and the international exchange of knowledge.

These achievements indicate a positive trend toward recognizing and incorporating CAM into healthcare education and practice. However, continued efforts are needed to address remaining challenges, promote equitable access to CAM education, and ensure that healthcare professionals are equipped to meet the diverse needs of their patients with an informed and open-minded approach to CAM.

ACCESS/EQUITABLE ACCESS TO COMPLEMENTARY AND ALTERNATIVE MEDICINE (CAM) EDUCATION: CHALLENGES

The journey toward ensuring access and equitable access to CAM education is fraught with challenges that span educational, cultural, regulatory, and financial domains. These challenges reflect the complexity of integrating CAM into a healthcare education system that has historically been rooted in conventional medicine. Understanding these challenges is crucial for developing targeted strategies to overcome them. Some of the key challenges include:

1. **Cultural and Professional Bias**: There remains a significant cultural and professional bias within parts of the healthcare and academic communities against CAM practices. This skepticism can lead to resistance to fully integrating CAM education into mainstream healthcare curricula, affecting the depth and breadth of CAM education offered.

2. **Lack of Standardization**: CAM encompasses a wide range of practices with varying degrees of evidence and methodologies. The lack of standardization in CAM education, including varying degrees of rigor, scope, and content across programs, makes it challenging to ensure a consistent and high-quality educational experience for all students.

3. **Regulatory and Accreditation Hurdles**: Navigating the regulatory and accreditation standards for CAM programs can be complex, given that these standards may not always align with the unique aspects of CAM practices. This can limit the establishment and recognition of CAM educational programs and restrict their integration into conventional healthcare training.

4. **Limited Research and Evidence Base**: While research into CAM is growing, gaps in the evidence base for certain CAM modalities can pose challenges to their acceptance and inclusion in education. The lack of robust, large-scale studies makes it difficult for educators to incorporate certain CAM topics into curricula with the same confidence as evidence-based conventional medicine.

5. **Resource Constraints**: Developing and implementing CAM education requires substantial resources, including faculty with expertise in CAM, appropriate teaching materials,

and clinical training opportunities. Financial constraints can limit an institution's ability to offer comprehensive CAM education, particularly in regions or settings with limited healthcare funding.

6. **Access to Clinical Training**: Providing students with hands-on clinical experience in CAM is essential for their learning. However, finding sufficient and varied clinical placement opportunities that encompass a broad range of CAM practices can be challenging, limiting students' practical exposure to CAM.

7. **Diverse Student Needs and Backgrounds**: Students interested in CAM come from diverse backgrounds, with varying levels of prior exposure to and understanding of CAM practices. Meeting the educational needs of a diverse student body, while ensuring equity and inclusivity in CAM education, poses a significant challenge.

8. **Integration into Practice**: Even with CAM education, healthcare professionals may find it challenging to integrate CAM into their practice due to regulatory, legal, and insurance-related barriers. This can discourage professionals from pursuing CAM education or limit the application of their CAM knowledge in clinical settings.

9. **Global Variability in CAM Acceptance and Regulation**: The acceptance and regulation of CAM practices vary widely around the world, impacting the global mobility of CAM practitioners and the international standardization of CAM education. This variability can create challenges for students and professionals seeking to practice or further their education in different countries.

10. **Ethical and Safety Considerations**: Ensuring that CAM education addresses ethical considerations and safety concerns associated with CAM practices is essential. Educators must navigate these issues carefully, balancing the teaching of CAM modalities with a commitment to evidence-based practice and patient safety.

11. **Evolving Legal and Regulatory Frameworks**: The legal and regulatory frameworks governing CAM practices and education are continuously evolving. Keeping educational programs updated with these changes requires constant vigilance and adaptability, which can be resource-intensive for educational institutions.

12. **Interdisciplinary Collaboration Barriers**: Effective CAM education often requires collaboration across various disciplines within healthcare. However, silos within the healthcare education system can impede interdisciplinary learning and collaboration, limiting students' exposure to a holistic view of patient care that includes CAM.

13. **Quality Assurance in CAM Education**: Ensuring the quality of CAM education is a significant challenge, given the diverse range of practices and the varying levels of scientific evidence supporting them. Establishing universally accepted quality standards and metrics for CAM education programs remains a complex issue.

14. **Balancing Traditional Knowledge with Scientific Inquiry**: CAM practices often stem from traditional knowledge systems that may not align with conventional scientific paradigms. Balancing respect for traditional knowledge while promoting rigorous scientific inquiry and evidence-based practice in CAM education is a delicate endeavor.

15. **Technological Access and Digital Divide**: While online education has the potential to increase access to CAM education, the digital divide can limit this access for students in under-resourced areas or those with limited technological literacy. Ensuring equitable access to digital learning resources is a challenge that needs addressing.

16. **Intellectual Property Issues**: Some CAM practices are based on indigenous or traditional knowledge that may have cultural and intellectual property considerations. Navigating these issues while developing educational content can be complex and requires sensitivity and adherence to ethical guidelines.

17. **Public Perception and Demand**: Shifting public perceptions and demand toward more holistic and integrative care models can be both a driver and a challenge for CAM education.

Aligning educational offerings with public interest while ensuring they are evidence-based requires nuanced planning and communication.

18. **Faculty Recruitment and Retention**: Recruiting and retaining faculty members who are both knowledgeable in CAM and skilled in education can be challenging. The niche nature of some CAM practices may limit the pool of qualified educators, impacting the quality and diversity of CAM education.

19. **Financial Sustainability of CAM Programs**: The financial sustainability of CAM educational programs, particularly those in smaller, specialized institutions, can be precarious. Without sufficient enrollment, funding, and institutional support, maintaining high-quality CAM programs can be difficult.

20. **Incorporating CAM into Clinical Guidelines**: Integrating CAM education that aligns with clinical guidelines and best practices is challenging. Educators must navigate how to incorporate CAM in a way that complements conventional care, ensuring students are prepared to make evidence-based decisions in their practice.

21. **Ethical Recruitment and Marketing Practices**: Ethically marketing CAM educational programs, particularly in a landscape with diverse opinions on CAM's efficacy, requires careful consideration. Ensuring that prospective students receive accurate information about the potential benefits and limitations of CAM practices is essential.

22. **Navigating the Diversity of CAM Modalities**: The sheer diversity of CAM practices, each with its theoretical foundations, traditions, and methodologies, presents a significant challenge for curriculum development. Educators must decide which modalities to include, how deeply to cover them, and how to provide a balanced view that respects the diversity of CAM while maintaining academic rigor.

23. **Ensuring Cultural Competence**: CAM practices are often deeply rooted in specific cultural or traditional backgrounds. Providing education that respects and accurately represents these cultural perspectives requires cultural competence. Educators and institutions must work to ensure that CAM education does not appropriate or misrepresent traditional knowledge, but rather, honors and learns from it.

24. **Adapting to Rapid Changes in CAM Practices**: The field of CAM is dynamic, with new therapies emerging and existing practices evolving. Keeping educational content up-to-date with the latest developments and research findings in CAM can be challenging for institutions, requiring ongoing curriculum review and faculty development.

25. **International Standards and Recognition**: With CAM practices and education varying widely across different countries, the challenge of establishing international standards and mutual recognition of CAM qualifications is significant. This affects the mobility of CAM practitioners and the global exchange of CAM knowledge and practices.

26. **Measuring Educational Outcomes**: Evaluating the effectiveness of CAM education is complex due to the subjective nature of many CAM outcomes and the individualized approach to CAM therapies. Developing robust metrics for assessing educational outcomes and the impact of CAM training on clinical practice requires innovative approaches and methodologies.

27. **Integration into Health Systems**: Beyond education, the integration of CAM into health systems poses challenges. Healthcare professionals with CAM training may face obstacles in applying their knowledge within settings that are predominantly oriented toward conventional medicine, including issues related to scope of practice, reimbursement, and interprofessional respect.

28. **Addressing Misinformation**: The proliferation of misinformation about CAM on social media and other platforms poses a challenge to CAM education. Educators must equip students with critical thinking skills and evidence-based knowledge to navigate and counteract misinformation, ensuring they can provide accurate advice to patients.

29. **Sustainability and Environmental Considerations**: Some CAM practices involve natural resources or traditional medicines that may be threatened by overharvesting or environmental degradation. Incorporating sustainability and ethical sourcing considerations into CAM education is crucial for promoting responsible practice.
30. **Accessibility for People with Disabilities**: Ensuring that CAM education is accessible to students and practitioners with disabilities is a challenge that requires thoughtful curriculum design and delivery methods. This includes providing accessible learning materials, accommodating diverse learning needs, and ensuring that clinical training sites are accessible.
31. **Ethical Use of Animals in Education and Practice**: Certain CAM practices may involve the use of animals or animal-derived products. Addressing ethical concerns related to animal welfare and promoting alternatives or ethical sourcing in CAM education is an important consideration.

Addressing these challenges requires a multifaceted approach that involves collaboration among educators, healthcare professionals, regulatory bodies, and CAM practitioners. Efforts must focus on increasing research into CAM, improving the standardization and quality of CAM education, and fostering an environment of openness and respect for the value that CAM can bring to holistic patient care.

ACCESS/EQUITABLE ACCESS TO COMPLEMENTARY AND ALTERNATIVE MEDICINE (CAM) EDUCATION: RECOMMENDATIONS

To address the challenges of access and equitable access to CAM education and to further integrate CAM into the healthcare system, several recommendations can be put forward. These recommendations aim to enhance the quality, availability, and inclusivity of CAM education, ensuring that all healthcare professionals are equipped to meet the diverse needs of their patients with a holistic approach. To enhance access and promote equitable access to CAM education, several recommendations can be considered (Al-Worafi, 2020a,b, 2022a,b, 2023a–n, 2024a–c; Hasan et al., 2019):

1. **Develop and Standardize CAM Curricula**: Educational institutions should work toward developing and standardizing CAM curricula that are evidence-based and comprehensive. This includes establishing core competencies for CAM practices that are recognized across healthcare professions, ensuring a consistent foundation of knowledge.
2. **Increase Interdisciplinary Collaboration**: Foster interdisciplinary collaboration among healthcare faculties to integrate CAM education into existing healthcare programs. This could involve joint lectures, cross-disciplinary workshops, and shared clinical rotations that expose students to a variety of healthcare perspectives, including CAM.
3. **Expand Research and Evidence-Based Practice**: Enhance funding and support for research into CAM practices to strengthen the evidence base for their efficacy and safety. Incorporating findings from this research into educational programs will ensure that CAM practices taught are evidence-based and reflect current knowledge.
4. **Leverage Technology for Accessibility**: Utilize online learning platforms and digital resources to make CAM education more accessible to a broader audience. Online courses, webinars, and virtual reality simulations can provide flexible learning opportunities, especially for those in remote areas or with limited access to traditional educational institutions.
5. **Promote Cultural Competence and Diversity**: Ensure that CAM education programs incorporate cultural competence training and reflect the diversity of CAM practices and their cultural origins. This includes respecting traditional knowledge and practices while integrating them ethically into healthcare.

6. **Advocate for Policy and Regulatory Support**: Work with healthcare policymakers and regulatory bodies to advocate for the inclusion of CAM in healthcare policy and practice guidelines. This could help in standardizing CAM practices, ensuring quality and safety, and integrating CAM more fully into healthcare systems.

7. **Enhance Continuing Education and Professional Development**: Provide ongoing continuing education and professional development opportunities in CAM for healthcare professionals. This ensures that practicing clinicians remain informed about the latest CAM research, practices, and integration strategies.

8. **Facilitate Clinical Training Opportunities**: Develop partnerships with CAM practitioners, clinics, and centers to offer clinical training opportunities for students. Hands-on experience is crucial for understanding the practical application of CAM therapies and for developing competency in their use.

9. **Address Financial Barriers**: Offer scholarships, grants, and financial aid targeted specifically at students pursuing CAM education, especially those from underrepresented or economically disadvantaged backgrounds. This can help to ensure that financial barriers do not prevent interested students from accessing CAM education.

10. **Improve Public and Professional Awareness**: Conduct awareness campaigns to educate the public and healthcare professionals about the benefits and evidence base of CAM practices. Increasing awareness can drive demand for CAM services and, by extension, for CAM education among healthcare professionals.

11. **Establish International Collaboration and Standards**: Work toward establishing international collaborations and standards for CAM education and practice. This can facilitate the exchange of knowledge, enhance the mobility of CAM practitioners, and ensure that CAM education and practice meet global quality standards.

12. **Incorporate Ethics and Sustainability**: Ensure that CAM education programs incorporate discussions on ethics, sustainability, and the responsible use of natural resources. Teaching future practitioners about ethical considerations and sustainability in CAM practices is essential for promoting responsible healthcare practices.

13. **Encourage Patient-Practitioner Communication Training**: Incorporate training modules that emphasize effective communication between healthcare practitioners and patients regarding CAM. This includes discussing CAM options, understanding patient preferences, and integrating these preferences into care plans, which is essential for patient-centered care.

14. **Create an Inclusive Environment for CAM Education**: Educational institutions should strive to create an inclusive environment that values and respects the diversity of CAM practices and the students who seek to learn them. This can involve creating support networks, mentorship programs, and student organizations focused on CAM.

15. **Utilize Competency-Based Education Models**: Shift toward competency-based education models that focus on the attainment of specific skills and knowledge in CAM, rather than solely on traditional academic metrics. This approach can ensure that students are fully prepared to apply CAM principles and practices in their professional lives.

16. **Promote Collaborative Research Initiatives**: Encourage collaborative research initiatives between CAM and conventional medical researchers. This can help bridge gaps in understanding, foster mutual respect, and integrate evidence-based CAM practices into mainstream healthcare more effectively.

17. **Strengthen Regulatory Frameworks for CAM Education**: Work with accrediting agencies and regulatory bodies to strengthen the regulatory frameworks surrounding CAM education. This ensures that programs meet high standards of quality and safety, increasing their legitimacy and integration into healthcare.

18. **Increase Access to CAM Resources in Libraries and Online**: Enhance the availability of CAM resources in academic libraries and online databases accessible to healthcare students

and professionals. Providing easy access to reputable CAM literature, research papers, and clinical guidelines can support self-directed learning and evidence-based practice.

19. **Facilitate Access to Interprofessional Education (IPE) Opportunities**: Encourage access to IPE opportunities that include CAM as part of the curriculum. IPE fosters a team-based approach to patient care, which is crucial for the integration of CAM into holistic patient management.

20. **Support Student Initiatives and Clubs Focused on CAM**: Provide support and resources for student-led initiatives, clubs, and organizations focused on CAM. These groups can offer peer learning opportunities, promote awareness, and create a community of practice that supports CAM education.

21. **Implement Feedback and Continuous Improvement Processes**: Establish mechanisms for collecting feedback from students, faculty, and CAM practitioners on CAM education programs. Use this feedback for continuous curriculum improvement, ensuring that CAM education remains relevant, responsive, and high-quality.

22. **Promote Ethical Practice and Patient Safety**: Emphasize the importance of ethical practice and patient safety in CAM education. This includes teaching students about informed consent, recognizing the limits of their expertise, and knowing when to refer patients to other healthcare professionals.

23. **Encourage Local and Global CAM Exchanges**: Develop programs that support local and global exchange opportunities for students and practitioners interested in CAM. These exchanges can provide exposure to different CAM practices and cultural perspectives, enriching the educational experience.

24. **Advocate for Inclusive Healthcare Policies**: Engage in advocacy to influence healthcare policies that recognize and integrate CAM practices. Policies that support the inclusion of CAM in healthcare delivery and insurance coverage can significantly impact the demand for and integration of CAM education.

25. **Develop Global Competencies for CAM Practices**: Work toward the development of global competencies for CAM practices that can be universally recognized and adopted. This effort would facilitate a standardized approach to CAM education and practice, ensuring that healthcare professionals are equipped with a core set of skills and knowledge that are relevant and applicable across different countries and healthcare systems.

26. **Leverage Community Health Settings for CAM Education**: Encourage CAM education programs to partner with community health centers and clinics to provide students with real-world learning experiences. These partnerships can offer valuable insights into how CAM practices are integrated into community health and wellness programs, providing students with a broader understanding of CAM's role in healthcare.

27. **Incorporate CAM into Public Health Education**: Integrate CAM topics into public health education to highlight the role of CAM in promoting wellness, preventing disease, and addressing public health challenges. This can encourage future public health professionals to consider CAM strategies in their work and advocate for their inclusion in public health programs.

28. **Promote Access to Specialized CAM Training for Healthcare Professionals**: Establish pathways for healthcare professionals to access specialized training in CAM modalities of interest. This could include postgraduate certificates, continuing education courses, and workshops that allow professionals to deepen their CAM knowledge and skills, catering to the growing interest in specific CAM practices.

29. **Use Case Studies and Clinical Scenarios in Education**: Incorporate case studies and clinical scenarios that involve CAM into healthcare education. This method can help students critically evaluate the use of CAM in different contexts, understand the decision-making process, and appreciate the complexities of integrating CAM into patient care.

30. **Establish a CAM Education Research Agenda**: Identify and promote a research agenda focused on CAM education. This could involve studying the effectiveness of different educational strategies, the impact of CAM education on clinical outcomes, and the barriers and facilitators to integrating CAM into healthcare curricula. Such research can provide evidence-based guidance for improving CAM education.

31. **Promote Leadership and Advocacy Training**: Include leadership and advocacy training within CAM education programs to empower students to become advocates for CAM integration within their professional spheres and the wider healthcare system. This training can equip them with the skills needed to navigate political, regulatory, and institutional barriers to CAM integration.

32. **Encourage Ethical and Critical Engagement with CAM Literature**: Teach students to engage ethically and critically with CAM literature, recognizing both the potential benefits and limitations of CAM research. This includes understanding how to evaluate the quality of CAM research and differentiate between evidence-based practices and those lacking scientific support.

33. **Strengthen Support Networks for CAM Students and Professionals**: Develop support networks for students and professionals pursuing CAM education and practice. These networks can provide mentorship, career guidance, and professional development opportunities, helping to build a strong community of CAM practitioners.

34. **Address Language and Communication Barriers**: Recognize and address language and communication barriers that may affect access to CAM education for non-native speakers or culturally diverse student populations. Offering resources and support in multiple languages can make CAM education more inclusive and accessible.

35. **Evaluate and Address Environmental Impacts of CAM Practices**: Educate students about the environmental impacts of certain CAM practices and the importance of sustainable and ethical sourcing of CAM materials. This awareness can foster responsible use of natural resources and promote sustainability within CAM practices.

36. **Implement Mentorship Programs**: Establish mentorship programs linking students and new practitioners with experienced CAM professionals. These relationships can provide guidance, support career development, and encourage the practical application of CAM in clinical settings.

37. **Foster International Exchange Programs**: Develop international exchange programs focused on CAM, allowing students and practitioners to experience CAM practices within different cultural and healthcare contexts. Such exchanges can broaden perspectives, deepen understanding, and facilitate global networks of CAM professionals.

38. **Enhance Access to CAM Journals and Publications**: Increase access to scientific journals and publications related to CAM for students and practitioners. Providing free or subsidized access to these resources can support ongoing education and keep practitioners informed about the latest research and developments.

39. **Incorporate Health Equity into CAM Education**: Integrate health equity themes into CAM education to address disparities in healthcare access and outcomes. This includes teaching about the social determinants of health and how CAM can be used to support underserved and marginalized populations.

40. **Promote Patient Education and Engagement**: Encourage healthcare professionals to educate and engage patients about CAM options, empowering patients to make informed choices about their healthcare. This can improve patient satisfaction, adherence to treatment plans, and overall health outcomes.

41. **Develop Specialized CAM Research Centers**: Establish or support specialized research centers focused on CAM to conduct high-quality research, provide education, and serve as a resource for healthcare professionals and the public. These centers can play a crucial role in advancing the field of CAM.

42. **Incorporate CAM into Health Technology Assessments**: Include CAM practices in health technology assessments to evaluate their effectiveness, cost-effectiveness, and impact on patient outcomes. This can provide evidence-based data to support the integration of CAM into healthcare systems.

43. **Leverage Social Media for CAM Education**: Utilize social media platforms to disseminate accurate information about CAM, promote CAM education opportunities, and engage with a broader audience. This approach can help combat misinformation and raise awareness about the benefits of CAM.

44. **Enhance Library and Information Services**: Improve library and information services related to CAM by expanding collections of CAM resources, providing training on how to find and evaluate CAM information, and offering information literacy workshops for students and practitioners.

45. **Support CAM Student Organizations**: Encourage and support the formation of student organizations focused on CAM within healthcare education institutions. These organizations can provide a platform for student advocacy, education, and community building around CAM interests.

46. **Integrate CAM into Emergency and Disaster Response Training**: Include CAM approaches in emergency and disaster response training programs, recognizing the role CAM can play in managing stress, trauma, and community health in these settings.

47. **Promote Multidisciplinary Research Teams**: Encourage the formation of multidisciplinary research teams that include CAM practitioners, conventional healthcare providers, and researchers to study the integration of CAM into healthcare and its impact on health outcomes.

48. **Develop Policy Briefs and Position Papers**: Create policy briefs and position papers on the role of CAM in healthcare to inform policymakers, healthcare administrators, and the public. These documents can advocate for the inclusion of CAM in healthcare policies and practices.

49. **Host CAM Awareness Events**: Organize CAM awareness events, such as conferences, seminars, and public lectures, to educate healthcare professionals, students, and the public about CAM practices and their benefits.

50. **Implement Quality Assurance Programs for CAM Providers**: Develop quality assurance programs for CAM providers to ensure high standards of practice, ethical conduct, and patient safety. This can increase trust in CAM practices among healthcare professionals and patients.

51. **Facilitate Access to CAM for Healthcare Workers**: Provide healthcare workers with easy access to CAM treatments and education as part of their wellness programs. This can improve healthcare workers' well-being and increase their personal experience with CAM, potentially influencing their professional practice.

52. **Develop Guidelines for Integrating CAM into Clinical Practice**: Create comprehensive guidelines for healthcare professionals on integrating CAM into clinical practice, addressing practical considerations, ethical issues, and communication strategies.

53. **Encourage CAM Practice within Hospital Settings**: Advocate for the inclusion of CAM services within hospital settings, allowing patients to access a broader range of treatment options and promoting the integration of CAM into conventional healthcare.

54. **Support Language and Cultural Training**: Offer language and cultural training for CAM practitioners and students to improve communication with patients from diverse backgrounds and enhance the cultural competence of CAM providers.

55. **Promote Sustainability Practices in CAM**: Encourage and educate about sustainability practices in the sourcing, production, and use of CAM therapies and products, aligning CAM practices with environmental stewardship.

By implementing these recommendations, stakeholders in healthcare education and policy can work toward creating a more inclusive, accessible, and comprehensive approach to CAM education. This will not only benefit future healthcare professionals but also contribute to a more holistic and patient-centered healthcare system that embraces the full spectrum of healing practices.

CONCLUSION

In conclusion, enhancing access and equitable access to CAM education is a multifaceted challenge that requires concerted efforts from various stakeholders, including educational institutions, healthcare providers, policymakers, and the CAM community. The recommendations outlined address key areas such as curriculum development, interdisciplinary collaboration, research, policy advocacy, and the use of technology, all aimed at integrating CAM more fully into the healthcare landscape. Achieving equitable access to CAM education is not just about incorporating diverse healing practices into healthcare training; it's about fostering a healthcare system that is inclusive, patient-centered, and responsive to the evolving needs and values of the population it serves. By embracing the diversity of CAM modalities and ensuring that all healthcare professionals have the knowledge and skills to utilize these practices effectively, we can move toward a more holistic approach to health and wellness. Moreover, the integration of CAM into healthcare education and practice has the potential to address critical issues such as health disparities, patient satisfaction, and the sustainability of healthcare systems. It encourages a broader view of health that includes physical, emotional, spiritual, and environmental well-being, aligning with the growing demand for healthcare that is personalized, preventive, and focused on the whole person. As we move forward, it is essential to continue advocating for the inclusion of CAM in healthcare curricula, support research that builds the evidence base for CAM practices, and develop policies that facilitate the integration of CAM into mainstream healthcare. Collaboration across disciplines and cultures, along with a commitment to equity and quality, will be key to overcoming the challenges and realizing the full potential of CAM in enhancing health and healing. Ultimately, access and equitable access to CAM education represent not just an opportunity to expand the toolkit of healthcare professionals but also a commitment to a more diverse, equitable, and comprehensive approach to health and wellness. Through continued efforts to integrate CAM into the healthcare system, we can aspire to meet the diverse needs of patients and communities, promoting health and well-being in its fullest sense.

REFERENCES

Al-Worafi, Y.M. (Ed.). (2020a). *Drug Safety in Developing Countries: Achievements and Challenges.* Academic Press.

Al-Worafi, Y.M. (2020b). Herbal medicines safety issues. In: Al-Worafi, Y.M. (ed), *Drug Safety in Developing Countries* (pp. 163–178). Academic Press.

Al-Worafi, Y.M. (2022a). *A Guide to Online Pharmacy Education: Teaching Strategies and Assessment Methods.* CRC Press.

Al-Worafi, Y.M. (2022b). Competencies and learning outcomes. In: Al-Worafi, Y.M. (ed), *A Guide to Online Pharmacy Education: Teaching Strategies and Assessment Methods.* CRC Press.

Al-Worafi, Y.M. (2023a). *Patient Safety in Developing Countries: Education, Research, Case Studies.* CRC Press.

Al-Worafi, Y.M. (2023b). *Technology for Drug Safety: Current Status and Future Developments.* Springer Nature.

Al-Worafi, Y.M. (2023c). Patient safety-related issues: Patient care errors and related problems. In: Al-Worafi, Y.M. (ed), *Patient Safety in Developing Countries: Education, Research, Case Studies.* CRC Press.

Al-Worafi, Y.M. (2023d). Patient care errors and related problems: Preventive medicine errors & related problems. In: Al-Worafi, Y.M. (ed), *Patient Safety in Developing Countries: Education, Research, Case Studies.* CRC Press.

Al-Worafi, Y.M. (2023e). Patient care errors and related problems: Patient assessment and diagnostic errors & related problems. In: Al-Worafi, Y.M. (ed), *Patient Safety in Developing Countries: Education, Research, Case Studies*. CRC Press.

Al-Worafi, Y.M. (2023f). Patient care errors and related problems: Non-pharmacological errors & related problems. In: Al-Worafi, Y.M. (ed), *Patient Safety in Developing Countries: Education, Research, Case Studies*. CRC Press.

Al-Worafi, Y.M. (2023g). Patient care errors and related problems: Medical errors & related problems. In: Al-Worafi, Y.M. (ed), *Patient Safety in Developing Countries: Education, Research, Case Studies*. CRC Press.

Al-Worafi, Y.M. (2023h). Patient care errors and related problems: Monitoring errors & related problems. In: Al-Worafi, Y.M. (ed), *Patient Safety in Developing Countries: Education, Research, Case Studies*. CRC Press.

Al-Worafi, Y.M. (2023i). Patient care errors and related problems: Patient education and counselling errors and related problems. In: Al-Worafi, Y.M. (ed), *Patient Safety in Developing Countries: Education, Research, Case Studies*. CRC Press.

Al-Worafi, Y.M. (2023j). Patient safety-related issues: Other medication safety issues. In: Al-Worafi, YM. (eds) *Patient Safety in Developing Countries: Education, Research, Case Studies*. CRC Press.

Al-Worafi, Y.M. (2023k). Patient safety culture. In: Al-Worafi, Y.M. (ed), *Patient Safety in Developing Countries: Education, Research, Case Studies*. CRC Press.

Al-Worafi, Y.M. (2023l). Patient safety education: Competencies and learning outcomes. In: Al-Worafi, Y.M. (ed), *Patient Safety in Developing Countries: Education, Research, Case Studies*. CRC Press.

Al-Worafi, Y.M. (Ed.). (2023m). *Clinical Case Studies on Medication Safety*. Academic Press.

Al-Worafi, Y.M. (Ed.). (2023n). *Comprehensive Healthcare Simulation: Pharmacy Education, Practice and Research*. Springer Nature.

Al-Worafi, Y.M. (Ed.). (2024a). *Handbook of Medical and Health Sciences in Developing Countries*. Springer, Cham.

Al-Worafi, Y.M. (2024b). Complementary and alternative medicine (CAM) in developing countries. In: Al-Worafi, Y.M. (ed), *Handbook of Medical and Health Sciences in Developing Countries*. Springer, Cham. https://doi.org/10.1007/978-3-030-74786-2_301-1

Al-Worafi, Y.M. (2024c). Access/equitable access to medical and health sciences in developing countries. In: Al-Worafi, Y.M. (ed), *Handbook of Medical and Health Sciences in Developing Countries*. Springer, Cham. https://doi.org/10.1007/978-3-030-74786-2_147-1

Hasan, S., Al-Omar, M.J., AlZubaidy, H., and Al-Worafi, Y.M. (2019). Use of medications in Arab countries. In Laher, I. (ed), *Handbook of Healthcare in the Arab World* (p. 42). Springer, Cham.

11 Complementary and Alternative Medicine (CAM) Education

Continuous Medical Education

INTRODUCTION

Complementary and Alternative Medicine (CAM) refers to a broad range of medical practices that fall outside the realm of conventional medicine. These practices can include herbal medicine, acupuncture, chiropractic care, naturopathy, yoga, and meditation, among others. The interest in CAM has grown significantly over the years, both among patients seeking holistic approaches to health and wellness and among healthcare professionals looking to broaden their treatment options. This growing interest has underscored the need for continuous medical education (CME) in the field of CAM. Continuous Medical Education in CAM is designed to keep healthcare providers updated on the latest advancements, research, and best practices in the field. It is crucial because CAM practices evolve over time, with new therapies being introduced and existing ones being refined based on emerging evidence. CME in CAM enables healthcare professionals to gain a deeper understanding of the benefits and limitations of various CAM therapies, ensuring they can offer informed advice to their patients and integrate CAM practices into their treatment plans safely and effectively. Educational programs in CAM for healthcare professionals vary widely in format, content, and duration. They can range from short workshops and seminars to more extensive certificate or degree programs. These educational opportunities are offered through various platforms, including universities, professional associations, and online courses. The content typically covers a broad spectrum of CAM modalities, ethical considerations, integration strategies with conventional medicine, and critical evaluation of CAM research. One of the key challenges in CAM education is the variability in the quality and rigor of educational offerings, given the diverse nature of CAM practices and the varying levels of evidence supporting their use. Therefore, accreditation and standardization of CAM education programs become vital to ensure healthcare professionals receive high-quality, evidence-based training. Professional bodies and educational institutions play a crucial role in developing standardized curricula and accreditation processes for CAM education. Continuous Medical Education in CAM not only enhances the competency of healthcare providers in delivering holistic care but also promotes a more integrative approach to healthcare. This integrative approach recognizes the value of combining conventional medical treatments with CAM therapies to address the physical, emotional, and spiritual aspects of health. By staying informed about CAM through continuous education, healthcare professionals can better guide their patients in making informed decisions about their health and wellness, ultimately leading to improved patient outcomes and satisfaction.

RATIONALITY AND IMPORTANCE OF CONTINUOUS MEDICAL EDUCATION IN COMPLEMENTARY AND ALTERNATIVE MEDICINE (CAM) EDUCATION

The rationale for continuous medical education (CME) in Complementary and Alternative Medicine (CAM) for educators hinges on the evolving nature of healthcare and the growing demand for

DOI: 10.1201/9781003327202-11

holistic treatment approaches. The importance of CME in CAM, particularly for educators, can be dissected into several key aspects:

1. **Keeping Pace with Evolving CAM Practices**

   CAM practices are dynamic, with continuous research and development leading to new discoveries and the refinement of existing therapies. Educators must stay abreast of these changes to provide the most current, evidence-based information to healthcare professionals and students. Continuous education ensures that educators are knowledgeable about the latest CAM therapies, their mechanisms of action, benefits, and potential risks.

2. **Enhancing Educational Quality and Relevance**

   For educators, CME in CAM is vital to ensure the quality and relevance of the education they provide. As CAM practices gain popularity and acceptance in the mainstream healthcare system, there is a growing need to integrate CAM content into medical and health sciences curricula. Educators need to be well-versed in CAM to develop and deliver curricula that reflect the current landscape of healthcare, blending conventional and alternative approaches seamlessly.

3. **Promoting Evidence-Based Practice**

   The field of CAM includes a wide range of practices, some of which are supported by robust scientific evidence while others are not. Educators play a crucial role in promoting evidence-based practice by teaching healthcare professionals how to critically evaluate the evidence supporting various CAM therapies. This critical evaluation is essential for safe and effective patient care, ensuring that healthcare professionals are equipped to recommend CAM practices based on solid evidence.

4. **Fostering an Integrative Healthcare Approach**

   Educators are instrumental in fostering an integrative approach to healthcare. By incorporating CAM into continuous medical education, they encourage future healthcare professionals to consider all aspects of patient care, including physical, emotional, and spiritual health. This holistic approach can lead to more personalized and patient-centered care, improving patient outcomes and satisfaction.

5. **Meeting Patient Expectations and Needs**

   Patients increasingly seek CAM therapies alongside conventional medicine. Educators with expertise in CAM can better prepare healthcare professionals to meet these patient expectations. By understanding the nuances of CAM, healthcare providers can engage in meaningful conversations with their patients about the use of CAM therapies, helping patients make informed decisions about their health and treatment options.

6. **Professional Development and Interdisciplinary Collaboration**

   Continuous education in CAM fosters professional development for educators and healthcare professionals alike. It encourages interdisciplinary collaboration among practitioners from different backgrounds, promoting a more comprehensive and collaborative approach to patient care. This collaboration is essential for integrating CAM into conventional medical settings effectively.

In summary, the rationality and importance of CME in CAM for educators lie in its potential to enhance the quality of healthcare education, promote evidence-based practice, support the integration of holistic treatment approaches, and ultimately improve patient care. Continuous education ensures that educators remain at the forefront of healthcare education, preparing the next generation of healthcare professionals to meet the challenges and opportunities presented by the integration of CAM into mainstream healthcare.

COMPLEMENTARY AND ALTERNATIVE MEDICINE (CAM) CONTINUOUS MEDICAL EDUCATION: FACILITATORS

Facilitators of Continuous Medical Education (CME) in Complementary and Alternative Medicine (CAM) play a crucial role in the dissemination, understanding, and integration of CAM practices within the healthcare system. These facilitators can vary widely, encompassing individuals, institutions, and platforms, each contributing uniquely to the education of healthcare professionals about CAM. Here's an overview of the primary facilitators in this field:

1. **Academic Institutions**

    Universities and colleges that offer degrees in health sciences often include CAM topics within their curricula or offer specific courses and programs dedicated to CAM. These institutions are pivotal in providing foundational knowledge to future healthcare professionals and may also offer postgraduate CME courses in CAM for practicing clinicians. Through academic research, these institutions also contribute to the evidence base that supports CAM practices.

2. **Professional Associations and Organizations**

    Many professional healthcare organizations recognize the importance of CAM and offer CME opportunities through workshops, seminars, and conferences. Organizations such as the American Holistic Medical Association or the National Center for Complementary and Integrative Health (NCCIH) provide resources and learning opportunities for healthcare professionals interested in incorporating CAM into their practice. These organizations often set standards for CAM education and practice, ensuring quality and reliability.

3. **Online Education Platforms**

    With the rise of digital learning, online platforms have become significant facilitators of CAM education. Websites offering MOOCs (Massive Open Online Courses), webinars, and virtual workshops enable healthcare professionals to access CAM education regardless of their geographic location. These platforms can offer flexibility in learning, allowing for self-paced study that can fit around the demands of clinical practice.

4. **Continuing Education Providers**

    Dedicated CME providers that specialize in CAM offer courses and certifications for healthcare professionals seeking to expand their knowledge and skills in this area. These providers often collaborate with experts in various CAM modalities to deliver high-quality, evidence-based education. They may also offer specialized training for integrating CAM into clinical practice.

5. **Healthcare Facilities and Hospitals**

    Some healthcare facilities and integrative health centers offer in-house training and workshops on CAM for their staff. This training can be particularly relevant for facilities that aim to incorporate CAM therapies into their patient care offerings. Such settings provide a practical learning environment where healthcare professionals can observe and practice CAM integration firsthand.

6. **Peer Networks and Collaborative Groups**

    Informal learning through peer networks and collaborative groups can also facilitate CAM education. Healthcare professionals may share experiences, insights, and resources related to CAM through professional networks, online forums, and collaborative interest groups. Peer learning can provide valuable practical advice and support for integrating CAM into practice.

7. **Research Institutions**

    Institutions focused on CAM research contribute to continuous education by generating new knowledge and evidence about the efficacy, safety, and mechanisms of CAM therapies. Research findings are disseminated through scientific journals, conferences, and symposiums, providing a basis for evidence-based CAM education and practice.

8. **Government and Regulatory Bodies**

Government health departments and regulatory agencies can facilitate CAM education by funding research, developing guidelines for CAM practice, and accrediting CAM educational programs. For instance, the National Institutes of Health (NIH) in the United States, through the National Center for Complementary and Integrative Health (NCCIH), supports research and disseminates information on CAM. Such bodies ensure that CAM practices integrated into healthcare are safe, ethical, and based on scientific evidence.

9. **Patient Advocacy Groups**

Patient advocacy groups and patient-led organizations often play a role in educating healthcare professionals about CAM. These groups can provide insights into patient experiences, preferences, and outcomes related to CAM therapies. By sharing these perspectives at conferences, in educational materials, and through direct engagement with healthcare providers, they help highlight the value and importance of patient-centered approaches in CAM.

10. **Libraries and Information Centers**

Medical libraries and specialized CAM information centers offer access to a vast array of resources, including books, journals, and databases dedicated to CAM research and practice. These resources are invaluable for healthcare professionals seeking to deepen their understanding of CAM modalities, evidence-based practices, and the latest research findings. Libraries also often host educational events, such as lectures and workshops, featuring experts in the field of CAM.

11. **Interprofessional Collaboration Projects**

Projects that bring together professionals from various healthcare disciplines to work on integrating CAM into healthcare systems can serve as powerful facilitators of CAM education. These projects provide opportunities for hands-on learning, sharing best practices, and developing collaborative care models that include CAM therapies. Interprofessional education initiatives can break down silos between different healthcare professions and promote a more unified approach to patient care.

12. **Cultural and Community Centers**

Cultural and community centers can be vital in educating healthcare professionals about traditional and indigenous health practices, many of which overlap with CAM modalities. These centers can offer workshops, cultural competency training, and opportunities for healthcare providers to learn directly from practitioners of traditional medicine. This education is crucial for providing culturally sensitive care and for understanding the historical and cultural contexts of certain CAM practices.

13. **Technology and Innovation Hubs**

Technology and innovation hubs that focus on healthcare advancements can facilitate CAM education through the development of new learning tools, such as virtual reality (VR) simulations, apps for learning CAM therapies, and online platforms for collaborative learning. These technologies can make CAM education more interactive, engaging, and accessible to a broader audience of healthcare professionals.

14. **Mentorship Programs**

Mentorship programs that pair experienced CAM practitioners with healthcare professionals interested in CAM can provide personalized learning experiences. Mentors can offer guidance, share their expertise and insights, and provide practical advice on incorporating CAM into clinical practice. This one-on-one learning can be particularly effective for developing specific skills and knowledge in CAM.

Each of these facilitators contributes to the ecosystem of CAM education, offering resources, knowledge, and support for healthcare professionals. The diversity of facilitators ensures that education in CAM can reach a wide audience, accommodate different learning styles and professional needs, and continually adapt to the evolving landscape of healthcare.

COMPLEMENTARY AND ALTERNATIVE MEDICINE (CAM) CONTINUOUS MEDICAL EDUCATION: BARRIERS

While Continuous Medical Education (CME) in Complementary and Alternative Medicine (CAM) presents numerous opportunities for healthcare professionals to expand their knowledge and skills, several barriers can hinder the effective delivery and uptake of CAM education. Understanding these challenges is crucial for developing strategies to overcome them and ensure that healthcare professionals can fully engage with CAM education. The main barriers include:

1. **Lack of Standardization and Regulation**

   CAM encompasses a wide range of practices, not all of which are regulated or standardized. This lack of standardization can make it challenging to develop comprehensive educational programs that cover the breadth of CAM practices in a consistent manner. Moreover, the variability in the quality and content of CAM education programs can lead to skepticism among healthcare professionals regarding their validity and usefulness.

2. **Limited Research and Evidence Base**

   Although research into CAM is growing, there remains a significant gap in high-quality, evidence-based studies for many CAM modalities. This lack of robust evidence can be a barrier to incorporating CAM topics into mainstream medical education, as healthcare professionals and educators may prioritize evidence-based practices. The skepticism toward CAM modalities not supported by strong evidence can limit interest and engagement in CAM education.

3. **Cultural and Institutional Resistance**

   There can be cultural and institutional resistance within the medical community toward integrating CAM into conventional medical education and practice. Some healthcare professionals may view CAM with skepticism or consider it outside the realm of conventional medicine. This resistance can stem from a lack of familiarity with CAM, concerns about efficacy and safety, or philosophical differences regarding healthcare approaches.

4. **Resource Constraints**

   Developing and implementing CAM education programs requires resources, including funding, expert faculty, and educational materials. Limited resources can be a significant barrier, particularly for smaller institutions or in regions where CAM is not widely recognized. Additionally, healthcare professionals often face time constraints, which can limit their ability to pursue additional education in CAM alongside their clinical duties and other continuing education requirements.

5. **Regulatory and Accreditation Challenges**

   The integration of CAM into mainstream medical education can be complicated by regulatory and accreditation standards. Accrediting bodies for medical and healthcare education may have specific requirements that do not readily accommodate CAM education. Furthermore, varying legal and professional regulations across regions regarding the practice of CAM can complicate the development of standardized education programs.

6. **Interprofessional Collaboration**

   Effective CAM education often requires collaboration between professionals from various disciplines, including those traditionally trained in CAM and those from conventional medical backgrounds. However, differences in language, practice philosophies, and educational backgrounds can hinder effective interprofessional collaboration and learning.

7. **Lack of Awareness and Interest**

   Some healthcare professionals may lack awareness of the potential benefits of CAM or may not have an interest in CAM due to preconceived notions or a lack of exposure to CAM practices. This lack of interest can reduce the demand for CAM education and limit opportunities for integration into clinical practice.

8. **Evaluation and Measurement Challenges**

   Evaluating the impact of CAM education on healthcare practice and patient outcomes can be challenging due to the diverse nature of CAM practices and the individualized approach often taken in CAM therapies. This difficulty in measuring outcomes can hinder the development of evidence-based CAM education programs and the ability to demonstrate their value.

9. **Integration into Clinical Practice**

   Even when healthcare professionals are educated in CAM, integrating these practices into conventional clinical settings can be challenging. There may be operational, regulatory, or logistical barriers that prevent the practical application of CAM therapies. For example, insurance coverage for CAM therapies is not universally available, which can limit the ability of patients to access recommended treatments and reduce the incentive for healthcare professionals to integrate CAM into their practice.

10. **Quality Assurance and Competency Evaluation**

    Ensuring the quality of CAM education and evaluating the competency of healthcare professionals in applying CAM practices pose significant challenges. Unlike more traditional areas of medicine, CAM encompasses a broad range of practices with varying levels of scientific support and standardization. Developing reliable methods for assessing competency in CAM requires clear benchmarks and standards, which are still evolving in many CAM disciplines.

11. **Ethical Considerations**

    Ethical issues related to patient safety, informed consent, and the potential for unproven treatments to replace evidence-based conventional care are significant concerns. Healthcare professionals must navigate these ethical considerations carefully, which requires a nuanced understanding of both CAM and conventional medicine ethics. Education programs need to address these ethical challenges directly, equipping professionals with the skills to make informed, patient-centered decisions.

12. **Cultural Competence and Sensitivity**

    CAM practices often originate from diverse cultural traditions. A barrier to effective CAM education is ensuring that it is delivered with cultural competence and sensitivity. Healthcare professionals need to understand the cultural contexts of CAM practices to respect patient beliefs and preferences. This aspect of CAM education requires a nuanced approach that may not be adequately covered in more general medical education programs.

13. **Technological Adaptation and Online Learning**

    While online learning platforms offer opportunities to expand access to CAM education, they also present challenges in terms of ensuring the quality and interactivity of educational content. Practical skills, which are crucial for many CAM modalities, may be difficult to teach and assess through online platforms. Developing effective online and hybrid models of CAM education that incorporate hands-on learning experiences is an ongoing challenge.

14. **Interdisciplinary Communication**

    Effective CAM education requires fostering communication and understanding across various healthcare disciplines. However, differences in terminology, diagnostic approaches, and treatment philosophies between conventional medicine and CAM can impede effective interdisciplinary communication. Educators must address these barriers by promoting a common language and mutual respect among all healthcare providers.

15. **Financial Incentives and Support**

    The lack of financial incentives for healthcare professionals to pursue CAM education and integrate CAM practices into their work is a significant barrier. Without financial support or reimbursement mechanisms for CAM services, healthcare providers may be less motivated to invest time and resources into CAM education. Advocacy for insurance coverage and reimbursement for CAM services is crucial to overcoming this barrier.

Overcoming these barriers requires concerted efforts from educators, healthcare institutions, professional organizations, and regulatory bodies. Strategies may include enhancing the evidence base for CAM through research, promoting cultural competency and interprofessional respect, securing resources for CAM education, and advocating for regulatory changes to accommodate CAM in healthcare education and practice.

COMPLEMENTARY AND ALTERNATIVE MEDICINE (CAM) CONTINUOUS MEDICAL EDUCATION: ACHIEVEMENTS

Despite the barriers to integrating CAM into CME, there have been significant achievements in this area, demonstrating progress and the growing acceptance of CAM within the healthcare community. These achievements contribute to a more inclusive, holistic approach to patient care and reflect the evolving landscape of healthcare education. Key accomplishments include:

1. **Incorporation into Medical Curricula**

   Many medical schools and healthcare education institutions have begun to incorporate CAM topics into their curricula. This integration ranges from elective courses on specific CAM modalities to modules within required courses that introduce students to the principles of holistic and integrative medicine. This inclusion represents a significant shift toward recognizing the value of CAM in providing comprehensive patient care.

2. **Development of Accreditation Standards**

   There has been progress in developing accreditation standards for CAM education programs, particularly in fields such as acupuncture, naturopathy, and chiropractic care. These standards ensure that practitioners receive high-quality, evidence-based education in their respective CAM modalities. Accreditation bodies, such as the Accreditation Commission for Acupuncture and Oriental Medicine (ACAOM) and the Council on Chiropractic Education (CCE), play crucial roles in maintaining educational standards.

3. **Growth of Research and Evidence Base**

   The research base for many CAM practices has grown, supported by institutions like the National Center for Complementary and Integrative Health (NCCIH) in the United States. Increased funding for CAM research has led to a better understanding of the mechanisms, efficacy, and safety of various CAM modalities. This growing body of evidence supports the integration of CAM into clinical practice and education.

4. **Expansion of Online and Distance Learning**

   The availability of online and distance learning options for CAM education has expanded significantly, making CAM education more accessible to healthcare professionals worldwide. These platforms offer courses ranging from introductory overviews to in-depth training in specific CAM modalities, allowing for flexible learning that can fit within the busy schedules of healthcare providers.

5. **Interprofessional Education and Collaborative Practice Models**

   There have been strides in promoting interprofessional education (IPE) that includes CAM, facilitating collaboration between conventional healthcare providers and CAM practitioners. This approach fosters a mutual understanding and respect for the contributions of different healthcare modalities, promoting a more integrated approach to patient care.

6. **Patient-Centered Care and Holistic Approaches**

   The achievements in CAM education reflect a broader shift toward patient-centered care and recognition of the importance of holistic approaches in healthcare. By educating healthcare professionals about CAM, patients are more likely to receive care that aligns with their values, preferences, and whole-person health needs.

7. **Professional Development and Continuing Education**

The development of continuing education programs in CAM has allowed healthcare professionals to deepen their understanding of CAM practices and integrate them into their clinical practice. These programs often include practical components, ensuring that participants can apply what they learn in a clinical setting.

8. **Global Recognition and Integration**

CAM education and practice have gained recognition and integration not just in Western countries but globally. Many health systems around the world now recognize the value of integrating CAM with conventional medicine to provide comprehensive healthcare.

These achievements illustrate the significant progress in CAM education within the medical and healthcare fields. They reflect a growing recognition of the importance of diverse, holistic approaches to health and wellness, and the need to equip healthcare professionals with the knowledge and skills to meet the evolving preferences and needs of patients. As CAM continues to evolve, ongoing efforts to enhance the quality of CAM education, research, and integration into healthcare practice will be essential to maximizing its contributions to health and well-being.

COMPLEMENTARY AND ALTERNATIVE MEDICINE (CAM) CONTINUOUS MEDICAL EDUCATION: CHALLENGES

The integration of CAM into CME faces several challenges that can affect the scope, quality, and effectiveness of education in this area. These challenges reflect broader issues within healthcare education and practice, as well as specific concerns related to the nature of CAM itself. Addressing these challenges is crucial for the advancement and acceptance of CAM within the healthcare system. Key challenges include:

1. **Evidence and Research Gaps**

One of the most significant challenges is the variability in the quality and quantity of scientific evidence supporting various CAM modalities. While some CAM practices are well-researched and have a strong evidence base, others lack rigorous scientific studies to validate their efficacy and safety. This discrepancy can lead to skepticism among healthcare professionals and educators, making it difficult to integrate CAM into evidence-based educational programs.

2. **Cultural and Institutional Resistance**

There can be significant cultural and institutional resistance within the medical community toward CAM. This resistance may stem from a perceived lack of scientific rigor, concerns about professionalism, or philosophical differences between conventional medicine and CAM. Overcoming these biases requires a concerted effort to demonstrate the value of CAM practices through evidence-based outcomes and to promote a more integrative approach to healthcare.

3. **Standardization and Accreditation**

The wide range of CAM practices, each with its own set of theories, diagnostic methods, and treatments, poses a challenge for standardization and accreditation. Developing standardized curricula that accurately reflect the diversity within CAM while ensuring high educational quality is a complex task. Additionally, the lack of uniform standards for CAM education and practice makes it difficult to assess and ensure the competence of CAM practitioners.

4. **Integration into Clinical Practice**

Even when CAM is included in CME, integrating these practices into clinical settings can be challenging. Healthcare professionals may face institutional barriers, lack of support for CAM therapies, or uncertainty about how to combine CAM with conventional treatments. There's also the challenge of ensuring that CAM practices are used appropriately and safely, considering patient-specific factors and evidence-based guidelines.

5. **Funding and Resources**

Funding for CAM education and research is often limited compared to conventional medicine. This lack of financial support can hinder the development of comprehensive CAM educational programs, research projects to build the evidence base for CAM practices, and initiatives to integrate CAM into healthcare systems. Securing resources is critical for advancing CAM education and practice.

6. **Regulatory and Legal Considerations**

Regulatory and legal frameworks for CAM vary significantly across regions, which can complicate the integration of CAM education and practice. Navigating these regulations requires a deep understanding of local laws and standards, which must be incorporated into CAM education programs to ensure that healthcare professionals are aware of their legal and ethical obligations when practicing CAM.

7. **Educational Material Development**

There is a need for high-quality, evidence-based educational materials on CAM. Developing these resources requires expertise in both CAM and educational theory, as well as ongoing updates to reflect the latest research findings. The dynamic nature of CAM, with new evidence and practices emerging regularly, makes this a continuous challenge.

8. **Professional Acceptance and Public Perception**

Gaining wider acceptance among healthcare professionals and changing public perception about the legitimacy and value of CAM are ongoing challenges. Misconceptions and misinformation about CAM can affect both professional and public willingness to consider CAM as a valid component of healthcare. Education plays a key role in addressing these perceptions by providing accurate, evidence-based information about CAM practices.

Overcoming these challenges requires collaborative efforts among educators, researchers, healthcare professionals, and policymakers. Strategies include increasing research funding for CAM, developing standardized curricula and accreditation processes, promoting interprofessional education, and advocating for regulatory changes that support the integration of CAM into healthcare systems. As these challenges are addressed, CAM's potential to contribute to comprehensive, patient-centered care becomes increasingly recognized and realized.

COMPLEMENTARY AND ALTERNATIVE MEDICINE (CAM) CONTINUOUS MEDICAL EDUCATION: RECOMMENDATIONS FOR THE BEST PRACTICE

To enhance the effectiveness and integration of CAM within CME, several best practice recommendations can be implemented (Al-Worafi, 2020a,b, 2022a–e, 2023a–n, 2024a–c; Hasan et al., 2019). These recommendations aim to address the challenges associated with CAM education, improve the quality and relevance of CAM training for healthcare professionals, and ultimately support the integration of CAM into patient care. Here are key recommendations for best practices in CAM CME:

1. **Evidence-Based Curriculum Development**
   - Develop and continuously update CAM curricula based on the latest evidence and research findings.
   - Include critical appraisal skills in the curriculum to enable healthcare professionals to evaluate the quality of CAM research and understand the evidence base behind different CAM modalities.

2. **Interprofessional Education**
   - Promote interprofessional education (IPE) programs that include both CAM and conventional healthcare professionals. This approach fosters mutual respect, understanding, and collaboration between different healthcare disciplines, enhancing patient care.
   - Encourage dialogue and learning across professions to share insights and best practices in integrating CAM into patient care.

3. **Standardization and Accreditation**
   - Work toward the standardization of CAM education and practice by developing clear competencies, educational standards, and accreditation processes for CAM programs.
   - Engage with professional bodies and regulatory agencies to ensure that CAM education meets high-quality and ethical standards.

4. **Practical and Clinical Training**
   - Incorporate practical training and clinical placements in CAM education programs to provide hands-on experience with CAM modalities.
   - Facilitate mentorship and learning opportunities with experienced CAM practitioners to enhance clinical skills and knowledge in real-world settings.

5. **Cultural Competence**
   - Integrate cultural competence training into CAM education to ensure healthcare professionals are sensitive to the cultural dimensions of CAM practices and can provide culturally appropriate care.
   - Educate on the historical and cultural origins of various CAM practices to enhance understanding and respect for their use among diverse patient populations.

6. **Patient-Centered Care**
   - Emphasize the importance of patient-centered care in CAM education, teaching healthcare professionals to consider the whole person, including their physical, emotional, and spiritual needs.
   - Train healthcare professionals to communicate effectively about CAM options and to involve patients in shared decision-making processes.

7. **Research and Innovation**
   - Encourage involvement in CAM research to contribute to the evidence base and understand the mechanisms, efficacy, and safety of CAM modalities.
   - Promote innovation in CAM education through the use of technology, such as online learning platforms, simulation-based learning, and virtual reality, to enhance accessibility and engagement.

8. **Ethical Practice and Legal Awareness**
   - Include training on ethical considerations and legal aspects of CAM practice to ensure healthcare professionals understand their responsibilities and the regulatory environment surrounding CAM.
   - Educate on the importance of informed consent and ethical patient care when recommending or practicing CAM therapies.

9. **Continual Professional Development**
   - Support ongoing professional development in CAM by providing access to advanced courses, workshops, and seminars for healthcare professionals to deepen their CAM knowledge and skills.
   - Encourage lifelong learning and critical thinking about the role of CAM in health and wellness.

10. **Collaboration and Networking**
    - Foster collaboration and networking opportunities among CAM and conventional healthcare professionals through conferences, forums, and professional associations.
    - Create platforms for sharing knowledge, research findings, and clinical experiences related to CAM to build a supportive community of practice.

11. **Utilize Technology and Innovation for Education**
    - Leverage digital technologies to create immersive and interactive CAM learning experiences, such as virtual reality (VR) simulations of CAM practices or augmented reality (AR) for anatomy and physiology education related to CAM therapies.
    - Develop online forums and communities of practice where healthcare professionals can discuss CAM case studies, share experiences, and update each other on the latest research and clinical guidelines.

12. **Promote Public Health Integration**
    - Include public health perspectives in CAM education to emphasize the role of CAM in preventive medicine, wellness, and community health.
    - Teach healthcare professionals how CAM can be used to address public health challenges, such as chronic disease management, mental health, and health disparities.

13. **Implement Outcome-Based Education**
    - Focus CAM education on achieving specific competencies and outcomes, such as the ability to make informed CAM referrals, integrate CAM into treatment plans effectively, and communicate about CAM with patients and other healthcare professionals.
    - Use outcome-based assessments to evaluate the competence of healthcare professionals in CAM, ensuring that they meet predefined standards of knowledge, skills, and attitudes.

14. **Encourage Reflective Practice**
    - Encourage reflective practice as part of CAM education, allowing healthcare professionals to reflect on their experiences, biases, and the impact of CAM on their clinical practice. This reflective process can foster a deeper understanding of the holistic nature of CAM and its potential benefits and limitations.
    - Integrate journals, portfolios, or discussion groups focused on reflective practice into CAM CME programs.

15. **Strengthen Patient Education and Engagement**
    - Train healthcare professionals to educate and engage patients effectively in discussions about CAM, including the evidence for and against specific CAM modalities, potential risks and benefits, and how CAM can fit into their overall health and wellness plans.
    - Develop patient education materials and resources as part of CAM CME programs to support healthcare professionals in this role.

16. **Cross-Cultural Exchange and Learning**
    - Facilitate cross-cultural exchange programs and partnerships that allow healthcare professionals to learn about CAM practices from different cultural and international perspectives. This can enrich understanding and foster a global approach to integrative medicine.
    - Organize international conferences, workshops, and seminars that bring together CAM and conventional medicine practitioners from around the world to share knowledge and experiences.

17. **Advocacy and Policy Engagement**
    - Encourage healthcare professionals to engage in advocacy and policy discussions related to CAM, aiming to influence health policy, insurance coverage for CAM therapies, and integration of CAM services into healthcare systems.
    - Educate healthcare professionals on how to navigate the policy and regulatory environments affecting CAM practice and how to advocate for changes that support integrative healthcare models.

18. **Sustainability and Environmental Health**
    - Integrate principles of sustainability and environmental health into CAM education, highlighting the connections between natural resources, herbal medicine, and holistic health.
    - Teach healthcare professionals about sustainable practices within CAM modalities and the importance of considering environmental impacts in healthcare decisions.

19. **Ethnobotanical and Traditional Wisdom**
   - Incorporate ethnobotanical knowledge and traditional wisdom into CAM education to provide a deeper understanding of the origins and uses of herbal medicines and natural therapies. This approach respects and preserves indigenous knowledge systems, while also examining their relevance and application in contemporary healthcare.
   - Facilitate partnerships with traditional healers and communities as part of educational programs, enabling direct learning from the source of much traditional medicinal knowledge.

20. **Global Health Perspectives**
   - Integrate global health perspectives into CAM CME to highlight the role of CAM in different health systems around the world. Understanding the global context can enrich healthcare professionals' perspectives on the diversity of health beliefs and practices, as well as the challenges and opportunities in integrating CAM into healthcare.
   - Encourage study abroad programs or international rotations that expose healthcare professionals to CAM practices in different cultural and healthcare settings.

21. **Personal Wellness and Self-Care for Healthcare Professionals**
   - Emphasize the importance of personal wellness and self-care among healthcare professionals within CAM education. By experiencing the benefits of CAM practices first-hand, healthcare providers can gain a more profound appreciation for these modalities and be better advocates for their integration into patient care.
   - Offer workshops and courses on self-care practices, such as mindfulness, yoga, or nutritional counseling, as part of CME programs.

22. **Interdisciplinary Research Collaborations**
   - Foster interdisciplinary research collaborations that bring together CAM practitioners, conventional healthcare professionals, and researchers. Such collaborations can generate innovative research questions, methodologies, and evidence that advance the understanding and integration of CAM.
   - Support the development of research networks focused on CAM to facilitate knowledge exchange, mentorship, and resource sharing among researchers.

23. **Healthcare System Integration**
   - Develop educational programs that specifically address the practical aspects of integrating CAM therapies into existing healthcare systems, including workflow, electronic health records, and multidisciplinary team collaboration.
   - Teach healthcare professionals about the logistics of referring patients to CAM practitioners, understanding insurance coverage for CAM therapies, and communicating with CAM providers.

24. **Ethics of CAM Practice and Research**
   - Deepen the focus on ethics within CAM education, addressing issues such as informed consent, patient autonomy, and the ethical considerations unique to CAM research and practice. This includes discussing the ethical use of natural resources and respecting cultural ownership of traditional medical knowledge.
   - Incorporate case studies and ethical dilemmas related to CAM into educational programs to stimulate critical thinking and discussion.

25. **Technological Innovation in CAM**
   - Leverage technological innovations to enhance CAM education and practice. This could include the development of apps for tracking the efficacy of CAM treatments, virtual reality simulations for learning acupuncture techniques, or AI-driven platforms for personalizing herbal medicine recommendations.
   - Encourage healthcare professionals to stay informed about technological advances that can support CAM practice and research.

26. **Patient Advocacy and Empowerment**
   - Train healthcare professionals in patient advocacy and empowerment, emphasizing the role of CAM in supporting patients to take an active role in their health and wellness. This includes educating patients on how to access reliable information about CAM and make informed choices about their health care.
   - Include communication skills training that enables healthcare professionals to effectively discuss CAM options with patients, respecting their values, beliefs, and preferences.

27. **Specialized Pathways for CAM Professions**
   - Develop specialized educational pathways for healthcare professionals interested in deepening their expertise in specific CAM modalities, leading to certification or advanced degrees in areas like herbalism, acupuncture, or mind-body therapies.

28. **Quality Assurance Mechanisms**
   - Implement quality assurance mechanisms for CAM CME programs, including peer review of educational content, feedback systems for participants, and ongoing program evaluation to ensure high standards.

29. **Collaborative Patient Care Models**
   - Teach collaborative care models that include CAM practitioners in the healthcare team, focusing on how to communicate effectively across disciplines and integrate care plans for the benefit of patients.

30. **Leadership and Change Management**
   - Offer courses on leadership and change management within healthcare settings to empower healthcare professionals to advocate for and implement integrative health-care models that include CAM.

31. **Ecotherapy and Environmental Medicine**
   - Integrate ecotherapy and environmental medicine principles into CAM education, emphasizing the connection between the environment, health, and well-being, and how to incorporate these principles into practice.

32. **Innovative Delivery Methods**
   - Utilize innovative delivery methods for CAM CME, such as gamification, interactive e-learning modules, and mobile learning apps, to engage learners and enhance the educational experience.

33. **Community Health Integration**
   - Focus on integrating CAM into community health initiatives, teaching healthcare professionals how to use CAM approaches in public health campaigns, community wellness programs, and population health management.

34. **Holistic Nutrition Education**
   - Expand education on holistic nutrition, covering the role of diet in preventing and treating illness and how to integrate nutritional counseling into clinical practice.

35. **Spiritual Care in Healthcare**
   - Include education on spiritual care within healthcare, recognizing the role of spirituality in healing and wellness and how to address spiritual needs as part of a holistic approach to patient care.

36. **Regenerative Medicine and CAM**
   - Explore the intersection of regenerative medicine and CAM, including how natural and traditional therapies can support tissue healing, regeneration, and wellness.

37. **Healthcare Policy and CAM**
   - Educate healthcare professionals on healthcare policy related to CAM, including regulation, insurance coverage, and the role of CAM in healthcare reform efforts.

38. **Data Science and CAM Research**
    - Encourage education in data science and biostatistics as applied to CAM research, enabling healthcare professionals to contribute to and critically evaluate research on CAM efficacy and safety.
39. **Emergency Preparedness and CAM**
    - Include training on the role of CAM in emergency preparedness and disaster response, such as stress management techniques, first aid using natural remedies, and supporting community resilience.
40. **Aging and Geriatric Care**
    - Focus on the application of CAM in aging and geriatric care, teaching healthcare professionals how to use CAM modalities to support healthy aging and manage age-related conditions.
41. **Pediatric CAM Education**
    - Develop specialized training on the safe and effective use of CAM in pediatric care, including considerations for dosage, safety, and communicating with parents about CAM.
42. **Mental Health and CAM**
    - Integrate education on the use of CAM in mental health care, covering modalities that support mental and emotional well-being, such as mindfulness, yoga, and herbal supplements for mood regulation.
43. **Cultural Sensitivity and Humility**
    - Emphasize the importance of cultural sensitivity and humility in CAM education, preparing healthcare professionals to respect and understand the cultural contexts of CAM practices and patient beliefs.
44. **Advanced Diagnostic Techniques in CAM**
    - Offer training in advanced diagnostic techniques specific to CAM, such as pulse diagnosis in Traditional Chinese Medicine or Ayurvedic tongue diagnosis, enhancing the clinical skills of healthcare professionals.
45. **Veterinary CAM**
    - Address the growing interest in CAM for animals by providing resources and education on CAM modalities suitable for veterinary practice.
46. **Sustainable Practice Management**
    - Educate on sustainable practice management, including ethical sourcing of CAM materials, environmentally friendly practices, and sustainability in healthcare business models.
47. **Telehealth and CAM**
    - Integrate education on the use of telehealth technologies for CAM consultations and treatments, addressing the unique challenges and opportunities of delivering CAM services remotely.
48. **Pain Management and CAM**
    - Provide specialized training on CAM approaches to pain management, emphasizing non-pharmacological interventions for chronic pain, such as acupuncture, massage therapy, and mind-body techniques.
49. **Integrative Oncology**
    - Offer courses in integrative oncology, teaching healthcare professionals how CAM can support conventional cancer treatments, help manage side effects, and improve the quality of life for cancer patients.
50. **Sports Medicine and CAM**
    - Develop educational programs on the role of CAM in sports medicine, including injury prevention, performance enhancement, and recovery strategies using CAM modalities.

51. **Neuroscience and CAM**
   - Incorporate the latest neuroscience research into CAM education, exploring how CAM practices such as meditation, yoga, and acupuncture affect brain function and mental health.

52. **Phytotherapy and Herbal Pharmacology**
   - Enhance education on phytotherapy and herbal pharmacology, providing health-care professionals with a deep understanding of the medicinal properties of plants, herb-drug interactions, and the pharmacokinetics of herbal medicines.

53. **Ethnomedicine and Global Healing Traditions**
   - Explore ethnomedicine and global healing traditions in CAM education, providing insights into the healing practices and medicinal plants used by various cultures around the world.

54. **Mindfulness and Stress Reduction**
   - Emphasize training in mindfulness and stress reduction techniques, highlighting their application in clinical practice for improving patient outcomes in various health conditions.

55. **Naturopathic Medicine Integration**
   - Educate on the principles of naturopathic medicine and its integration into conventional healthcare settings, including the use of natural remedies, lifestyle interventions, and holistic patient assessments.

56. **Biofeedback and Neurofeedback**
   - Offer training in biofeedback and neurofeedback as CAM modalities, teaching health-care professionals how these techniques can be used for stress management, mental health conditions, and neurological disorders.

57. **Detoxification Therapies**
   - Provide education on detoxification therapies within CAM, including their theoretical basis, methods of implementation, and the scientific evidence regarding their efficacy and safety.

58. **Energy Medicine**
   - Include courses on energy medicine, such as Reiki, qi gong, and therapeutic touch, focusing on their theoretical foundations, clinical applications, and integration into patient care plans.

59. **Holistic Skin Care**
   - Develop educational content on holistic skin care, teaching the role of CAM in treating dermatological conditions and promoting skin health through natural products and therapies.

60. **Disability and Rehabilitation**
   - Address the application of CAM in disability and rehabilitation, exploring how CAM modalities can support physical therapy, occupational therapy, and rehabilitation from injuries and surgeries.

61. **Women's Health and CAM**
   - Focus on women's health, offering CME on CAM approaches to managing menstrual disorders, fertility issues, pregnancy, childbirth, and menopause.

62. **Environmental Toxins and Health**
   - Educate healthcare professionals on the impact of environmental toxins on health and how CAM approaches can support detoxification and recovery from exposure.

63. **Addiction and Recovery**
   - Provide training on using CAM as part of addiction treatment and recovery programs, including modalities that support detoxification, reduce cravings, and address the underlying emotional and psychological issues.

64. **Pedagogical Skills for CAM Educators**
   - Enhance the pedagogical skills of CAM educators, focusing on effective teaching strategies, curriculum development, and assessment methods tailored to CAM education.
65. **Healthcare Design and CAM**
   - Explore the role of healthcare design in supporting CAM practices, including the creation of healing environments that facilitate relaxation, meditation, and other CAM therapies.
66. **Financial and Business Aspects of CAM Practice**
   - Include education on the financial and business aspects of running a CAM practice, covering topics such as billing, insurance, marketing, and regulatory compliance.

Implementing these best practices requires a collaborative effort among educators, healthcare institutions, professional bodies, and regulatory agencies. By adhering to these recommendations, the integration of CAM into CME can be enhanced, leading to more comprehensive, holistic, and patient-centered healthcare.

CONCLUSION

In conclusion, the expansive recommendations for Continuous Medical Education (CME) in Complementary and Alternative Medicine (CAM) underscore a dynamic and comprehensive approach to integrating CAM into healthcare education and practice. These recommendations reflect a multifaceted strategy aimed at addressing the complexities of CAM, including its diverse modalities, evidence base, and integration challenges within the conventional healthcare system. By focusing on evidence-based curriculum development, interprofessional education, practical training, and global health perspectives, among others, these guidelines aim to enhance the competency of healthcare professionals in CAM. Furthermore, they emphasize the importance of patient-centered care, cultural competence, and ethical practice in CAM education. The inclusion of advanced topics such as integrative oncology, pain management, neuroscience, and the use of technology in CAM education highlights the evolving nature of healthcare and the need for a healthcare workforce that is knowledgeable, skilled, and adaptable in the use of CAM therapies. Addressing the challenges of standardization, quality assurance, and research gaps in CAM is crucial for its legitimate integration into healthcare systems globally. Ultimately, the goal of these comprehensive CME recommendations in CAM is to foster an integrative healthcare environment where CAM and conventional medicine work synergistically to improve patient outcomes, promote wellness, and enhance the quality of life. By embracing these recommendations, healthcare educators, practitioners, and policymakers can contribute to a more holistic, inclusive, and patient-centered approach to health and wellness, recognizing the valuable role CAM plays in addressing the diverse needs of populations worldwide.

REFERENCES

Al-Worafi, Y.M. (Ed.). (2020a). *Drug Safety in Developing Countries: Achievements and Challenges*. Academic Press.

Al-Worafi, Y.M. (2020b). Herbal medicines safety issues. In: Al-Worafi, Y.M. (ed), *Drug Safety in Developing Countries* (pp. 163–178). Academic Press.

Al-Worafi, Y.M. (2022a). *A Guide to Online Pharmacy Education: Teaching Strategies and Assessment Methods*. CRC Press.

Al-Worafi, Y.M. (2022b). Competencies and learning outcomes. In: Al-Worafi, Y.M. (ed), *A Guide to Online Pharmacy Education: Teaching Strategies and Assessment Methods*. CRC Press.

Al-Worafi, Y.M. (2022c). Self-learning and self-directed learning. In: Al-Worafi, Y.M. (ed), *A Guide to Online Pharmacy Education: Teaching Strategies and Assessment Methods*. CRC Press.

Al-Worafi, Y.M. (2022d). Pharmacy education: Learning styles. In: Al-Worafi, Y.M. (ed), *A Guide to Online Pharmacy Education: Teaching Strategies and Assessment Methods*. CRC Press.

Al-Worafi, Y.M. (2022e). Continuous pharmacy education and professional development for pharmacy educators. In: Al-Worafi, Y.M. (ed), *A Guide to Online Pharmacy Education: Teaching Strategies and Assessment Methods*. CRC Press.

Al-Worafi, Y.M. (2023a). *Patient Safety in Developing Countries: Education, Research, Case Studies*. CRC Press.

Al-Worafi, Y.M. (2023b). *Technology for Drug Safety: Current Status and Future Developments*. Springer Nature.

Al-Worafi, Y.M. (2023c). Patient safety-related issues: Patient care errors and related problems. In: Al-Worafi, Y.M. (ed), *Patient Safety in Developing Countries: Education, Research, Case Studies*. CRC Press.

Al-Worafi, Y.M. (2023d). Patient care errors and related problems: Preventive medicine errors & related problems. In: Al-Worafi, Y.M. (ed), *Patient Safety in Developing Countries: Education, Research, Case Studies*. CRC Press.

Al-Worafi, Y.M. (2023e). Patient care errors and related problems: Patient assessment and diagnostic errors & related problems. In: Al-Worafi, Y.M. (ed), *Patient Safety in Developing Countries: Education, Research, Case Studies*. CRC Press.

Al-Worafi, Y.M. (2023f). Patient care errors and related problems: Non-pharmacological errors & related problems. In: Al-Worafi, Y.M. (ed), *Patient Safety in Developing Countries: Education, Research, Case Studies*. CRC Press.

Al-Worafi, Y.M. (2023g). Patient care errors and related problems: Medical errors & related problems. In: Al-Worafi, Y.M. (ed), *Patient Safety in Developing Countries: Education, Research, Case Studies*. CRC Press.

Al-Worafi, Y.M. (2023h). Patient care errors and related problems: Monitoring errors & related problems. In: Al-Worafi, Y.M. (ed), *Patient Safety in Developing Countries: Education, Research, Case Studies*. CRC Press.

Al-Worafi, Y.M. (2023i). Patient care errors and related problems: Patient education and counselling errors and related problems. In: Al-Worafi, Y.M. (ed), *Patient Safety in Developing Countries: Education, Research, Case Studies*. CRC Press.

Al-Worafi, Y.M. (2023j). Patient safety-related issues: Other medication safety issues. In: Al-Worafi, Y.M. (ed), *Patient Safety in Developing Countries: Education, Research, Case Studies*. CRC Press.

Al-Worafi, Y.M. (2023k). Patient safety culture. In: Al-Worafi, Y.M. (ed), *Patient Safety in Developing Countries: Education, Research, Case Studies*. CRC Press.

Al-Worafi, Y.M. (2023l). Patient safety education: Competencies and learning outcomes. In: Al-Worafi, Y.M. (ed), *Patient Safety in Developing Countries: Education, Research, Case Studies*. CRC Press.

Al-Worafi, Y.M. (Ed.). (2023m). *Clinical Case Studies on Medication Safety*. Academic Press.

Al-Worafi, Y.M. (Ed.). (2023n). *Comprehensive Healthcare Simulation: Pharmacy Education, Practice and Research*. Springer Nature.

Al-Worafi, Y.M. (Ed.). (2024a). *Handbook of Medical and Health Sciences in Developing Countries*. Springer, Cham.

Al-Worafi, Y.M. (2024b). Complementary and alternative medicine (CAM) in developing countries. In: Al-Worafi, Y.M. (ed), *Handbook of Medical and Health Sciences in Developing Countries*. Springer, Cham. https://doi.org/10.1007/978-3-030-74786-2_301-1

Al-Worafi, Y.M. (2024c). Access/equitable access to medical and health sciences in developing countries. In: Al-Worafi, Y.M. (ed), *Handbook of Medical and Health Sciences in Developing Countries*. Springer, Cham. https://doi.org/10.1007/978-3-030-74786-2_147-1

Hasan, S., Al-Omar, M.J., AlZubaidy, H., and Al-Worafi, Y.M. (2019). Use of medications in Arab countries. In: Laher, I. (ed), *Handbook of Healthcare in the Arab World* (p. 42). Springer, Cham.

12 Complementary and Alternative Medicine (CAM) Education
Cost of Education and Training

INTRODUCTION

Complementary and alternative medicine (CAM) encompasses a broad range of healthcare practices, products, and therapies that are not generally considered part of conventional medicine. CAM education and training vary significantly due to the diverse nature of CAM therapies, which include herbal medicine, acupuncture, chiropractic care, yoga, and meditation, among others. These therapies are often rooted in historical and cultural traditions rather than the scientific evidence that forms the basis of conventional medical education. Education and training in CAM are tailored to the specific type of therapy and the level of expertise required in practicing it safely and effectively. For instance, acupuncture training typically involves several years of study, including both theoretical coursework and hands-on clinical practice. Similarly, chiropractic education requires a Doctor of Chiropractic degree, which involves extensive study in anatomy, physiology, and chiropractic techniques, followed by a clinical internship. Unlike conventional medical training, which follows a standardized curriculum across medical schools, CAM education can be more variable. Some CAM practitioners may receive their training through specialized schools or institutions dedicated to a particular CAM therapy. Others may complete courses or workshops that supplement their existing healthcare credentials. Additionally, there are certification programs and continuing education courses for healthcare professionals interested in integrating CAM therapies into their practice. The regulatory framework for CAM education and training also varies by country and region, reflecting differing approaches to CAM integration into the healthcare system. In some countries, specific CAM practices are regulated, and practitioners are required to meet certain education and licensure requirements. In contrast, other countries have less formal regulatory structures, leading to a wider variation in the quality and consistency of CAM education. Despite these challenges, there is a growing interest in CAM education within conventional healthcare training programs. This is partly driven by an increasing demand from patients for CAM therapies and a growing recognition of the value of holistic and patient-centered approaches to care. As a result, some medical, nursing, and pharmacy schools are incorporating CAM topics into their curricula, providing future healthcare professionals with a broader understanding of the range of therapies available to patients. Overall, CAM education and training are evolving fields that reflect the dynamic and diverse nature of CAM therapies. As interest in and acceptance of CAM continue to grow, it is likely that education and training programs will become more standardized and integrated into the broader healthcare education system.

COMPLEMENTARY AND ALTERNATIVE MEDICINE (CAM) EDUCATION AND TRAINING: COST COMPONENTS

The cost components of CAM education and training can vary widely depending on the type of CAM practice, the length and depth of the program, the geographic location, and whether the

DOI: 10.1201/9781003327202-12

training is part of a degree program or a standalone course. Below are several key cost components associated with CAM education and training:

1. **Tuition Fees**: This is often the most significant expense for CAM education and training. Tuition can vary greatly depending on the institution, the type of program (e.g., certificate, diploma, or degree), and its duration. For instance, degree programs in chiropractic or naturopathic medicine, which can last several years, tend to have higher tuition fees compared to shorter certificate or diploma courses in massage therapy or herbalism.
2. **Materials and Supplies**: Students may need to purchase textbooks, equipment, and other supplies relevant to their field of study. For example, acupuncture students might need to buy needles and anatomical models, while herbal medicine students might need to purchase herbs and compounding supplies. The cost of materials can add significantly to the overall expense of the program.
3. **Facility Fees**: Some CAM programs may charge additional fees for the use of clinical or laboratory facilities, especially if the training involves hands-on practice with patients or the preparation of herbal remedies.
4. **Licensing and Certification Fees**: After completing their education, CAM practitioners often need to obtain licensure or certification to practice, which can involve additional costs. These may include fees for licensing exams, application fees, and the cost of maintaining licensure through continuing education.
5. **Travel and Accommodation**: For students attending CAM programs away from their home area, travel to and from the institution, as well as accommodation, can be significant expenses. Even for local students, commuting costs can add up over time.
6. **Insurance**: Some CAM students are required to have liability insurance, especially if they are involved in clinical training or practice. The cost of insurance can vary based on the type of CAM practice and the level of coverage needed.
7. **Continuing Education**: CAM practitioners often engage in continuing education to maintain their licensure and stay current in their field. The cost of attending workshops, seminars, and conferences, as well as enrolling in additional courses, can contribute to ongoing expenses.
8. **Technology Fees**: Online or hybrid CAM programs may require students to have access to specific technology or software, leading to additional costs for hardware, software, and possibly internet access.

The total cost of CAM education and training can range from a few thousand dollars for short-term certificate programs to tens of thousands of dollars for degree programs in practices like naturopathy or chiropractic medicine. Prospective CAM students should carefully research the costs associated with their chosen field of study, including indirect expenses and potential financial aid options, to fully understand the financial commitment required.

COMPLEMENTARY AND ALTERNATIVE MEDICINE (CAM) EDUCATION AND TRAINING COSTS: HIGH-INCOME COUNTRIES

In high-income countries, the costs associated with education and training in CAM can be substantial, reflecting both the comprehensive nature of the training required for certain CAM practices and the higher cost of education in these countries generally. Here's a breakdown of how these costs can manifest across different aspects of CAM education:

1. **Degree Programs**: For professions such as chiropractic, naturopathic medicine, and traditional Chinese medicine (TCM), which often require a degree, the tuition fees can be significant. Degree programs can last from 3 to 5 years or more, with annual tuition costs

ranging from $10,000 to $30,000 or higher. For example, a full program in naturopathic medicine or chiropractic care in the United States or Canada can cost upwards of $100,000 to $250,000 in total tuition.

2. **Accreditation and Quality of Institutions**: In high-income countries, there's a greater emphasis on attending accredited institutions that meet specific educational standards, which can also drive up costs. These institutions often provide comprehensive training that includes clinical hours, which are essential for obtaining licensure.

3. **Licensing and Examination Fees**: Post-graduation, CAM practitioners in high-income countries usually need to pass licensing or certification exams to practice legally. These exams can be expensive, with fees ranging from a few hundred to over a thousand dollars. Additionally, maintaining licensure often requires ongoing education, which incurs further costs.

4. **Supplementary Costs**: High-income countries may see higher costs for textbooks, equipment, and other necessary supplies. For example, acupuncture needles, anatomical models, or specialized software for practice management can represent a significant investment.

5. **Living Expenses**: The cost of living in high-income countries can significantly affect the overall expenses of CAM education. This includes accommodation, food, transportation, and personal expenses, which can add thousands of dollars to annual educational costs.

6. **Online and Hybrid Learning Options**: The rise of online and hybrid programs has made CAM education more accessible in some respects, but these programs still require technology investments and, in some cases, travel for in-person intensives or clinical training, adding to the overall cost.

7. **Financial Aid and Scholarships**: While there are opportunities for financial aid, scholarships, and grants, these can be competitive and may not cover the full cost of education. Students often rely on loans, which add to the future burden of loan repayment.

In summary, pursuing CAM education and training in high-income countries represents a significant financial investment. Prospective students must consider not only the direct costs of tuition and fees but also indirect expenses like living costs and the financial impact of the time commitment to education and training. Careful planning and research into financial aid and flexible learning options are crucial for managing these costs effectively.

COMPLEMENTARY AND ALTERNATIVE MEDICINE (CAM) EDUCATION AND TRAINING COSTS: MIDDLE-INCOME COUNTRIES

In middle-income countries, the landscape of CAM education and training costs can be quite distinct from that in high-income countries. While there are variations depending on the specific country, type of CAM practice, and educational institution, several general factors influence the cost of CAM education in these regions:

1. **Tuition Fees**: Generally, tuition fees for CAM education in middle-income countries may be lower than in high-income countries. However, the cost can still be significant relative to the average income in these countries. Degree programs in areas such as TCM, Ayurveda, or chiropractic care might be more affordable but can still represent a considerable investment for many students.

2. **Government Support and Regulation**: In some middle-income countries, CAM practices are more integrated into the healthcare system and may receive more government support. This can include subsidies for CAM education or more affordable training programs at public institutions. However, the level of support and regulation varies widely between countries, affecting both the cost and the quality of education.

3. **Quality and Accreditation of Educational Institutions**: The availability of accredited and high-quality CAM education programs can vary. In countries where CAM is more mainstream and regulated, there may be a higher number of reputable institutions offering accredited programs. However, in countries where CAM is less regulated, students may need to be more cautious in selecting programs to ensure they receive quality education that will be recognized by employers and professional bodies.

4. **Availability of Programs and Specialties**: The range of CAM programs and specialties available in middle-income countries may differ from those in high-income countries, reflecting local cultural practices and healthcare needs. This can influence both the cost and the type of CAM education pursued. For example, Ayurveda in India or TCM in China may be more accessible and affordable due to their widespread acceptance and integration into the healthcare system.

5. **Materials and Supplies**: Costs for materials and supplies needed for CAM training, such as herbs, acupuncture needles, or massage tables, can vary. These costs might be lower in countries where these supplies are locally produced or more readily available.

6. **Cost of Living**: The overall cost of living in middle-income countries can be lower than in high-income countries, potentially making it more affordable for students to cover living expenses while studying. However, this can vary greatly depending on the city or region.

7. **Online and Distance Learning Options**: With the increasing availability of online and distance learning options, students in middle-income countries may have more affordable paths to CAM education. However, the effectiveness and recognition of these programs can vary, and students may still face costs related to technology and internet access.

8. **Scholarships and Financial Aid**: Opportunities for scholarships, grants, and financial aid may be more limited in middle-income countries, though some institutions and organizations do offer support for CAM students. The availability and amount of financial aid can significantly affect the accessibility of CAM education for many students.

In conclusion, while CAM education and training in middle-income countries may be more accessible and affordable compared to high-income countries, students still face significant financial considerations. The variability in program quality, regulation, and government support across different countries and regions means that prospective CAM students must carefully research and consider their options to ensure they are making a viable investment in their education and future careers.

COMPLEMENTARY AND ALTERNATIVE MEDICINE (CAM) EDUCATION AND TRAINING COSTS: LOW-INCOME COUNTRIES

In low-income countries, the landscape of CAM education and training costs reflects a mix of challenges and opportunities unique to these settings. Various factors influence the affordability and accessibility of CAM education in these countries:

1. **Tuition Fees**: Relative to the local economy, the cost of CAM education in low-income countries can be prohibitive for many individuals. While tuition fees might be lower in absolute terms compared to higher-income countries, they can still represent a significant burden when considered relative to the average household income. For local and traditional forms of CAM (such as traditional herbal medicine or indigenous healing practices), formal education programs might be less common, and training might occur more informally or through apprenticeship models, potentially reducing direct costs.

2. **Government Support and Regulation**: The extent to which CAM is integrated into the national healthcare system and supported or regulated by the government varies widely. In some low-income countries, there may be limited formal recognition or support for CAM

education, impacting the development of standardized educational programs and influencing costs. However, where traditional medicine is an integral part of cultural heritage, governments may offer some support for its preservation and transmission, possibly through subsidized training programs.

3. **Quality and Availability of Education**: Access to quality CAM education can be a significant challenge. There might be fewer educational institutions offering CAM training, and those that do may have limited resources, affecting the quality of education. This scarcity can make it difficult for prospective students to find reputable programs within their own country, potentially increasing the need to seek education abroad, which is often cost-prohibitive.

4. **Materials and Infrastructure**: Costs associated with materials, textbooks, and other educational resources can be a barrier. Additionally, the infrastructure for delivering CAM education, such as clinical training facilities, may be lacking or under-resourced in low-income countries, potentially compromising the quality of hands-on training.

5. **Cost of Living and Indirect Costs**: While the cost of living in low-income countries may be lower than in more affluent nations, students may still face significant indirect costs related to their education, such as transportation, housing, and living expenses, which can add a substantial burden.

6. **Online and Distance Learning**: The expansion of online education offers a potential avenue for more affordable CAM training. However, challenges related to internet access, digital literacy, and the recognition of online qualifications can limit the effectiveness and appeal of this option.

7. **Scholarships and Financial Aid**: Opportunities for scholarships and financial aid for CAM education in low-income countries are limited. When available, such support can play a critical role in enabling students to pursue CAM training.

8. **Cultural and Traditional Knowledge Transmission**: In many low-income countries, CAM practices are deeply rooted in local culture and tradition, and knowledge is often passed down through generations informally. This traditional mode of learning may not involve formal tuition costs but could include other forms of investment, such as time spent in apprenticeship or contributions to the community or mentor.

In summary, while CAM education and training in low-income countries face significant challenges related to affordability, quality, and accessibility, there are also unique opportunities related to the cultural integration and transmission of traditional knowledge. Efforts to improve the formalization and recognition of CAM education in these contexts could help mitigate some of the barriers to access, ensuring that CAM practices continue to be a valuable and accessible component of healthcare systems in low-income countries.

COMPLEMENTARY AND ALTERNATIVE MEDICINE (CAM) EDUCATION AND TRAINING COSTS: AFFORDABILITY

Affordability is a significant concern for students interested in pursuing education and training in CAM. Given the varied nature of CAM disciplines, the costs and financial implications of pursuing such education can be considerable. Here are some factors affecting affordability and strategies to manage these costs:

FACTORS AFFECTING AFFORDABILITY

1. **Tuition and Fees**: CAM programs, especially those leading to licensure in fields like naturopathy, acupuncture, and chiropractic, can be expensive. The cost of tuition varies widely depending on the institution and the length of the program.

2. **Lack of Financial Support**: There is often less financial aid available for CAM students compared to those in conventional medical programs. Scholarships, grants, and loans specifically designated for CAM studies are less common, which can make these programs less accessible.

3. **Supplementary Costs**: Beyond tuition, students may face additional expenses for materials, equipment, and supplementary courses or workshops. These costs can add up, especially for programs that require specialized tools or resources.

4. **Living Expenses**: Like any higher education endeavor, CAM education requires consideration of living expenses. Students often need to factor in housing, food, transportation, and other personal expenses for the duration of their studies.

STRATEGIES FOR MANAGING COSTS

1. **Seek Out Scholarships and Grants**: Some organizations and institutions offer scholarships specifically for CAM studies. It's important to research and apply for any available financial aid tailored to CAM students.

2. **Explore Part-Time Programs**: Some CAM programs offer part-time study options, allowing students to work and earn an income while completing their education. This can help mitigate the financial burden of full-time study.

3. **Attend Community Colleges or Public Institutions**: Some community colleges and public institutions offer courses or programs in CAM at a lower cost than private institutions. These options can provide a more affordable pathway into the field.

4. **Consider Online Learning**: Online courses and programs can sometimes offer a more affordable alternative to traditional, in-person CAM education. Additionally, online learning can reduce or eliminate the costs associated with relocation and commuting.

5. **Budgeting and Financial Planning**: Effective financial planning is crucial for managing the costs associated with CAM education. Creating a detailed budget that accounts for all expected expenses can help students make informed decisions and seek out additional financial resources if necessary.

6. **Income-Based Repayment Plans for Loans**: For students who take out loans to finance their CAM education, exploring income-based repayment plans can offer a manageable way to handle debt after graduation.

7. **Work-Study Programs**: Some institutions may offer work-study programs that provide students with employment opportunities as part of their financial aid package. These positions can help offset education costs and provide valuable experience.

8. **Participate in Assistantships or Fellowships**: Some educational institutions offer assistantships or fellowships to graduate students. These positions typically involve research, teaching, or administrative duties in exchange for tuition waivers or stipends. Although more common in conventional higher education settings, exploring such opportunities within CAM programs can provide financial relief and valuable professional experience.

9. **Community and Professional Organization Scholarships**: Beyond institutional financial aid, numerous community organizations, professional associations, and foundations offer scholarships and grants to students pursuing careers in CAM fields. Membership in professional organizations related to CAM disciplines can also provide access to exclusive financial aid opportunities.

10. **Crowdfunding and Community Support**: Some students have turned to crowdfunding platforms to solicit financial support for their education from a broader community. This approach can be particularly effective if the student has a compelling personal story or professional mission related to CAM.

11. **Income Share Agreements**: An emerging funding model for education is the income share agreement (ISA), where students agree to pay a percentage of their future income for

a fixed period after graduation in exchange for funding their education. While not widely available for CAM programs yet, ISAs represent an innovative approach to education funding that could increase in popularity.

12. **Tax Credits and Deductions**: In some jurisdictions, students can take advantage of tax credits or deductions for education expenses. These tax benefits can help reduce the overall cost of education by lowering the amount of tax owed.

13. **Military or Public Service Programs**: Individuals with military service may have access to education benefits that can cover the cost of CAM education. Additionally, some public service programs offer loan forgiveness or educational benefits in exchange for a commitment to work in underserved areas or in specific roles that benefit the community.

14. **Negotiating Payment Plans**: Some educational institutions may offer flexible payment plans that allow students to spread the cost of tuition over several months or years. Negotiating a payment plan that aligns with one's financial capacity can make tuition payments more manageable.

15. **Leveraging Professional Networks**: Networking with professionals already working in CAM fields can provide insights into funding opportunities and strategies for cost-effective education. Mentorship and professional guidance can also open doors to scholarships, internships, and employment opportunities that can help offset education costs.

Long-Term Considerations

- **Career Pathways and Earning Potential**: Understanding the career pathways and potential earning power in various CAM disciplines is crucial. Researching the demand for specific CAM practices, potential settings (private practice, integrative health centers, etc.), and geographical areas can help in making an informed decision about which CAM field to pursue.

- **Regulatory Environment**: The legal and regulatory environment for CAM practitioners varies by location and discipline. Awareness of licensure requirements, scope of practice, and insurance reimbursement policies can influence the decision on which CAM field to pursue and how to navigate one's career post-graduation.

- **Continued Professional Development**: The field of CAM is continuously evolving, with new research, techniques, and practices emerging. Committing to lifelong learning and continued professional development is essential for success and can also involve additional costs over time.

Affordability in CAM education and training is a multifaceted issue that requires a comprehensive approach to address. By exploring a variety of funding sources, and financial strategies, and considering the long-term implications of their educational choices, students can navigate the challenges of affordability while pursuing their passion in the field of CAM.

COMPLEMENTARY AND ALTERNATIVE MEDICINE (CAM) EDUCATION AND TRAINING COSTS: CHALLENGES

The education and training in CAM present unique challenges, particularly in terms of costs, which can be broadly categorized into direct and indirect factors:

1. **Direct Costs**
   - **Tuition Fees**: CAM education programs vary widely in their cost structure. Programs can range from short certificate courses to multiyear degree programs, such as

naturopathy, acupuncture, or chiropractic degrees. The tuition for these programs can be comparable to or even exceed those of conventional medical schools in some cases.
- **Materials and Equipment**: CAM practices often require specialized equipment, tools, or materials (e.g., herbs for herbal medicine, acupuncture needles, etc.), which can add to the overall cost of education.
- **Certification and Licensing Fees**: After completing their education, CAM practitioners often need to obtain certification or licensure to practice, which may involve additional fees for exams and licensing.

2. **Indirect Costs**
- **Opportunity Cost**: The time spent in training and education for CAM could have been used to earn an income or pursue other educational opportunities. For some, the length of training required for certain CAM practices can represent a significant opportunity cost.
- **Insurance and Liability Coverage**: CAM practitioners may need to secure professional liability insurance, which can be costly and vary significantly depending on the type of CAM practice and the geographical location.
- **Continuing Education**: To maintain licensure and stay current in their field, CAM practitioners often need to engage in ongoing professional development and continuing education, which can involve further costs.

CHALLENGES

- **Accreditation and Recognition:** Not all CAM education programs are accredited or recognized by national or international bodies, which can affect graduates' ability to practice legally and limit their career prospects.
- **Evidence-based Curriculum:** Integrating evidence-based medicine into CAM education and ensuring that curricula reflect the latest research findings can be challenging. This integration is crucial for the legitimacy and efficacy of CAM practices.
- **Financial Aid and Support:** Financial aid and scholarships for CAM education may be less available compared to conventional medical programs. This lack of financial support can make CAM education less accessible to some students.
- **Market Saturation:** In some areas, the market for certain CAM practitioners is saturated, making it difficult for new graduates to establish successful practices. This concern can deter potential students from pursuing CAM education due to fears of inadequate return on investment.

Addressing these challenges requires concerted efforts from educators, professional organizations, regulatory bodies, and the CAM community to improve the accessibility, affordability, and quality of CAM education and training. Strategies could include increasing public and private funding for CAM education, enhancing the rigor and recognition of CAM programs, and promoting research into the efficacy and cost-effectiveness of CAM practices.

COMPLEMENTARY AND ALTERNATIVE MEDICINE (CAM) EDUCATION AND TRAINING COSTS: RECOMMENDATIONS

To address the varied challenges associated with the costs of CAM education and training across different economic settings, several recommendations can be made to improve accessibility, affordability, and the overall quality of CAM education (Al-Worafi, 2020a,b, 2022a,b, 2023a–n, 2024a–c; Hasan et al., 2019). These suggestions aim to support prospective students, educational institutions, and policymakers:

FOR PROSPECTIVE STUDENTS

1. **Research and Compare Programs**: Carefully investigate CAM programs, including tuition fees, accreditation status, and the curriculum. Compare various educational paths to find one that offers a balance of quality and affordability.
2. **Seek Scholarships and Financial Aid**: Look for scholarships, grants, and financial aid opportunities specifically targeted at CAM students. Some institutions, professional associations, and nonprofit organizations offer financial support to students pursuing careers in CAM.
3. **Consider Online and Hybrid Learning Options**: Online and hybrid programs can be more cost-effective than traditional on-campus programs. They can reduce travel and living expenses and offer greater flexibility to work while studying.
4. **Explore Apprenticeship and Mentorship Opportunities**: In fields where it's applicable, seek out apprenticeship or mentorship programs that offer hands-on training at a reduced cost. This is particularly relevant for CAM practices that are traditionally learned through apprenticeship.
5. **Plan for Additional Expenses**: Budget for indirect costs such as books, supplies, and any required insurance or licensing fees. Understanding the full financial commitment can help in planning and managing resources effectively.

FOR EDUCATIONAL INSTITUTIONS

1. **Develop Flexible Payment Options**: Institutions could offer flexible payment plans, sliding scale fees based on income, or other financing options to make CAM education more accessible.
2. **Increase Online and Hybrid Offerings**: By expanding online and hybrid education options, institutions can reduce overhead costs and pass those savings on to students, making CAM education more affordable.
3. **Strengthen Scholarship Programs**: Establish or expand scholarship and financial aid programs for students pursuing CAM studies, particularly for those from underserved communities.
4. **Engage in Community Partnerships**: Partner with healthcare providers, CAM practitioners, and community organizations to offer practical training opportunities to students, potentially reducing the cost of clinical education.

FOR POLICYMAKERS AND PROFESSIONAL ASSOCIATIONS

1. **Support Accreditation and Quality Standards**: Develop and enforce accreditation standards for CAM education to ensure quality and support the establishment of more standardized training pathways, which can help in controlling costs.
2. **Foster Public-Private Partnerships**: Encourage partnerships between governments, educational institutions, and the private sector to fund CAM education programs, scholarships, and research.
3. **Promote Recognition and Integration of CAM**: Work toward the integration of CAM practices into the broader healthcare system, which can increase the demand for CAM practitioners and justify investment in CAM education.
4. **Facilitate International Cooperation**: Support international exchanges and cooperation in CAM education to share resources, and best practices, and reduce costs through shared online platforms and joint programs.

By implementing these recommendations, stakeholders can work toward making CAM education and training more accessible and affordable, ensuring a diverse and well-prepared CAM workforce to meet the growing demand for holistic and alternative healthcare options worldwide.

CONCLUSION

The landscape of CAM education and training is complex and diverse, reflecting the wide range of practices and traditions that fall under the CAM umbrella. Across high-income, middle-income, and low-income countries, the costs associated with CAM education and training present significant challenges but also opportunities for innovation and improvement. To make CAM education more accessible and affordable, a multifaceted approach is needed. Prospective students should be proactive in researching programs, seeking financial aid, and considering alternative learning paths such as online education or apprenticeships. Educational institutions can play a crucial role by developing flexible payment options, expanding online and hybrid offerings, and strengthening scholarship programs. Moreover, policymakers and professional associations must support quality standards, foster public-private partnerships, and promote the integration of CAM into the healthcare system. As the demand for CAM continues to grow, alongside an increasing recognition of its value in achieving holistic health and well-being, addressing the challenges of CAM education and training costs becomes ever more critical. By implementing thoughtful strategies and recommendations, stakeholders can ensure that CAM education is both high-quality and accessible, preparing a skilled workforce ready to meet the diverse healthcare needs of populations around the world. This concerted effort can help to bridge the gap between traditional and conventional medicine, ultimately leading to more integrated and patient-centered healthcare systems.

REFERENCES

Al-Worafi, Y.M. (Ed.). (2020a). *Drug Safety in Developing Countries: Achievements and Challenges.* Academic Press.

Al-Worafi, Y.M. (2020b). Herbal medicines safety issues. In: Al-Worafi, Y.M. (ed), *Drug Safety in Developing Countries* (pp. 163–178). Academic Press.

Al-Worafi, Y.M. (2022a). *A Guide to Online Pharmacy Education: Teaching Strategies and Assessment Methods.* CRC Press.

Al-Worafi, Y.M. (2022b). Competencies and learning outcomes. In: Al-Worafi, Y.M. (ed), *A Guide to Online Pharmacy Education: Teaching Strategies and Assessment Methods.* CRC Press.

Al-Worafi, Y.M. (2023a). *Patient Safety in Developing Countries: Education, Research, Case Studies.* CRC Press.

Al-Worafi, Y.M. (2023b). *Technology for Drug Safety: Current Status and Future Developments.* Springer Nature.

Al-Worafi, Y.M. (2023c). Patient safety-related issues: Patient care errors and related problems. In: Al-Worafi, Y.M. (ed), *Patient Safety in Developing Countries: Education, Research, Case Studies.* CRC Press.

Al-Worafi, Y.M. (2023d). Patient care errors and related problems: Preventive medicine errors & related problems. In: Al-Worafi, Y.M. (ed), *Patient Safety in Developing Countries: Education, Research, Case Studies.* CRC Press.

Al-Worafi, Y.M. (2023e). Patient care errors and related problems: Patient assessment and diagnostic errors & related problems. In: Al-Worafi, Y.M. (ed), *Patient Safety in Developing Countries: Education, Research, Case Studies.* CRC Press.

Al-Worafi, Y.M. (2023f). Patient care errors and related problems: Non-pharmacological errors & related problems. In: Al-Worafi, Y.M. (ed), *Patient Safety in Developing Countries: Education, Research, Case Studies.* CRC Press.

Al-Worafi, Y.M. (2023g). Patient care errors and related problems: Medical errors & related problems. In: Al-Worafi, Y.M. (ed), *Patient Safety in Developing Countries: Education, Research, Case Studies.* CRC Press.

Al-Worafi, Y.M. (2023h). Patient care errors and related problems: Monitoring errors & related problems. In: Al-Worafi, Y.M. (ed), *Patient Safety in Developing Countries: Education, Research, Case Studies.* CRC Press.

Al-Worafi, Y.M. (2023i). Patient care errors and related problems: Patient education and counselling errors and related problems. In: Al-Worafi, Y.M. (ed), *Patient Safety in Developing Countries: Education, Research, Case Studies.* CRC Press.

Al-Worafi, Y.M. (2023j). Patient safety-related issues: Other medication safety issues. In: Al-Worafi, Y.M. (ed), *Patient Safety in Developing Countries: Education, Research, Case Studies.* CRC Press.

Al-Worafi, Y.M. (2023k). Patient safety culture. In: Al-Worafi, Y.M. (ed), *Patient Safety in Developing Countries: Education, Research, Case Studies.* CRC Press.

Al-Worafi, Y.M. (2023l). Patient safety education: Competencies and learning outcomes. In: Al-Worafi, Y.M. (ed), *Patient Safety in Developing Countries: Education, Research, Case Studies.* CRC Press.

Al-Worafi, Y.M. (Ed.). (2023m). *Clinical Case Studies on Medication Safety.* Academic Press.

Al-Worafi, Y.M. (Ed.). (2023n). *Comprehensive Healthcare Simulation: Pharmacy Education, Practice and Research.* Springer Nature.

Al-Worafi, Y.M. (Ed.). (2024a). *Handbook of Medical and Health Sciences in Developing Countries.* Springer, Cham.

Al-Worafi, Y.M. (2024b). Complementary and alternative medicine (CAM) in developing countries. In: Al-Worafi, Y.M. (ed), *Handbook of Medical and Health Sciences in Developing Countries.* Springer, Cham. https://doi.org/10.1007/978-3-030-74786-2_301-1

Al-Worafi, Y.M. (2024c). Access/equitable access to medical and health sciences in developing countries. In: Al-Worafi, Y.M. (ed), *Handbook of Medical and Health Sciences in Developing Countries.* Springer, Cham. https://doi.org/10.1007/978-3-030-74786-2_147-1

Hasan, S., Al-Omar, M.J., AlZubaidy, H., and Al-Worafi, Y.M. (2019). Use of medications in Arab countries. In: Laher, I. (ed), *Handbook of Healthcare in the Arab World* (p. 42). Springer, Cham.

13 Trends in Research & Graduation Projects in Complementary and Alternative Medicine (CAM)

INTRODUCTION

The field of complementary and alternative medicine (CAM) has seen significant growth and evolution over the past few decades, both in practice and in academic research. This expansion is reflected in the diversity and innovation of research and graduation projects within this field. CAM encompasses a wide range of practices, therapies, and products that are not generally considered part of conventional medicine, including herbal medicine, acupuncture, chiropractic, naturopathy, and mind-body interventions, among others. The trends in research and graduation projects in CAM illustrate a dynamic interplay between traditional knowledge and contemporary scientific inquiry, aiming to validate and integrate CAM therapies into mainstream healthcare. Initially, research in CAM was primarily exploratory, focusing on documenting traditional practices and their potential health benefits. Graduation projects often involved case studies or literature reviews that aimed to provide a scientific basis for traditional therapies. Over time, the focus has shifted toward more rigorous scientific research methodologies, including randomized controlled trials, systematic reviews, and meta-analyses, to evaluate the efficacy, safety, and mechanisms of action of CAM therapies. A significant trend in CAM research and graduation projects is the emphasis on evidence-based practice. There is a growing demand for high-quality evidence to support the integration of effective CAM therapies into conventional healthcare settings. This has led to an increase in interdisciplinary projects that combine CAM with conventional medical research, aiming to offer holistic and patient-centered care approaches. Another notable trend is the exploration of the biological mechanisms underlying CAM therapies. This includes research into the pharmacological properties of herbal medicines, the physiological effects of acupuncture, and the neurobiological mechanisms of meditation and other mind-body interventions. Such projects aim to bridge the gap between traditional knowledge and modern science, enhancing the credibility and acceptance of CAM practices.

Technological advancements have also influenced CAM research and graduation projects. Digital health technologies, including mobile health apps and wearable devices, are being used to monitor the effects of CAM therapies, improve patient adherence, and collect large-scale data for research purposes. This convergence of technology and CAM opens new avenues for personalized medicine and remote healthcare delivery. Furthermore, there is an increasing focus on the holistic and preventive aspects of CAM, with research projects exploring how these therapies can contribute to overall well-being, stress reduction, and the prevention of chronic diseases. This aligns with a broader shift in healthcare toward preventive medicine and the promotion of health and wellness. In conclusion, the trends in research and graduation projects in CAM reflect a growing recognition of the value of integrating traditional and alternative therapies with conventional medicine. Through rigorous scientific research, interdisciplinary collaboration, and the embrace of technological innovations, the field of CAM continues to evolve, offering new insights and approaches to health and wellness. The emphasis on evidence-based practice, understanding the mechanisms of action, and

DOI: 10.1201/9781003327202-13

exploring preventive and holistic health strategies are key themes that will likely continue to shape the future of CAM research and practice.

RATIONALITY AND IMPORTANCE OF GRADUATION PROJECTS IN COMPLEMENTARY AND ALTERNATIVE MEDICINE (CAM) PROGRAMS

Graduation projects in CAM programs hold significant rationality and importance for multiple reasons, intertwining academic pursuits with practical implications for healthcare. These projects not only contribute to the body of knowledge in CAM but also play a critical role in the broader healthcare ecosystem. Here's an exploration of their rationality and importance:

1. **Bridging Traditional Knowledge and Modern Healthcare**

   Graduation projects in CAM are vital for bridging the gap between traditional healing practices and contemporary medical science. They provide a structured way to investigate, document, and validate the efficacy and safety of traditional remedies and practices, many of which have been used for centuries. By applying rigorous scientific methodologies to study these practices, students can help integrate beneficial CAM therapies into mainstream healthcare, enhancing patient care options.

2. **Expanding Evidence-Based Practice**

   A core rationale for CAM graduation projects is the expansion of evidence-based practice within the field. Despite the widespread use of CAM therapies, many remain under-researched or lack robust scientific evidence supporting their effectiveness. Graduation projects can address this gap by generating high-quality evidence through clinical trials, systematic reviews, and meta-analyses. This evidence is crucial for healthcare professionals, patients, and policymakers to make informed decisions about incorporating CAM therapies into treatment plans.

3. **Innovating Healthcare Solutions**

   CAM programs encourage innovation by allowing students to explore novel healthcare solutions that encompass holistic and preventive approaches to health and wellness. Graduation projects can lead to the development of new therapies, products, or intervention programs that contribute to patient care, wellness, and disease prevention. This innovation is especially important in areas where conventional medicine may not offer adequate solutions, such as chronic pain management, mental health, and lifestyle diseases.

4. **Promoting Interdisciplinary Collaboration**

   Graduation projects in CAM also foster interdisciplinary collaboration, bringing together insights from various fields such as biology, pharmacology, psychology, and engineering. This collaborative approach enriches the research process and outcomes, facilitating the development of comprehensive and integrated healthcare solutions that address complex health issues from multiple angles.

5. **Empowering Patient-Centered Care**

   By focusing on CAM, graduation projects inherently promote a patient-centered approach to healthcare. CAM therapies often emphasize the holistic well-being of the individual, considering physical, emotional, and spiritual health. Research in this area can lead to more personalized and compassionate care strategies, empowering patients to take an active role in their health and treatment processes.

6. **Addressing Public Interest and Demand**

   The growing public interest and demand for CAM therapies underscore the importance of graduation projects in this field. As more individuals seek alternative and complementary approaches to health and wellness, there is a pressing need for qualified professionals who can provide evidence-based guidance and care. These projects prepare students to meet this demand, equipping them with the knowledge and skills necessary to practice safely and effectively within the CAM field.

7. **Global Health Contributions**

Finally, CAM graduation projects have the potential to make significant contributions to global health challenges. By exploring diverse medical traditions and practices from around the world, these projects can uncover valuable knowledge and approaches that could be adapted and applied globally, offering new strategies for disease prevention, health promotion, and culturally sensitive care.

In essence, graduation projects in CAM programs are fundamental to advancing the field, promoting health innovation, and enhancing patient care. They offer a unique opportunity to explore the vast potential of integrating traditional and alternative medical practices with conventional healthcare, aiming for a more holistic, evidence-based, and patient-centered approach to health and wellness.

RATIONALITY AND IMPORTANCE OF RESEARCH ABOUT COMPLEMENTARY AND ALTERNATIVE MEDICINE (CAM) EDUCATION

Research into CAM education holds critical importance and rationality, driven by the growing integration of CAM practices into the broader healthcare landscape and the increasing consumer demand for such therapies. This research area not only seeks to understand and improve how CAM is taught and learned but also aims to ensure that future healthcare providers are well-equipped to make informed decisions about the integration of CAM into patient care. Here's a deeper look into the rationale and importance of this research focus:

1. **Ensuring Educational Excellence and Standards**

Research in CAM education is essential for establishing and maintaining high educational standards and competencies. With CAM practices becoming more prevalent, there is a need for rigorous educational frameworks that ensure practitioners have a thorough understanding of both the theoretical and practical aspects of CAM. Research can help identify core competencies, develop standardized curricula, and assess educational outcomes, thereby enhancing the quality of CAM education and practice.

2. **Promoting Evidence-Based CAM Practices**

One of the primary goals of researching CAM education is to promote evidence-based practices. By integrating research findings into the educational content, future CAM practitioners can be trained to critically evaluate the evidence surrounding CAM therapies and apply this knowledge in their practice. This is crucial for patient safety, the efficacy of treatment, and the professional credibility of CAM practitioners.

3. **Facilitating Interdisciplinary Integration**

As healthcare moves toward more integrative models, understanding how CAM is taught and how it can be effectively integrated with conventional medical education becomes increasingly important. Research in CAM education can identify effective strategies for interdisciplinary learning, encouraging collaboration and understanding between conventional and CAM practitioners. This interdisciplinary approach can lead to more holistic and patient-centered care.

4. **Addressing Public Demand and Expectations**

Public interest in and the use of CAM therapies are on the rise. Research in CAM education can help align educational programs with the public's health beliefs, practices, and expectations, ensuring that healthcare providers are prepared to discuss and advise on CAM therapies. This alignment is essential for patient satisfaction, engagement, and overall health outcomes.

5. **Improving Patient Care and Safety**

   Research into CAM education directly impacts patient care and safety. By ensuring that CAM practitioners and healthcare professionals receive proper training, including knowledge of potential interactions between CAM therapies and conventional treatments, patient safety can be enhanced. Furthermore, well-educated CAM practitioners are better equipped to provide care that is evidence-based, culturally sensitive, and aligned with patients' values and beliefs.

6. **Advancing Global Health Perspectives**

   CAM practices often draw from global traditions and knowledge systems. Research in CAM education can foster a greater understanding and appreciation of these diverse health perspectives, contributing to more culturally competent healthcare providers. This global outlook is particularly important in increasingly multicultural societies and can improve the accessibility and acceptability of healthcare services.

7. **Supporting Regulatory and Policy Development**

   Finally, research in CAM education can inform policy and regulatory frameworks. By providing evidence about the effectiveness of different educational strategies and the competencies of CAM practitioners, research can guide policy decisions that support the safe, effective, and integrated use of CAM therapies in healthcare systems.

In summary, research in CAM education is pivotal for advancing the field, ensuring high standards of practice, and meeting the health needs of the public. It supports the development of a healthcare workforce that is knowledgeable, skilled, and capable of integrating CAM therapies into patient care in an evidence-based and patient-centered manner.

GRADUATION PROJECTS AND RESEARCH IN COMPLEMENTARY AND ALTERNATIVE MEDICINE (CAM) EDUCATION: FACILITATORS

Graduation projects and research in CAM education benefit significantly from various facilitators that support and enhance these academic endeavors. These facilitators play a crucial role in ensuring that research and projects are conducted efficiently, effectively, and with the highest standards of academic rigor and relevance. Here are some key facilitators for CAM education research and graduation projects:

1. **Interdisciplinary Collaboration**
   - *Rationale*: Encourages the integration of diverse perspectives and expertise, enriching the research and educational content.
   - *Implementation*: Partnerships between CAM institutions, conventional medical schools, and other academic departments can foster interdisciplinary research projects and educational programs.

2. **Access to Quality CAM Resources**
   - *Rationale*: Comprehensive access to CAM literature, traditional knowledge, and modern research findings is essential for informed research.
   - *Implementation*: Libraries, online databases, and subscriptions to CAM journals and publications provide necessary resources for students and researchers.

3. **Expert Guidance and Mentorship**
   - *Rationale*: Experienced faculty and practitioners can guide students through the complexities of CAM research and practice.
   - *Implementation*: Incorporating mentorship programs and inviting CAM practitioners as guest lecturers or advisors for projects.

4. **Research Funding and Grants**
   - *Rationale*: Financial support is crucial for conducting high-quality research, especially for projects requiring specialized equipment, travel, or extensive resources.
   - *Implementation*: Seeking funding from institutions, governmental bodies, and private organizations dedicated to CAM research and education.
5. **Ethical and Cultural Sensitivity Training**
   - *Rationale*: Understanding and respecting the cultural roots and ethical considerations of CAM practices are vital for conducting responsible research.
   - *Implementation*: Workshops and courses on research ethics, cultural competency, and the historical context of CAM therapies.
6. **Technological Support**
   - *Rationale*: Technology facilitates data collection, analysis, and the dissemination of research findings.
   - *Implementation*: Utilizing software for statistical analysis, online platforms for surveys, and digital tools for project management and collaboration.
7. **Community and Clinical Partnerships**
   - *Rationale*: Engagement with local communities and clinical settings can enhance the relevance and applicability of research findings.
   - *Implementation*: Establishing partnerships with healthcare providers, CAM practitioners, and community organizations for fieldwork and practical projects.
8. **Professional Development Opportunities**
   - *Rationale*: Participation in conferences, workshops, and seminars helps students and researchers stay updated on the latest CAM research and trends.
   - *Implementation*: Support for attending professional gatherings and opportunities to present research findings to a wider audience.
9. **Regulatory Guidance**
   - *Rationale*: Navigating the regulatory landscape is essential for research involving human subjects or novel therapies.
   - *Implementation*: Access to experts in healthcare law, ethics boards, and regulatory bodies to ensure compliance with all research regulations.
10. **Peer Support and Collaborative Networks**
    - *Rationale*: Building a supportive community of peers engaged in CAM research fosters a collaborative environment.
    - *Implementation*: Creating forums, study groups, and online communities for students and researchers to share insights, challenges, and successes.

These facilitators, when effectively leveraged, can significantly enhance the quality and impact of graduation projects and research in CAM education, contributing to the advancement of the field and the integration of CAM into broader healthcare practices.

GRADUATION PROJECTS AND RESEARCH IN COMPLEMENTARY AND ALTERNATIVE MEDICINE (CAM) EDUCATION: BARRIERS

While there are many facilitators that can enhance graduation projects and research in CAM education, several barriers also exist that can impede progress in these areas. Overcoming these challenges is essential for advancing CAM research and education. Here are some of the main barriers faced by students and researchers in this field:

1. **Limited Recognition and Acceptance**
   - *Issue*: CAM practices often face skepticism and limited acceptance within the main-stream medical community. This skepticism can extend to academic and research settings, affecting the credibility and support for CAM projects.
   - *Impact*: Difficulty in securing funding, partnerships, and institutional support for CAM research and education projects.
2. **Insufficient Funding**
   - *Issue*: CAM research and education programs frequently struggle to obtain funding, as they may not be considered a priority by traditional funding agencies focused on conventional medicine.
   - *Impact*: Limited resources to conduct high-quality research, access to necessary materials, and implementation of innovative CAM educational programs.
3. **Regulatory and Ethical Complexities**
   - *Issue*: Navigating the regulatory and ethical approval processes for CAM research can be challenging, particularly for projects involving traditional remedies or practices not well understood by regulatory bodies.
   - *Impact*: Delays in project approval, limitations on research scope, and challenges in conducting clinical trials involving CAM therapies.
4. **Lack of Standardized Curriculum**
   - *Issue*: The vast diversity of CAM practices and the absence of standardized curricula can create challenges in developing comprehensive and coherent educational programs.
   - *Impact*: Variability in the quality and content of CAM education, making it difficult to ensure that graduates have met consistent competency standards.
5. **Research Methodology Challenges**
   - *Issue*: The holistic and individualized nature of many CAM therapies can pose difficulties in applying traditional randomized controlled trial (RCT) methodologies, which are the gold standard in medical research.
   - *Impact*: Challenges in proving the efficacy and safety of CAM modalities to the standards expected by the scientific community, affecting the integration of CAM practices into mainstream healthcare.
6. **Access to Quality Resources**
   - *Issue*: There can be a lack of access to high-quality, peer-reviewed research resources and databases dedicated to CAM, making it difficult for students and researchers to conduct comprehensive literature reviews.
   - *Impact*: Limitations on the scope and depth of research projects, affecting the quality and reliability of CAM education and research findings.
7. **Cultural and Ethical Sensitivities**
   - *Issue*: CAM practices often have deep cultural roots, and research in this area requires sensitivity to cultural beliefs and ethical considerations, which may not be fully understood or respected by all in the academic community.
   - *Impact*: Potential for cultural appropriation, ethical breaches, and misunderstanding of CAM practices, undermining the validity and respect for CAM research.
8. **Interdisciplinary Collaboration Barriers**
   - *Issue*: Effective CAM research and education often require interdisciplinary collaboration, which can be hindered by differences in terminology, methodology, and professional culture between CAM and conventional medical fields.
   - *Impact*: Challenges in communication and collaboration, affecting the development of integrative health approaches and the comprehensive education of students.
9. **Professional Isolation**
   - *Issue*: CAM practitioners and researchers may experience professional isolation due to the marginal status of CAM within the broader healthcare and academic systems.

- *Impact*: Limited professional networking opportunities, mentorship, and career development paths for individuals in the CAM field.
10. **Public Perception and Demand**
    - *Issue*: While public interest in CAM is growing, misconceptions and varying levels of awareness about CAM practices can affect the demand for and perceived value of CAM education and research.
    - *Impact*: Difficulty in engaging a wider audience and securing the support necessary for the advancement of CAM research and educational programs.

Overcoming these barriers requires concerted efforts from educators, researchers, policymakers, and the CAM community to advocate for greater recognition, develop robust research methodologies, secure funding, and promote interdisciplinary collaboration. By addressing these challenges, the field of CAM can advance, contributing valuable insights and approaches to the broader healthcare landscape.

GRADUATION PROJECTS AND RESEARCH IN COMPLEMENTARY AND ALTERNATIVE MEDICINE (CAM) EDUCATION: RECOMMENDATIONS

To enhance the quality and impact of graduation projects and research in CAM education, adopting best practices is crucial. These recommendations are designed to address common barriers and leverage facilitators, aiming to foster rigorous, impactful, and innovative research within the CAM field. Here are key recommendations for best practices in CAM education and research (Al-Worafi, 2020a,b, 2022a,b, 2023a–n, 2024a–c; Hasan et al., 2019):

1. **Foster Interdisciplinary Partnerships**
   - *Action*: Create collaborative partnerships between CAM and conventional medicine programs, as well as other relevant disciplines such as pharmacology, psychology, and public health. This encourages the exchange of knowledge and methodologies, enriching the research and educational experience.
2. **Secure Diverse Funding Sources**
   - *Action*: Explore a variety of funding sources, including government grants, private foundations, and nonprofit organizations dedicated to CAM research. Diversifying funding sources can provide stability and support for innovative projects.
3. **Enhance Curriculum Development**
   - *Action*: Develop standardized, evidence-based curricula that include both the theory and practice of CAM, ensuring that students receive a comprehensive education. Incorporating feedback from CAM practitioners, educators, and students can help in refining and updating the curriculum.
4. **Utilize Mixed Research Methodologies**
   - *Action*: Encourage the use of mixed methods research that combines qualitative and quantitative approaches. This allows for a more holistic understanding of CAM practices and their effects, accommodating the individualized and holistic nature of many CAM therapies.
5. **Promote Ethical and Cultural Competency**
   - *Action*: Integrate ethical and cultural competency training into CAM education programs. This prepares students to conduct research and practice in a manner that is respectful of diverse cultures and ethical standards, particularly when working with traditional knowledge and practices.

6. **Implement Rigorous Research Training**
   - *Action*: Ensure that CAM education includes strong research methodology training, emphasizing the importance of evidence-based practice. Providing students with the skills to critically evaluate research and conduct high-quality studies is essential for advancing the field.
7. **Expand Access to CAM Resources**
   - *Action*: Improve access to CAM research databases, libraries, and journals for students and researchers. Making comprehensive resources available supports thorough literature reviews and informed research projects.
8. **Enhance Public Engagement and Communication**
   - *Action*: Develop programs and initiatives that engage the public and healthcare professionals in discussions about CAM research and education. Effective communication strategies can improve the understanding and acceptance of CAM practices.
9. **Support Professional Development**
   - *Action*: Offer ongoing professional development opportunities for CAM educators and researchers, including workshops, seminars, and conferences. Staying updated on the latest research and educational strategies is key to maintaining high standards in CAM education.
10. **Advocate for CAM Integration in Healthcare Systems**
    - *Action*: Work toward the integration of evidence-based CAM practices into conventional healthcare systems. Advocacy and policy engagement are important for creating a healthcare environment that values holistic and patient-centered care approaches.
11. **Monitor and Evaluate Program Effectiveness**
    - *Action*: Regularly monitor and evaluate the effectiveness of CAM education programs and research projects. Feedback mechanisms and outcome assessments can help identify areas for improvement and demonstrate the impact of CAM education.
12. **Cultivate a Supportive CAM Community**
    - *Action*: Build a supportive community of CAM researchers, practitioners, and students. Networking and mentorship programs can foster a collaborative environment that supports the professional and personal development of those in the CAM field.
13. **Leverage Technology and Digital Platforms**
    - *Action*: Utilize digital platforms and technologies for both education and research in CAM. Online learning platforms can enhance access to CAM education, while digital tools and mobile apps can facilitate innovative research methodologies and improve data collection and analysis.
14. **Encourage Patient and Community Involvement**
    - *Action*: Involve patients and local communities in CAM research and education projects. community-based participatory research (CBPR) approaches can ensure that research is relevant, respectful, and beneficial to the communities involved, fostering trust and enhancing the applicability of findings.
15. **Strengthen International Collaboration**
    - *Action*: Establish international collaborations to share knowledge, resources, and best practices in CAM education and research. Global partnerships can enrich the educational experience, broaden research perspectives, and facilitate the exchange of traditional and innovative CAM practices.
16. **Develop Specialized Research Centers**
    - *Action*: Advocate for the creation of specialized research centers focused on CAM. These centers can serve as hubs for interdisciplinary research, education, and practice, providing leadership, resources, and advocacy for the CAM field.

17. **Incorporate Real-World Clinical Experience**
   - *Action*: Ensure that CAM education programs include substantial clinical experience, allowing students to apply their knowledge in real-world settings under the guidance of experienced practitioners. This hands-on experience is critical for developing practical skills and understanding the complexities of patient care.

18. **Promote Research Literacy among Practitioners**
   - *Action*: Develop initiatives aimed at enhancing research literacy among CAM practitioners who may not be actively involved in academia. Understanding how to critically appraise research findings and apply them in practice is essential for evidence-based CAM practice.

19. **Advocate for Policy Support**
   - *Action*: Engage in advocacy efforts to gain policy support for CAM research and education. Policy changes can provide the necessary framework and resources for integrating CAM into health systems, ensuring regulatory support, and enhancing public trust in CAM practices.

20. **Ensure Sustainability in CAM Initiatives**
   - *Action*: Plan for the long-term sustainability of CAM education and research initiatives. This includes securing ongoing funding, developing leadership succession plans, and embedding CAM principles into institutional cultures.

21. **Foster a Culture of Continuous Improvement**
   - *Action*: Cultivate a culture of continuous improvement within CAM education and research organizations. Encourage feedback, experimentation, and the willingness to adapt based on new evidence, ensuring that CAM practices remain dynamic, relevant, and effective.

22. **Emphasize the Importance of Mental and Emotional Well-Being**
   - *Action*: Integrate the study of mental and emotional well-being into CAM education and research. Recognizing the interconnectedness of the mind, body, and spirit is fundamental to many CAM practices and is increasingly relevant in addressing holistic health in the modern world.

23. **Embrace Digital Health Innovations**
   - *Action*: Integrate digital health innovations into CAM research and practice. This could involve telehealth services for CAM practices, the use of health apps to support wellness and mindfulness, and wearable technology for monitoring health outcomes. Embracing these technologies can expand the reach and efficacy of CAM interventions.

24. **Cultivate Resilience and Adaptability**
   - *Action*: Prepare students and practitioners for the dynamic nature of healthcare by emphasizing resilience and adaptability in CAM education. Developing skills to navigate changes in healthcare policies, patient needs, and medical technologies ensures that CAM professionals remain relevant and effective.

25. **Implement Personalized Medicine Approaches**
   - *Action*: Encourage the integration of personalized medicine approaches in CAM research and education. This includes understanding genetic, environmental, and lifestyle factors that influence health, allowing for more tailored and effective CAM interventions.

26. **Enhance Data Analytics Skills**
   - *Action*: Incorporate data analytics and bioinformatics into CAM curricula to analyze complex datasets from research studies and clinical trials. This skill set is crucial for identifying patterns, efficacy, and safety profiles of CAM therapies, contributing to evidence-based practice.

27. **Promote Environmental Sustainability**
    - *Action*: Embed principles of environmental sustainability into CAM education and practice. Since many CAM practices rely on natural resources (e.g., herbal medicines), advocating for sustainable sourcing and conservation is vital. This approach aligns with the holistic philosophy of CAM, emphasizing harmony with nature.
28. **Strengthen Leadership and Management Training**
    - *Action*: Provide leadership and management training within CAM education programs to prepare practitioners for roles in healthcare leadership, policy development, and the management of CAM practices or wellness centers.
29. **Expand Community Health Initiatives**
    - *Action*: Develop and support community health initiatives that incorporate CAM approaches. Projects that focus on wellness, prevention, and holistic health can demonstrate the practical value of CAM in public health and encourage community engagement.
30. **Foster Ethical Innovation**
    - *Action*: Encourage ethical innovation in CAM research and product development. This involves not only adhering to ethical guidelines in research but also considering the ethical implications of CAM innovations on society, including issues of accessibility and equity.
31. **Build Global Health Networks**
    - *Action*: Establish global health networks focusing on CAM to facilitate the exchange of knowledge, resources, and best practices across borders. This can enhance the understanding of CAM's role in different cultural and healthcare systems and identify universal principles that can be applied globally.
32. **Advocate for Inclusive Healthcare Policies**
    - *Action*: Work toward the development of healthcare policies that are inclusive of CAM, advocating for insurance coverage of CAM therapies and the integration of CAM services in mainstream healthcare facilities. This requires demonstrating the value and efficacy of CAM through rigorous research and education.
33. **Encourage Lifelong Learning**
    - *Action*: Promote a culture of lifelong learning among CAM practitioners and researchers. Continuous education in the latest CAM research findings, technologies, and practices ensures that professionals can provide the best possible care and contribute to the field's advancement.
34. **Utilize Social Media and Online Platforms**
    - *Action*: Leverage social media and online platforms to disseminate CAM research findings, educate the public about CAM benefits, and engage with a broader audience. Effective use of these platforms can enhance the visibility and acceptance of CAM.
35. **Promote AI and Machine Learning Applications**
    - *Action*: Encourage the exploration of artificial intelligence (AI) and machine learning in CAM research for predictive analytics, personalized treatment recommendations, and the analysis of large-scale health data. Training CAM researchers and practitioners in these technologies can lead to groundbreaking discoveries and innovations in patient care.
36. **Implement Virtual Reality (VR) and Augmented Reality (AR) Technologies**
    - *Action*: Utilize VR and AR technologies for educational purposes, such as simulating acupuncture techniques or visualizing the impact of stress reduction practices on the body. These technologies can enhance learning experiences and provide immersive training environments.

37. **Focus on Mental Health and Well-Being**
   - *Action*: Place a greater emphasis on mental health and well-being in CAM research and education, reflecting the increasing global awareness of mental health issues. This includes integrating mental health-focused CAM practices into curricula and research agendas.

38. **Develop Entrepreneurial Skills**
   - *Action*: Incorporate entrepreneurship training into CAM education to empower practitioners to launch their own practices, develop innovative CAM products, and navigate the business aspects of healthcare. This can lead to the expansion of CAM services and products available to the public.

39. **Strengthen Patient Advocacy and Empowerment**
   - *Action*: Teach CAM students and practitioners the principles of patient advocacy and empowerment, emphasizing the role of CAM in supporting patients' rights to choose their treatment paths. This approach can build stronger practitioner-patient relationships and promote patient-centered care.

40. **Enhance Global Health Literacy**
   - *Action*: Focus on increasing global health literacy within the CAM community, understanding the global implications of health trends, diseases, and treatments. This global perspective is essential for addressing health issues that cross national boundaries and affect diverse populations.

41. **Address Health Disparities**
   - *Action*: Undertake research and education initiatives aimed at understanding and addressing health disparities through CAM. This involves studying how CAM can be made more accessible and effective for underserved and marginalized populations.

42. **Foster Resilient Healthcare Systems**
   - *Action*: Research and promote the role of CAM in building resilient healthcare systems, especially in response to global health crises such as pandemics. This includes exploring how CAM practices can support public health measures, enhance individual resilience, and contribute to healthcare sustainability.

43. **Integrate Climate Change Considerations**
   - *Action*: Integrate the implications of climate change into CAM research and education, recognizing how environmental changes impact health and how CAM practices can adapt. This also involves advocating for sustainable practices within the CAM industry.

44. **Promote Collaborative Patient Data Platforms**
   - *Action*: Support the development of collaborative patient data platforms that include CAM interventions alongside conventional treatments. This facilitates holistic patient care planning and the integration of CAM into mainstream healthcare records.

45. **Incorporate Genomics and Personalized Medicine**
   - *Action*: Integrate genomics and personalized medicine into CAM research to explore how individual genetic variations affect responses to CAM therapies. This can lead to more personalized and effective CAM treatment plans.

Adhering to these recommendations can significantly improve the quality and impact of CAM education and research, facilitating the integration of CAM into broader healthcare practices and contributing to the advancement of holistic, patient-centered care.

CONCLUSION

The realm of CAM is vast and dynamically evolving, reflecting broader shifts within healthcare, societal attitudes toward wellness, and the integration of technological advancements. The recommendations provided for advancing CAM education and research span from enhancing interdisciplinary collaborations to leveraging cutting-edge technologies like AI, VR, and personalized medicine. These suggestions aim to address current barriers while capitalizing on facilitators to push the boundaries of what CAM can offer to healthcare professionals, patients, and the global community. In conclusion, the future of CAM education and research is bright, with immense potential to contribute to more holistic, patient-centered, and integrated healthcare systems. By embracing innovation, advocating for inclusive healthcare policies, and continuously striving for evidence-based practices, the CAM community can ensure that its contributions are recognized, respected, and integrated into mainstream healthcare. The focus on global health literacy, environmental sustainability, mental well-being, and addressing health disparities highlights the commitment of the CAM field to not only individual health but also to the well-being of communities and the planet. The successful implementation of these recommendations requires a concerted effort from educators, researchers, healthcare practitioners, policymakers, and the community at large. It involves overcoming existing barriers, such as limited funding and skepticism from the conventional medical community, and capitalizing on the growing interest in and acceptance of CAM practices. Ultimately, the goal is to foster an environment where CAM and conventional medicine complement each other, offering patients the best possible care and supporting the health and well-being of populations worldwide. As CAM continues to evolve, it is imperative to remain open to new ideas, research, and practices while grounding decisions in rigorous scientific inquiry and ethical considerations. The journey of integrating CAM more fully into the healthcare landscape is ongoing, but with dedication, collaboration, and innovation, it promises to enrich the ways we understand and approach health and wellness in the 21st century and beyond.

REFERENCES

Al-Worafi, Y.M. (Ed.). (2020a). *Drug Safety in Developing Countries: Achievements and Challenges.* Academic Press.

Al-Worafi, Y.M. (2020b). Herbal medicines safety issues. In: Al-Worafi, Y.M. (ed), *Drug Safety in Developing Countries* (pp. 163–178). Academic Press.

Al-Worafi, Y.M. (2022a). *A Guide to Online Pharmacy Education: Teaching Strategies and Assessment Methods.* CRC Press.

Al-Worafi, Y.M. (2022b). Competencies and learning outcomes. In: Al-Worafi, Y.M. (ed), *A Guide to Online Pharmacy Education: Teaching Strategies and Assessment Methods.* CRC Press.

Al-Worafi, Y.M. (2023a). *Patient Safety in Developing Countries: Education, Research, Case Studies.* CRC Press.

Al-Worafi, Y.M. (2023b). *Technology for Drug Safety: Current Status and Future Developments.* Springer Nature.

Al-Worafi, Y.M. (2023c). Patient safety-related issues: Patient care errors and related problems. In: *Patient Safety in Developing Countries: Education, Research, Case Studies.* CRC Press.

Al-Worafi, Y.M. (2023d). Patient care errors and related problems: Preventive medicine errors & related problems. In: Al-Worafi, Y.M. (ed), *Patient Safety in Developing Countries: Education, Research, Case Studies.* CRC Press.

Al-Worafi, Y.M. (2023e). Patient care errors and related problems: Patient assessment and diagnostic errors & related problems. In: Al-Worafi, Y.M. (ed), *Patient Safety in Developing Countries: Education, Research, Case Studies.* CRC Press.

Al-Worafi, Y.M. (2023f). Patient care errors and related problems: Non-pharmacological errors & related problems. In: Al-Worafi, Y.M. (ed), *Patient Safety in Developing Countries: Education, Research, Case Studies.* CRC Press.

Al-Worafi, Y.M. (2023g). Patient care errors and related problems: Medical errors & related problems. In: Al-Worafi, Y.M. (ed), *Patient Safety in Developing Countries: Education, Research, Case Studies.* CRC Press.

Al-Worafi, Y.M. (2023h). Patient care errors and related problems: Monitoring errors & related problems. In: Al-Worafi, Y.M. (ed), *Patient Safety in Developing Countries: Education, Research, Case Studies.* CRC Press.

Al-Worafi, Y.M. (2023i). Patient care errors and related problems: Patient education and counselling errors and related problems. In: Al-Worafi, Y.M. (ed), *Patient Safety in Developing Countries: Education, Research, Case Studies.* CRC Press.

Al-Worafi, Y.M. (2023j). Patient safety-related issues: Other medication safety issues. In: Al-Worafi, Y.M. (ed), *Patient Safety in Developing Countries: Education, Research, Case Studies.* CRC Press.

Al-Worafi, Y.M. (2023k). Patient safety culture. In: Al-Worafi, Y.M. (ed), *Patient Safety in Developing Countries: Education, Research, Case Studies.* CRC Press.

Al-Worafi, Y.M. (2023l). Patient safety education: Competencies and learning outcomes. In: Al-Worafi, Y.M. (ed), *Patient Safety in Developing Countries: Education, Research, Case Studies.* CRC Press.

Al-Worafi, Y.M. (Ed.). (2023m). *Clinical Case Studies on Medication Safety.* Academic Press.

Al-Worafi, Y.M. (Ed.). (2023n). *Comprehensive Healthcare Simulation: Pharmacy Education, Practice and Research.* Springer Nature.

Al-Worafi, Y.M. (Ed.). (2024a). *Handbook of Medical and Health Sciences in Developing Countries.* Springer, Cham.

Al-Worafi, Y.M. (2024b). Complementary and alternative medicine (CAM) in developing countries. In: Al-Worafi, Y.M. (ed), *Handbook of Medical and Health Sciences in Developing Countries.* Springer, Cham. https://doi.org/10.1007/978-3-030-74786-2_301-1

Al-Worafi, Y.M. (2024c). Access/equitable access to medical and health sciences in developing countries. In: Al-Worafi, Y.M. (ed), *Handbook of Medical and Health Sciences in Developing Countries.* Springer, Cham. https://doi.org/10.1007/978-3-030-74786-2_147-1

Hasan, S., Al-Omar, M.J., AlZubaidy, H., and Al-Worafi, Y.M. (2019). Use of medications in Arab countries. In: Laher, I. (ed), *Handbook of Healthcare in the Arab World* (p. 42). Springer, Cham.

14 Technology in Complementary and Alternative Medicine (CAM) Education

INTRODUCTION

Technology in CAM Education is an evolving field that integrates modern digital tools and platforms to enhance the teaching and learning experience in CAM disciplines. This integration of technology aims to provide a comprehensive understanding of alternative therapies and practices, which include, but are not limited to, herbal medicine, acupuncture, chiropractic, and yoga. As the demand for CAM increases, so does the need for innovative educational methods to train practitioners and inform the public. The use of online learning platforms is a significant aspect of technology in CAM education. These platforms offer access to a wide range of resources, including video tutorials, interactive modules, and digital textbooks, allowing learners to study at their own pace and convenience. virtual reality (VR) and augmented reality (AR) are also becoming increasingly popular in CAM education, providing immersive experiences that can simulate real-life scenarios for students. For example, VR can be used to teach anatomy in a more interactive way, or to simulate acupuncture treatment processes, giving students a closer understanding of the techniques involved before practicing on real patients. Another crucial component is the use of data analytics and artificial intelligence (AI) to personalize learning experiences and to analyze the effectiveness of different CAM treatments. AI algorithms can help in creating customized learning paths for students, based on their progress and areas of interest. Additionally, mobile apps are developed to support learning on-the-go, offering features such as flashcards, quizzes, and the ability to track progress, making learning more flexible and accessible. Moreover, social media and forums have become vital in CAM education, facilitating peer-to-peer learning and collaboration among students and professionals worldwide. These platforms provide a space for sharing knowledge, discussing cases, and networking, which is essential for the CAM community's growth and development. Despite the potential benefits, the integration of technology in CAM education faces challenges, including ensuring the quality and accuracy of online content, addressing the digital divide, and maintaining the hands-on and personalized nature of many CAM practices. However, as technology advances and becomes more integrated into our daily lives, its role in CAM education is expected to grow, offering new and innovative ways to learn and practice complementary and alternative medicine.

RATIONALITY AND IMPORTANCE OF TO THE TECHNOLOGY IN THE COMPLEMENTARY AND ALTERNATIVE MEDICINE (CAM) EDUCATION

The integration of technology into CAM education is a significant development that reflects broader trends in healthcare and education. The rationality behind this integration and its importance can be explored through several dimensions:

1. **Enhanced Educational Resources**
   - *Rationality*: Traditional CAM education may rely heavily on apprenticeship models and face-to-face instruction in techniques such as acupuncture, herbal medicine, and manual therapies. The integration of technology provides a wealth of digital resources,

DOI: 10.1201/9781003327202-14

including online databases of research articles, digital textbooks, and instructional videos. This access enhances the breadth and depth of information available to students and practitioners.

- *Importance*: It ensures that CAM practitioners have a comprehensive understanding of their field, including the latest research and evidence-based practices. This is crucial for ensuring the efficacy and safety of CAM therapies.

2. **Improved Accessibility**
   - *Rationality*: Technology enables remote learning opportunities, making CAM education more accessible to a wider audience. Students who might not have access to traditional CAM educational institutions due to geographical or financial constraints can now access quality education online.
   - *Importance*: This democratization of education can lead to a more diverse group of practitioners. It also helps spread CAM practices to regions where they might not have been traditionally available, potentially enriching healthcare options for those communities.

3. **Interdisciplinary Learning and Research**
   - *Rationality*: Technology facilitates interdisciplinary learning by allowing CAM students and practitioners to easily access information from other medical and scientific fields. This is important because CAM often intersects with conventional medicine, psychology, pharmacology, and other areas.
   - *Importance*: Encouraging an interdisciplinary approach is crucial for the development of integrative medicine practices that combine the best of conventional medicine and CAM for holistic patient care. It also fosters a culture of collaboration and continuous learning among healthcare professionals.

4. **Simulation and Virtual Reality (VR)**
   - *Rationality*: For certain CAM practices, like acupuncture or chiropractic techniques, technology such as simulation models and VR can offer a safe environment for students to practice and hone their skills before working with actual patients.
   - *Importance*: This can improve the quality of education and training, reduce the risk of harm to patients during the learning process, and increase the confidence and competence of CAM practitioners.

5. **Data Analysis and Research**
   - *Rationality*: Technology plays a crucial role in the collection, analysis, and interpretation of data related to CAM practices. This includes everything from the clinical outcomes of CAM therapies to patterns of use among different populations.
   - *Importance*: The ability to analyze large datasets is essential for advancing the evidence base of CAM practices, informing best practices, and integrating CAM therapies into mainstream healthcare where appropriate.

The rationality of integrating technology into CAM education is grounded in enhancing educational resources, improving accessibility, fostering interdisciplinary learning, and advancing research. Its importance cannot be overstated, as it contributes to the development of a well-informed, skilled, and adaptable CAM workforce capable of meeting the diverse health needs of the population. This integration also supports the ongoing dialogue between CAM and conventional medicine, encouraging evidence-based practice and patient-centered care.

TYPES OF TECHNOLOGIES IN THE COMPLEMENTARY AND ALTERNATIVE MEDICINE (CAM) EDUCATION

The integration of technology in CAM education encompasses a variety of tools and platforms designed to enhance learning, facilitate research, and improve clinical practice. Here are some of the key types of technologies being utilized:

1. **E-Learning Platforms and Online Courses**
   - *Description*: Digital platforms offering courses and comprehensive programs in various CAM disciplines. These platforms may include interactive modules, video lectures, and forums for discussion.
   - *Examples*: learning management systems (LMS) like Moodle or Blackboard, specialized platforms for CAM education, and massive open online courses (MOOC) platforms like Coursera or Udemy.
2. **Virtual Reality (VR) and Augmented Reality (AR)**
   - *Description*: VR and AR technologies are used to simulate real-life environments and procedures, allowing students to practice skills in a controlled, virtual space. This is particularly useful for techniques that require precise manual skills, such as acupuncture or chiropractic adjustments.
   - *Examples*: VR simulations for acupuncture point location, AR apps for anatomy education, and VR-based practice environments for manual therapies.
3. **Mobile Applications**
   - *Description*: Apps designed for smartphones and tablets that support learning and practice in CAM fields. These can range from reference guides and diagnostic tools to meditation and wellness tracking apps.
   - *Examples*: herbal medicine databases, point location guides for acupuncture, yoga and meditation apps, and interactive anatomy apps.
4. **Online Databases and Digital Libraries**
   - *Description*: Comprehensive online repositories that provide access to scholarly articles, research papers, clinical trial data, and traditional medicine texts. These resources are crucial for evidence-based practice and research in CAM.
   - *Examples*: PubMed, Cochrane Library, specialized databases focusing on herbal medicine or traditional Chinese medicine, and digital archives of ancient texts.
5. **Telehealth and Remote Monitoring Tools**
   - *Description*: Technologies that enable the remote delivery of healthcare services and monitoring of patients. In CAM education, these tools can also be used to teach students about remote patient interaction, consultation, and management.
   - *Examples*: video conferencing tools, remote patient monitoring devices, and apps for virtual consultations between practitioners and patients.
6. **Simulation Software and Tools**
   - *Description*: Software that simulates clinical scenarios or body systems for educational purposes. This includes 3D anatomical models and simulation software for practicing diagnostic or therapeutic techniques.
   - *Examples*: 3D anatomy software, simulation tools for bodywork and energy healing practices, and interactive models for studying the effects of various CAM therapies.
7. **Learning Analytics and AI**
   - *Description*: Advanced analytics and artificial intelligence tools that analyze learning data to provide insights into student performance, personalize learning experiences, and predict learning outcomes.
   - *Examples*: AI-driven tutoring systems, predictive analytics for student success, and personalized learning pathways based on individual student data.

The use of technology in CAM education is multifaceted, reflecting the diverse nature of CAM practices and the broad spectrum of learning needs. From immersive VR environments to sophisticated data analysis tools, technology enhances the ability of students to learn, practice, and apply CAM therapies effectively. As technology continues to evolve, its role in CAM education is likely to expand, offering new and innovative ways to support students and practitioners in their professional development and practice.

COMPLEMENTARY AND ALTERNATIVE MEDICINE (CAM) EDUCATION TECHNOLOGIES: FACILITATORS

The integration of technology into CAM education serves as a facilitator in various significant ways, enhancing both the learning experience and the effectiveness of CAM practices. Here are some key facilitators brought about by the use of technology in CAM education:

1. **Accessibility and Flexibility**

    Technology enables CAM education to be more accessible and flexible, breaking down geographical barriers and allowing students from all over the world to access quality education and resources. Online courses, digital textbooks, and webinars allow students to learn at their own pace and according to their own schedules, making education more inclusive and adaptable to individual needs.

2. **Enhanced Learning Experience**

    Digital platforms and tools such as VR and AR provide immersive learning experiences that traditional classroom settings cannot offer. These technologies can simulate real-life scenarios and procedures, allowing students to gain practical skills and deepen their understanding of complex concepts in a controlled, risk-free environment.

3. **Interactive and Engaging Content**

    Technological tools enable the creation of interactive and engaging educational content. From interactive e-books and apps that offer quizzes and flashcards to sophisticated simulation software, students can engage with the material in a more meaningful way, which can improve retention and understanding.

4. **Personalized Learning**

    AI and learning analytics offer personalized learning experiences by analyzing students' performance and learning habits. This technology can tailor educational content to match the student's pace and level of understanding, addressing individual learning needs and optimizing the educational process.

5. **Collaboration and Community Building**

    Technology facilitates collaboration and community building among students, educators, and practitioners. Online forums, social media, and collaborative platforms enable the sharing of knowledge and experiences, fostering a sense of community and support among individuals with similar interests and goals.

6. **Evidence-Based Practice and Research**

    Online databases and digital libraries provide access to a vast amount of research and clinical data, supporting evidence-based practice and research in CAM. This access is crucial for students and practitioners to stay updated with the latest findings, integrate scientific evidence into their practices, and contribute to the ongoing research in the field.

7. **Professional Development and Continuing Education**

    Technology offers numerous opportunities for professional development and continuing education. With the fast-paced advancements in both CAM and technology, practitioners need to continuously update their knowledge and skills. Online courses, webinars, and workshops make it easier for professionals to access ongoing education and meet their licensing and certification requirements.

The incorporation of technology in CAM education acts as a facilitator by making learning more accessible, engaging, and effective. It not only enhances the educational experience but also supports the professional development of CAM practitioners, ensuring they are well-equipped to meet the needs of their patients with evidence-based and informed practices. As technology continues to evolve, its role in facilitating CAM education is expected to grow, further transforming how CAM is taught, practiced, and integrated into the broader healthcare landscape.

COMPLEMENTARY AND ALTERNATIVE MEDICINE (CAM) EDUCATION TECHNOLOGIES: BARRIERS

While technology offers numerous advantages in CAM education, it also presents several barriers and challenges. Addressing these barriers is crucial for maximizing the benefits of technology in this field. Here are some of the key challenges associated with integrating technology into CAM education:

1. **Digital Divide and Access Issues**
   - *Barrier*: Not all students and practitioners have equal access to digital technologies due to economic, geographic, or infrastructural limitations. This digital divide can hinder the ability of some individuals to benefit from technological advancements in CAM education.
   - *Impact*: It can lead to disparities in educational quality and access, with some students gaining a more advanced and interactive learning experience than others.
2. **Quality and Credibility of Online Resources**
   - *Barrier*: The vast amount of information available online includes both high-quality, evidence-based resources and misleading or inaccurate information. Discerning credible sources can be challenging for students and practitioners.
   - *Impact*: This may lead to the dissemination of incorrect or harmful practices, undermining the effectiveness and safety of CAM therapies.
3. **Technical Skills and Literacy**
   - *Barrier*: The effective use of technology in CAM education requires a certain level of technical skills and literacy among both educators and students. There can be a steep learning curve for those not familiar with digital tools and platforms.
   - *Impact*: This barrier can limit the adoption and effective use of technology in CAM education, potentially hindering the learning process.
4. **Lack of Standardized Regulations and Accreditation**
   - *Barrier*: The online delivery of CAM education and the use of technology in these programs often lack standardized regulations and accreditation processes. This can make it difficult for students to determine the quality and legitimacy of educational programs.
   - *Impact*: Graduates may find their qualifications questioned or not recognized by professional bodies or employers, impacting their career opportunities.
5. **Integration with Traditional Learning**
   - *Barrier*: Effectively integrating technology with traditional, hands-on aspects of CAM education, such as acupuncture, massage, or herbal medicine preparation, can be challenging. These practices often require tactile and sensory skills that are difficult to replicate digitally.
   - *Impact*: There is a risk that students may not develop the practical skills needed to the same extent as in traditional learning environments.
6. **Privacy and Security Concerns**
   - *Barrier*: The use of digital platforms for teaching and patient consultations raises concerns about data privacy and security. Ensuring the confidentiality and integrity of patient information is crucial but can be challenging in digital environments.

- *Impact*: Concerns over data breaches and privacy violations may deter educators, students, and practitioners from fully embracing technology in CAM practices.

7. **Cost of Implementation**
   - *Barrier*: The cost of developing, implementing, and maintaining advanced technological tools and platforms can be prohibitive for some educational institutions, especially smaller or resource-limited organizations.
   - *Impact*: This can limit the ability of these institutions to offer advanced technological tools and learning resources, potentially affecting the quality of education.

While technology holds great promise for enhancing CAM education, addressing these barriers is essential to fully realize its benefits. Solutions may include investing in infrastructure, improving digital literacy, establishing quality standards and accreditation for online programs, and ensuring equitable access to technology. By overcoming these challenges, the integration of technology in CAM education can continue to evolve, providing high-quality, accessible, and effective training for future practitioners.

COMPLEMENTARY AND ALTERNATIVE MEDICINE (CAM) EDUCATION TECHNOLOGIES: ACHIEVEMENTS

The integration of technology into CAM education has led to significant achievements, transforming how CAM is taught, learned, and practiced. These advancements have facilitated a broader acceptance and integration of CAM into mainstream healthcare and education. Here are some notable achievements:

1. **Widespread Accessibility of CAM Education**
   - *Achievement*: Technology has dramatically increased the accessibility of CAM education, allowing students from all over the world to access quality education and resources online. This has contributed to the global spread and acceptance of CAM practices.
   - *Impact*: It has democratized education in CAM fields, enabling a diverse range of students to pursue their interest in CAM therapies, regardless of their geographical location.
2. **Interactive and Engaging Learning Experiences**
   - *Achievement*: The use of VR, AR, and simulation technologies has provided students with immersive and interactive learning experiences. These technologies simulate real-life scenarios, making complex concepts easier to understand and engage with.
   - *Impact*: Enhanced learning experiences have improved the competence and confidence of CAM practitioners, leading to better patient outcomes and increased safety in practice.
3. **Advancements in Online Collaborative Platforms**
   - *Achievement*: The development of online forums, social media groups, and collaborative platforms specifically for CAM education has fostered a sense of community among students, educators, and practitioners. These platforms enable the sharing of knowledge, experiences, and best practices.
   - *Impact*: The collaborative environment has enhanced the quality of CAM education and practice by promoting continuous learning and professional development.
4. **Integration of Evidence-Based Practice in CAM Education**
   - *Achievement*: Technology has facilitated easier access to research databases, online journals, and evidence-based resources, integrating scientific evidence into CAM education. This has encouraged a more rigorous and analytical approach to CAM practices.
   - *Impact*: The emphasis on evidence-based practice has helped legitimize CAM therapies in the eyes of conventional healthcare providers and the public, promoting interdisciplinary collaboration and integration into healthcare systems.

5. **Development of Specialized CAM Education Software and Apps**
   - *Achievement*: The creation of specialized software and mobile applications for CAM education, such as herbal medicine databases, acupuncture point locators, and meditation apps, has enriched the educational resources available to students and practitioners.
   - *Impact*: These tools have made it easier for learners to access information, practice skills, and integrate CAM practices into their daily lives and professional activities.
6. **Expansion of Telehealth in CAM Practices**
   - *Achievement*: The adoption of telehealth technologies has extended the reach of CAM practitioners, allowing them to consult with and treat patients remotely. This has been particularly beneficial in making CAM therapies accessible to underserved populations.
   - *Impact*: Telehealth has not only expanded the practice of CAM but also integrated it further into holistic patient care, offering patients a wider range of treatment options.
7. **Enhancement of Professional Development Opportunities**
   - *Achievement*: The availability of online courses, webinars, and workshops has facilitated ongoing professional development and lifelong learning among CAM practitioners. This ensures that they remain up-to-date with the latest research, techniques, and technologies.
   - *Impact*: Continuous professional development has raised the standards of CAM practice, ensuring that practitioners can provide the best possible care to their patients.

The achievements in integrating technology into CAM education are substantial, reflecting a significant shift toward more accessible, interactive, and evidence-based CAM practices. These advancements not only improve the quality of CAM education but also support the broader goal of integrating CAM into holistic healthcare delivery, promoting well-being, and expanding treatment options for patients worldwide. As technology continues to evolve, it promises to bring further innovations and improvements to the field of CAM education and practice.

COMPLEMENTARY AND ALTERNATIVE MEDICINE (CAM) EDUCATION TECHNOLOGIES: CHALLENGES

While the integration of technology in CAM education has brought significant achievements, it also presents several challenges that need addressing to maximize its potential benefits. These challenges span across various aspects of implementation, accessibility, and quality assurance. Here's a closer look at some of the key challenges:

1. **Technological Literacy and Resistance to Change**
   - *Challenge*: Both educators and students may face difficulties in adapting to new technologies due to varying levels of technological literacy. Additionally, there can be resistance to change from traditional teaching and learning methods to technology-based approaches.
   - *Implications*: This can hinder the effective integration of technology in CAM education, leading to underutilization of available tools and platforms.
2. **Quality Control and Accreditation**
   - *Challenge*: With the proliferation of online CAM courses and programs, ensuring the quality and credibility of these offerings becomes a significant challenge. The lack of standardized accreditation processes for online CAM education can make it difficult for students to discern reputable programs.
   - *Implications*: Students may end up with qualifications that are not recognized by professional bodies or healthcare institutions, affecting their career opportunities and professional development.

3. **Equity and Access**
   - *Challenge*: The digital divide remains a significant barrier, with students in resource-limited settings facing challenges in accessing technology-based education due to a lack of infrastructure, high costs, or limited internet connectivity.
   - *Implications*: This can exacerbate educational inequalities, leaving some students without access to the benefits of technology-enhanced CAM education.
4. **Data Privacy and Security**
   - *Challenge*: The use of digital platforms for education and patient care raises concerns about data privacy and security. Ensuring the protection of personal and health information in online environments is crucial but challenging.
   - *Implications*: Privacy breaches and security lapses could undermine trust in CAM education platforms and telehealth services, potentially deterring their use.
5. **Practical Skills Development**
   - *Challenge*: Certain CAM practices require hands-on experience and the development of tactile skills that are difficult to replicate through virtual simulations or online learning environments.
   - *Implications*: There's a risk that students may not acquire the necessary practical skills to the same extent as in traditional, in-person training environments, potentially impacting the quality of patient care.
6. **Interprofessional Integration**
   - *Challenge*: Integrating CAM education within the broader healthcare education system and promoting interdisciplinary learning between CAM and conventional medical fields through technology can be challenging due to differing curricula, educational standards, and professional biases.
   - *Implications*: This may limit the opportunities for CAM practitioners to work collaboratively with other healthcare professionals, affecting patient care and the holistic integration of healthcare services.
7. **Adapting to Rapid Technological Advances**
   - *Challenge*: The rapid pace of technological innovation requires continuous updates to educational content and teaching methodologies, which can be resource-intensive and challenging for educational institutions to keep up with.
   - *Implications*: There's a risk that CAM education may become outdated if it does not adapt to incorporate new technologies and scientific advancements, potentially hindering the relevance and effectiveness of CAM practices.

Addressing these challenges requires a multifaceted approach, including investing in digital infrastructure, enhancing digital literacy among educators and students, establishing quality standards for online CAM education, and ensuring robust data protection measures. Overcoming these obstacles is essential for fully realizing the potential of technology in enhancing CAM education, improving the quality of care, and supporting the integration of CAM practices into mainstream healthcare.

COMPLEMENTARY AND ALTERNATIVE MEDICINE (CAM) EDUCATION TECHNOLOGIES: RECOMMENDATIONS FOR THE BEST PRACTICE

Implementing technology in CAM education, while challenging, offers significant opportunities for enhancing teaching, learning, and practice. To navigate these challenges effectively and maximize the benefits of technology, several best practices are recommended (Al-Worafi, 2020a,b, 2022a,b, 2023a–n, 2024a–d; Al-Worafi et al., 2023a,b; 2024; Hasan et al., 2019). These recommendations aim to ensure that technology integration supports high-quality, accessible, and effective CAM education:

1. **Develop and Implement Quality Standards**
   • Establish clear quality standards and accreditation processes for online CAM courses and programs to ensure they meet educational objectives and professional requirements.
   • Encourage transparency and accountability by having courses reviewed and accredited by recognized professional bodies.
2. **Enhance Digital Literacy**
   • Provide training and resources for both educators and students to improve their digital literacy, ensuring they can effectively use technological tools and platforms.
   • Include modules on navigating online research databases and evaluating the credibility of online sources as part of the curriculum.
3. **Ensure Equitable Access**
   • Work toward reducing the digital divide by offering scholarships, subsidized technology access, or loan programs for students who may face financial or infrastructural barriers.
   • Consider hybrid models that combine online learning with local, in-person workshops or practical sessions to cater to students in areas with limited internet access.
4. **Prioritize Data Privacy and Security**
   • Adopt stringent data protection measures to safeguard personal and health information shared online, complying with relevant laws and regulations.
   • Educate students and faculty about best practices in data privacy and security, particularly when using telehealth platforms.
5. **Incorporate Hands-on Experience**
   • Use blended learning models that combine online theoretical study with hands-on practical experience, ensuring students develop the tactile and practical skills necessary for CAM practice.
   • Leverage technology such as AR and VR to simulate hands-on experiences as closely as possible, complementing these with real-world practice.
6. **Foster Interprofessional Collaboration**
   • Create interdisciplinary online forums and collaborative projects that bring together CAM and conventional medicine students and practitioners, promoting mutual understanding and learning.
   • Integrate case studies and content that highlight the integration of CAM practices within conventional healthcare settings to prepare students for collaborative practice environments.
7. **Stay Updated with Technological Advances**
   • Continuously evaluate and update the technological tools and platforms used in CAM education to ensure they remain current and effective.
   • Encourage a culture of innovation within CAM educational institutions, where educators and students are motivated to explore and adopt new technologies.
8. **Evaluate and Adapt Educational Approaches**
   • Implement regular feedback mechanisms to assess the effectiveness of technology-enhanced learning and identify areas for improvement.
   • Be flexible and willing to adapt educational strategies based on feedback from students, educators, and technological advancements.
9. **Promote Evidence-Based Practice**
   • Utilize technology to provide easy access to research databases and resources, encouraging students and practitioners to integrate evidence-based practices into their learning and clinical work.
   • Offer training in research methods and critical appraisal skills to empower students to engage with scientific literature effectively.

10. **Leverage Community Partnerships**
    - Build partnerships with healthcare facilities, CAM practitioners, and community organizations to provide students with real-world practice opportunities alongside their technology-enhanced learning.
    - These partnerships can facilitate internships, apprenticeships, or practical workshops, offering students valuable hands-on experience in a supervised setting.

11. **Promote Global and Cultural Perspectives**
    - Utilize technology to expose students to CAM practices from various cultures and regions, broadening their understanding and appreciation of global health perspectives.
    - Encourage cross-cultural exchanges through virtual seminars, guest lectures from international CAM practitioners, and collaborative online projects with students from other countries.

12. **Incorporate Patient-Centered Technologies**
    - Train students in the use of patient-centered technologies, such as telehealth platforms and health tracking apps, preparing them for the evolving landscape of healthcare delivery.
    - Include coursework on the ethical, legal, and practical considerations of using these technologies in patient care.

13. **Adopt Adaptive Learning Technologies**
    - Implement adaptive learning platforms that personalize the educational experience to the individual learner's pace, style, and mastery level. These technologies can adjust content difficulty and presentation based on real-time feedback and assessments.
    - This personalized approach can help address diverse learning needs and preferences, potentially increasing engagement and retention.

14. **Support Faculty Development and Innovation**
    - Offer ongoing professional development opportunities for educators to stay abreast of technological advances and pedagogical strategies in CAM education.
    - Encourage faculty to innovate in their teaching methods, experiment with new technologies, and share their experiences and best practices with colleagues.

15. **Utilize Analytics for Continuous Improvement**
    - Use learning analytics to gather data on student engagement, performance, and outcomes. This information can inform targeted interventions, curriculum adjustments, and the overall improvement of educational programs.
    - Regularly review analytics to identify trends, successes, and areas needing enhancement, ensuring that the curriculum remains relevant and effective.

16. **Ensure Scalability and Sustainability**
    - Design technology-enhanced CAM education programs with scalability and sustainability in mind, allowing for expansion and adaptation as demand grows and technology evolves.
    - Consider using open-source platforms and collaborative content creation to reduce costs and increase the program's reach and longevity.

17. **Engage in Research and Scholarship**
    - Encourage and support research on the effectiveness of technology in CAM education, contributing to the scholarship of teaching and learning in this field.
    - Share findings through academic publications, presentations, and practitioner forums, fostering a culture of evidence-based educational practice.

18. **Integrate Simulation-Based Learning for Complex Skills**
    - *Strategy*: Beyond VR and AR, develop high-fidelity simulation environments that mimic real-life patient interactions and complex CAM treatment scenarios. This could include patient communication, diagnosis, and treatment planning.
    - *Benefit*: Enhances critical thinking, decision-making, and clinical skills in a safe, controlled setting, preparing students for real-world practice.

19. **Promote Interdisciplinary and Integrative Medicine Education**
    - *Strategy*: Foster programs that bridge CAM and conventional medicine, promoting an integrative approach to health. This involves creating courses that include both CAM and conventional medical practitioners as educators and participants.
    - *Benefit*: Encourages a holistic view of patient care, enhances mutual respect among healthcare professionals, and prepares students for collaborative practice environments.
20. **Leverage Big Data and AI for Personalized Learning and Research**
    - *Strategy*: Utilize big data analytics and AI to analyze learning patterns, optimize educational content delivery, and personalize learning experiences. Also, integrate these technologies into CAM research methodologies for exploring large datasets on health outcomes and therapy effectiveness.
    - *Benefit*: Improves learning outcomes through customization while advancing CAM research and evidence-based practice.
21. **Incorporate Social Media and Digital Marketing Skills**
    - *Strategy*: Train students in the ethical and effective use of social media and digital marketing for CAM practice promotion, patient education, and community building.
    - *Benefit*: Prepares CAM practitioners to navigate the digital landscape for professional networking, practice growth, and public health advocacy.
22. **Develop Digital Ethics and Legal Frameworks**
    - *Strategy*: Include comprehensive modules on digital ethics, data protection, and the legal aspects of telehealth and online CAM practice.
    - *Benefit*: Ensures that CAM practitioners are aware of their ethical responsibilities and legal requirements in digital healthcare delivery.
23. **Implement Community-Engaged Learning Projects**
    - *Strategy*: Integrate service-learning projects that involve working with communities to address health issues using CAM approaches, facilitated through digital platforms for coordination and collaboration.
    - *Benefit*: Enhances student understanding of social determinants of health, cultural competency, and the practical impact of CAM in diverse communities.
24. **Foster Lifelong Learning and Professional Curiosity**
    - *Strategy*: Encourage an ethos of lifelong learning and professional curiosity among students and practitioners by providing access to a wide range of continuing education resources and forums for ongoing professional development.
    - *Benefit*: Prepares CAM professionals to adapt to changes in healthcare, maintain competency, and continuously improve their practice.
25. **Global Collaboration and Knowledge Exchange**
    - *Strategy*: Utilize digital platforms to facilitate global collaborations and knowledge exchanges among CAM institutions, practitioners, and students worldwide.
    - *Benefit*: Promotes a global perspective on CAM education and practice, enriching the curriculum with diverse cultural and clinical insights and fostering international research collaborations.
26. **Sustainability in CAM Practice**
    - *Strategy*: Incorporate sustainability principles into the curriculum, focusing on environmentally sustainable practices in CAM (e.g., sustainable sourcing of herbal medicines, eco-friendly clinic operations).
    - *Benefit*: Prepares students to practice CAM in a way that is mindful of environmental impact and sustainability, aligning with global health and sustainability goals.
27. **Embrace Emerging Technologies**
    - *Strategy*: Stay abreast of emerging technologies such as blockchain for secure patient records, wearable health technology for real-time health monitoring, and advanced

biometric systems. Evaluate their applicability in CAM education and practice for teaching, patient care, and research.

- *Benefit*: Prepares students for a future where these technologies become integral to healthcare delivery, ensuring CAM practices remain relevant and integrated within the broader healthcare ecosystem.

28. **Cultivate Digital Leadership and Entrepreneurship**
- *Strategy*: Develop courses that inspire digital leadership and entrepreneurship, focusing on how to lead digital transformation within healthcare settings and how to innovate CAM practices through technology.
- *Benefit*: Encourages the development of forward-thinking CAM practitioners who are not only healthcare providers but also leaders and innovators in the digital health space.

29. **Promote Ethical Use of Technology in Healthcare**
- *Strategy*: Embed discussions and case studies on the ethical considerations of using technology in healthcare, including issues related to artificial intelligence, data mining, and patient privacy within CAM contexts.
- *Benefit*: It ensures that future CAM practitioners are well-versed in the ethical complexities of modern healthcare, fostering responsible and patient-centered use of technology.

30. **Incorporate Technology in Patient Education and Engagement**
- *Strategy*: Train CAM students in leveraging technology to educate and engage patients, using tools such as patient portals, educational apps, and online support communities.
- *Benefit*: Enhances patient care by empowering patients with knowledge and engaging them actively in their health and wellness journey, aligning with holistic CAM principles.

31. **Foster Resilience and Adaptability**
- *Strategy*: Include modules that focus on building resilience and adaptability in the face of technological and healthcare changes, preparing students to navigate and thrive amidst the uncertainties of the healthcare landscape.
- *Benefit*: Cultivates a generation of CAM practitioners who are not only technically proficient but also emotionally and professionally resilient, capable of leading through change.

32. **Expand Research Methodologies with Technology**
- *Strategy*: Integrate training on utilizing technology in research methodologies, including digital data collection, online survey tools, and statistical software for analyzing large datasets, to enhance CAM research capabilities.
- *Benefit*: Strengthens the evidence base of CAM practices by equipping practitioners with the skills to conduct rigorous, technology-enabled research.

33. **Global Health and CAM**
- *Strategy*: Include a global health perspective in CAM education, examining how technology can bridge gaps in healthcare delivery worldwide and how CAM practices can be adapted and integrated into different cultural and healthcare systems.
- *Benefit*: Prepares CAM practitioners with a global outlook, understanding the role of CAM in international health contexts, and contributing to global health equity.

34. **Accessibility and Inclusive Design in Technology**
- *Strategy*: Educate students on the principles of accessibility and inclusive design in digital health tools, ensuring technologies are usable by people with a wide range of abilities and backgrounds.
- *Benefit*: Promotes the creation and adoption of CAM technologies that are accessible to all, ensuring equitable healthcare delivery.

35. **Interdisciplinary Collaboration Platforms**
    - *Strategy*: Create online platforms for interdisciplinary collaboration among students, educators, and practitioners from CAM and other healthcare disciplines, fostering an environment of mutual learning and innovation.
    - *Benefit*: Enhances the integrative healthcare model by building strong collaborative networks across disciplines, improving patient care through shared knowledge and practices.
36. **Innovative Delivery Models for CAM Education**
    - *Strategy*: Develop and experiment with new models of education delivery, such as microlearning modules, gamified learning experiences, and interactive e-books, tailored to the unique aspects of CAM practices.
    - *Benefit*: These innovative delivery models can increase engagement, improve knowledge retention, and cater to diverse learning preferences, making CAM education more adaptable and appealing to a broader audience.
37. **Sustainable Practices within CAM Education**
    - *Strategy*: Embed principles of sustainability and environmental stewardship into the curriculum, emphasizing the role of CAM practices in promoting health without compromising the planet's resources.
    - *Benefit*: Prepares CAM practitioners to be advocates for sustainable healthcare, aligning CAM education with global sustainability goals and ethical practices.
38. **Advanced Data Analytics in CAM Research**
    - *Strategy*: Incorporate advanced data analytics, machine learning, and AI methodologies in CAM research curricula to analyze complex datasets, predict trends, and uncover new insights into CAM effectiveness and mechanisms.
    - *Benefit*: By harnessing these advanced technologies, CAM research can leap forward, providing a stronger evidence base for CAM practices and therapies.
39. **Global Telemedicine Initiatives**
    - *Strategy*: Train CAM students in delivering care through telemedicine platforms, with a focus on global health challenges and cross-border healthcare delivery, addressing regulatory, cultural, and linguistic barriers.
    - *Benefit*: Expands the reach of CAM practitioners to underserved populations worldwide, promoting global health equity and the international exchange of CAM knowledge and practices.
40. **Blockchain for CAM Data Integrity**
    - *Strategy*: Explore the application of blockchain technology for securing patient records, CAM research data, and ensuring the integrity and traceability of herbal medicines and supplements.
    - *Benefit*: Enhances trust and transparency in CAM practices and products, safeguarding patient data and ensuring the quality and safety of CAM interventions.
41. **Virtual International Exchanges and Collaborations**
    - *Strategy*: Facilitate virtual exchange programs and collaborative research projects with CAM institutions worldwide, using digital platforms to share knowledge, cultural practices, and educational resources.
    - *Benefit*: Fosters a global CAM community, enriching education with diverse perspectives and promoting international collaborations in research and practice.
42. **Integration of Genomics and Personalized Medicine**
    - *Strategy*: Integrate genomics and personalized medicine into CAM education, exploring how individual genetic profiles can influence responses to CAM therapies and tailoring treatments to the individual.
    - *Benefit*: Positions CAM at the forefront of personalized healthcare, enhancing the effectiveness of CAM treatments through a deeper understanding of genetic influences.

43. **Ethical AI in CAM Practice**
   - *Strategy*: Address the ethical considerations of using AI in healthcare, particularly in CAM practices, focusing on the development and use of AI that respects patient autonomy, privacy, and cultural values.
   - *Benefit*: Ensures that the integration of AI into CAM practices enhances patient care while adhering to ethical standards and respecting human values.

44. **Eco-Technological Innovations in CAM**
   - *Strategy*: Promote the development and use of eco-friendly technologies in CAM practices, such as digital platforms that reduce the need for physical resources and travel, contributing to a lower carbon footprint.
   - *Benefit*: Aligns CAM education and practice with environmental sustainability, reinforcing the holistic ethos of CAM that includes the health of the planet.

45. *Leadership and Change Management in CAM*
   - *Strategy*: Equip CAM students with leadership and change management skills, focusing on how to lead technological innovation and digital transformation within healthcare organizations and the CAM profession.
   - *Benefit*: Prepares future CAM leaders to drive positive change, advocate for the integration of CAM in healthcare systems, and navigate the challenges of a rapidly evolving healthcare landscape.

46. **Cross-Disciplinary Tech-Innovation Labs**
   - *Strategy*: Establish tech-innovation labs that bring together CAM practitioners, technologists, and educators to co-create new educational technologies and digital health solutions. These labs could focus on developing apps, wearables, and diagnostic tools specific to CAM practices.
   - *Benefit*: Fosters a culture of innovation within CAM education, encouraging the development of cutting-edge, practical technologies that can enhance both learning and clinical practice.

47. **Holistic Health Data Ecosystems**
   - *Strategy*: Advocate for and participate in the development of holistic health data ecosystems that integrate CAM treatments and outcomes into broader health databases, allowing for comprehensive patient health profiles that include CAM interventions.
   - *Benefit*: Enables more nuanced, data-driven insights into patient health, promoting the integration of CAM into holistic healthcare plans and facilitating research into CAM's effectiveness and mechanisms of action.

48. **Inclusive Technology Design in CAM**
   - *Strategy*: Prioritize the inclusive design of educational technologies, ensuring they are accessible to users of all abilities and backgrounds, including those with disabilities and those from diverse cultural and linguistic backgrounds.
   - *Benefit*: It ensures equitable access to CAM education and promotes diversity within the CAM practitioner community, enriching the field with a wide range of perspectives and experiences.

49. **Sustainable and Ethical Sourcing in CAM Education**
   - *Strategy*: Incorporate principles of sustainable and ethical sourcing into CAM education, particularly concerning the use of natural resources, herbs, and traditional remedies. This could include digital traceability systems for ingredients and materials.
   - *Benefit*: It educates future CAM practitioners about the importance of environmental stewardship and ethical practices, ensuring the sustainability of CAM therapies and respect for indigenous knowledge and resources.

50. **Global Health Diplomacy and CAM**
    - *Strategy*: Integrate global health diplomacy into CAM education, preparing practitioners to navigate international health policies and collaborate on global health initiatives that recognize and integrate CAM practices.
    - *Benefit*: Positions CAM practitioners as key players in global health, capable of contributing to international health policy and practice with a focus on holistic, integrative care.

51. **Neuroscience and CAM**
    - *Strategy*: Leverage advances in neuroscience to explore the mechanisms behind CAM therapies, incorporating neuroscientific principles into CAM education to deepen understanding of how these therapies impact brain health and function.
    - *Benefit*: Enhances the scientific foundation of CAM practices, providing a stronger evidence base for their efficacy and facilitating integration into mainstream healthcare.

52. **Digital Storytelling and Patient Narratives**
    - *Strategy*: Use digital storytelling and patient narratives as educational tools, highlighting the personal and cultural dimensions of CAM therapies. This approach can deepen understanding and empathy, enriching the practitioner-patient relationship.
    - *Benefit*: Supports the development of compassionate, patient-centered CAM practitioners who value and incorporate patient experiences and cultural contexts into their practice.

53. **Climate Change and CAM Education**
    - *Strategy*: Address the impacts of climate change on health and CAM practices within the curriculum, exploring how CAM can contribute to both mitigation and adaptation strategies, and how environmental changes may affect medicinal plants and traditional healing practices.
    - *Benefit*: Prepares CAM practitioners to respond to the health challenges posed by climate change and to advocate for sustainable practices that protect both health and the environment.

54. **Digital Literacy for Health Empowerment**
    - *Strategy*: Beyond professional development, include digital literacy for patients and communities as part of CAM education, empowering individuals to make informed health decisions and engage with digital health resources effectively.
    - *Benefit*: Strengthens the role of CAM practitioners in public health education and advocacy, supporting healthier, more informed communities.

55. **Futuristic CAM Modalities**
    - *Strategy*: Anticipate and explore the potential integration of futuristic modalities such as biofeedback, neurofeedback, and advanced body-mapping technologies into CAM practices, preparing students for the next wave of health innovations.
    - *Benefit*: Keeps CAM education at the cutting edge of healthcare innovation, ensuring that CAM practices evolve alongside technological advancements to meet future health needs.

By adhering to these recommendations, CAM educational institutions can navigate the challenges of integrating technology into their programs and harness its potential to enhance CAM education. Emphasizing quality, accessibility, practical skill development, and interprofessional collaboration will ensure that CAM practitioners are well-prepared to meet the demands of modern healthcare environments.

CONCLUSION

The integration of technology in CAM education represents a dynamic and evolving frontier, offering profound opportunities to enhance teaching, learning, and practice within this diverse field. As we've explored, the strategic incorporation of digital tools, platforms, and innovative pedagogical approaches can significantly elevate the quality and accessibility of CAM education, preparing practitioners for a healthcare landscape that is increasingly digital, interconnected, and patient-centered. From leveraging virtual reality and simulation-based learning to embracing global telemedicine initiatives and advanced data analytics, technology enables CAM educators and students to access a wealth of resources, engage in immersive learning experiences, and participate in global health communities. These technological advancements not only facilitate a deeper understanding of CAM practices but also foster an environment of continuous innovation and collaboration. However, navigating the challenges—such as ensuring equitable access, maintaining data privacy, integrating practical skills, and upholding quality standards—requires a concerted effort from educators, practitioners, policymakers, and technologists alike. Addressing these challenges head-on is crucial for realizing the full potential of technology in CAM education and ensuring that CAM practices are informed by the latest evidence, innovations, and ethical considerations. Looking forward, the continuous evolution of technology presents an ongoing opportunity to reimagine and reshape CAM education. By embracing a forward-looking approach that anticipates future healthcare needs, ethical considerations, and environmental sustainability, CAM education can lead in preparing practitioners who are not only skilled in traditional and alternative therapies but are also adept at leveraging technology for better health outcomes and global wellness. This approach ensures that CAM remains a vital, dynamic component of the global healthcare ecosystem, ready to meet the challenges and opportunities of the future. In conclusion, the integration of technology in CAM education is more than a trend; it's a transformative movement toward creating a more accessible, effective, and holistic healthcare paradigm. By adopting innovative strategies and embracing the potential of technology, CAM education can continue to play a crucial role in shaping the future of healthcare, promoting holistic well-being, and contributing to the global health landscape in meaningful, sustainable ways.

REFERENCES

Al-Worafi, Y.M. (Ed.). (2020a). *Drug Safety in Developing Countries: Achievements and Challenges.* Academic Press.

Al-Worafi, Y.M. (2020b). Herbal medicines safety issues. In: Al-Worafi, Y.M. (ed), *Drug Safety in Developing Countries* (pp. 163–178). Academic Press.

Al-Worafi, Y.M. (2022a). *A Guide to Online Pharmacy Education: Teaching Strategies and Assessment Methods.* CRC Press.

Al-Worafi, Y.M. (2022b). Competencies and learning outcomes. In: Al-Worafi, Y.M. (ed), *A Guide to Online Pharmacy Education: Teaching Strategies and Assessment Methods.* CRC Press.

Al-Worafi, Y.M. (2023a). *Patient Safety in Developing Countries: Education, Research, Case Studies.* CRC Press.

Al-Worafi, Y.M. (2023b). *Technology for Drug Safety: Current Status and Future Developments.* Springer Nature.

Al-Worafi, Y.M. (2023c). Patient safety-related issues: Patient care errors and related problems. In: Al-Worafi, Y.M. (ed), *Patient Safety in Developing Countries: Education, Research, Case Studies.* CRC Press.

Al-Worafi, Y.M. (2023d). Patient care errors and related problems: Preventive medicine errors & related problems. In: Al-Worafi, Y.M. (ed), *Patient Safety in Developing Countries: Education, Research, Case Studies.* CRC Press.

Al-Worafi, Y.M. (2023e). Patient care errors and related problems: Patient assessment and diagnostic errors & related problems. In: Al-Worafi, Y.M. (ed), *Patient Safety in Developing Countries: Education, Research, Case Studies.* CRC Press.

Al-Worafi, Y.M. (2023f). Patient care errors and related problems: Non-pharmacological errors & related problems. In: Al-Worafi, Y.M. (ed), *Patient Safety in Developing Countries: Education, Research, Case Studies*. CRC Press.

Al-Worafi, Y.M. (2023g). Patient care errors and related problems: Medical errors & related problems. In: Al-Worafi, Y.M. (ed), *Patient Safety in Developing Countries: Education, Research, Case Studies*. CRC Press.

Al-Worafi, Y.M. (2023h). Patient care errors and related problems: Monitoring errors & related problems. In: Al-Worafi, Y.M. (ed), *Patient Safety in Developing Countries: Education, Research, Case Studies*. CRC Press.

Al-Worafi, Y.M. (2023i). Patient care errors and related problems: Patient education and counselling errors and related problems. In: Al-Worafi, Y.M. (ed), *Patient Safety in Developing Countries: Education, Research, Case Studies*. CRC Press.

Al-Worafi, Y.M. (2023j). Patient safety-related issues: Other medication safety issues. In: Al-Worafi, Y.M. (ed), *Patient Safety in Developing Countries: Education, Research, Case Studies*. CRC Press.

Al-Worafi, Y.M. (2023k). Patient safety culture. In: Al-Worafi, Y.M. (ed), *Patient Safety in Developing Countries: Education, Research, Case Studies*. CRC Press.

Al-Worafi, Y.M. (2023l). Patient safety education: Competencies and learning outcomes. In: Al-Worafi, Y.M. (ed), *Patient Safety in Developing Countries: Education, Research, Case Studies*. CRC Press.

Al-Worafi, Y.M. (Ed.). (2023m). *Clinical Case Studies on Medication Safety*. Academic Press.

Al-Worafi, Y.M. (Ed.). (2023n). *Comprehensive Healthcare Simulation: Pharmacy Education, Practice and Research*. Springer Nature.

Al-Worafi, Y.M., Hermansyah, A., Tan, C.S., Choo, C.Y., Bouyahya, A., Paneerselvam, G.S., Liew, K.B., Goh, K.W., and Ming, L.C. (2023a). Applications, benefits, and risks of ChatGPT in medical and health sciences research: An experimental study. *Progress in Microbes & Molecular Biology*, 6(1). https://doi.org/10.36877/pmmb.a0000337.

Al-Worafi, Y.M., Hermansyah, A., Goh, K.W., and Ming, L.C. (2023b). Artificial intelligence use in university: Should we ban ChatGPT? *Preprints*. https://doi.org/10.20944/preprints202302.0400.v1

Al-Worafi, Y.M. (Ed.). (2024a). *Handbook of Medical and Health Sciences in Developing Countries*. Springer, Cham.

Al-Worafi, Y.M. (2024b). Complementary and alternative medicine (CAM) in developing countries. In: Al-Worafi, Y.M. (ed), *Handbook of Medical and Health Sciences in Developing Countries*. Springer, Cham. https://doi.org/10.1007/978-3-030-74786-2_301-1

Al-Worafi, Y.M. (2024c). Access/equitable access to medical and health sciences in developing countries. In: Al-Worafi, Y.M. (ed), *Handbook of Medical and Health Sciences in Developing Countries*. Springer, Cham. https://doi.org/10.1007/978-3-030-74786-2_147-1

Al-Worafi, Y.M. (2024d). Artificial intelligence and machine learning for drug safety. In: *Technology for Drug Safety: Current Status and Future Developments* (pp. 69–80). Springer International Publishing, Cham.

Al-Worafi, Y.M., Goh, K.W., Hermansyah, A., Tan, C.S., Ming, L.C. (2024). The use of ChatGPT for education modules on integrated pharmacotherapy of infectious disease: Educators' perspectives. JMIR Medical Education, 10 (1):e47339. doi: 10.2196/47339.

Hasan, S., Al-Omar, M.J., AlZubaidy, H., and Al-Worafi, Y.M. (2019). Use of medications in Arab countries. In: Laher, I. (ed), *Handbook of Healthcare in the Arab World* (p. 42). Springer, Cham.

15 Community Services by Complementary and Alternative Medicine (CAM) Schools/Departments

INTRODUCTION

The field of CAM encompasses a diverse range of practices and therapies that fall outside the realm of conventional medicine. These include, but are not limited to, acupuncture, chiropractic care, herbal medicine, homeopathy, and yoga. CAM schools and departments have increasingly played a significant role in not only educating practitioners but also in serving their communities through various services. Many CAM institutions incorporate community service as a core component of their educational philosophy, emphasizing the importance of holistic health, wellness, and preventive care. These services often aim to reach underserved populations who might not have access to conventional healthcare or who seek alternatives to mainstream medical treatments. By offering low-cost or free clinics, CAM schools provide valuable healthcare options to the community while giving their students practical experience in their field of study. For example, acupuncture and traditional Chinese medicine (TCM) schools often run clinics where students, supervised by licensed practitioners, offer treatments for a variety of conditions. These clinics serve as a bridge between theoretical knowledge and practical application, allowing students to gain hands-on experience. Similarly, chiropractic colleges may provide community health centers where students assist in treating musculoskeletal issues, contributing to the community's overall health and well-being. Herbal medicine departments within CAM schools might engage in community outreach by organizing workshops and seminars to educate the public about the benefits and uses of medicinal plants. These initiatives often include the cultivation of medicinal gardens, providing a tangible connection to the natural resources used in practice. Moreover, departments focusing on mind-body practices such as yoga and meditation often hold free or donation-based classes for the community. These sessions aim to promote mental and physical well-being, stress reduction, and overall health. They are particularly beneficial in communities affected by high levels of stress and health disparities. Through these and other services, CAM schools and departments contribute significantly to the health and wellness of their communities. They offer alternative and complementary healthcare options, provide educational resources, and foster a holistic approach to health. In doing so, they not only enrich the educational experience of their students but also play a pivotal role in promoting health equity and access to diverse healthcare options.

RATIONALITY AND IMPORTANCE OF TO THE COMMUNITY SERVICES BY THE COMPLEMENTARY AND ALTERNATIVE MEDICINE (CAM) SCHOOLS AND DEPARTMENTS

The rationale behind and importance of community services provided by CAM schools and departments are multifaceted, reflecting a commitment to holistic health, education, and social responsibility. These services address several critical needs and principles, which together underscore their value to individuals and communities at large.

DOI: 10.1201/9781003327202-15

Holistic Health Approach

CAM emphasizes treating the whole person—body, mind, and spirit—rather than just the symptoms of disease. By offering community services, CAM schools practice and promote this holistic approach, addressing not only physical ailments but also contributing to mental and emotional well-being. This comprehensive care model is particularly appealing to individuals seeking alternatives to the more segmented approach of conventional medicine.

Access to Healthcare

CAM community services often provide low-cost or free healthcare options to underserved and marginalized populations. This accessibility is crucial in areas where conventional medical services are either unaffordable, unavailable, or mistrusted. By filling these gaps, CAM schools ensure broader access to health services, contributing to health equity and reducing disparities.

Preventive Care

Many CAM practices focus on prevention and maintaining balance and wellness before health issues become severe. Community services such as free workshops on nutrition, stress management, and preventive practices help educate the public on how to maintain health and prevent illness. This emphasis on prevention can lead to a healthier community overall, reducing the need for medical interventions and the burden on the healthcare system.

Educational Opportunities

For CAM students, community services offer invaluable hands-on experience and a chance to apply theoretical knowledge in real-world settings. This experiential learning is critical for developing competent, confident practitioners. Additionally, these services provide a platform for interprofessional collaboration and learning, as CAM students and practitioners often work alongside or in coordination with conventional healthcare providers.

Promoting CAM Awareness and Integration

Through community services, CAM schools play a vital role in increasing awareness and understanding of alternative and complementary health practices. This exposure can lead to greater acceptance and integration of CAM within the broader healthcare system, encouraging a more inclusive approach to health and wellness that respects patient preferences and cultural practices.

Empowerment and Self-Care

CAM community services empower individuals by providing them with tools and knowledge for self-care and self-management of their health. This empowerment aligns with the growing trend toward patient-centered care, where individuals take an active role in their health decisions and management. It fosters a sense of autonomy and responsibility for one's health, which is fundamental for sustainable health and wellness.

In summary, the rationality and importance of community services offered by CAM schools and departments lie in their commitment to holistic health, increased healthcare accessibility, preventive care, educational enrichment, and the promotion of a more integrated, inclusive healthcare system. These services not only benefit the recipients but also enrich the education of CAM practitioners and contribute to the overall health and well-being of the community.

COMMUNITY SERVICES BY COMPLEMENTARY AND ALTERNATIVE MEDICINE (CAM) SCHOOLS AND DEPARTMENTS: FACILITATORS

Community services provided by CAM schools and departments are supported by a range of facilitators that enable these institutions to effectively deliver their programs and outreach activities. These facilitators are essential for the planning, execution, and sustainability of community services, ensuring that CAM's holistic health benefits reach broader segments of the population. Here are some of the key facilitators:

INSTITUTIONAL COMMITMENT

A strong institutional commitment to community engagement and service is fundamental. This commitment is often reflected in the school's mission, values, and strategic plans, which prioritize community health and wellness as core components of their educational objectives. Institutions may allocate resources, including faculty time, facilities, and funding, specifically for community service initiatives.

COLLABORATIVE PARTNERSHIPS

Partnerships with local healthcare providers, community organizations, government agencies, and other educational institutions can significantly enhance the scope and impact of CAM community services. These collaborations can provide additional resources, expertise, and access to populations in need. They also facilitate interdisciplinary learning and integrative healthcare models, combining CAM practices with conventional medicine for comprehensive care.

STUDENT AND FACULTY INVOLVEMENT

The active involvement of students and faculty is crucial for the success of community service programs. Students gain practical experience and learn the value of service and holistic care, while faculty can guide clinical practices, ensure quality of care, and foster a culture of community engagement. Volunteer opportunities and service-learning components integrated into the curriculum can encourage participation.

FINANCIAL SUPPORT

Sustainable funding sources are essential for the longevity of community service programs. This support can come from a variety of sources, including university budgets, grants, donations, and fundraising activities. Financial support not only covers operational costs but also enables the expansion of services to meet community needs effectively.

COMMUNITY ENGAGEMENT AND OUTREACH

Effective communication and outreach strategies help to build awareness of CAM services within the community. Engaging with community leaders and members through workshops, health fairs, and social media can inform the public about available services and the benefits of CAM practices. This engagement also helps to tailor services to the specific needs and cultural contexts of the community.

REGULATORY SUPPORT AND ACCREDITATION

Compliance with regulatory standards and accreditation requirements ensures that community services provided by CAM schools meet high-quality and safety standards. Accreditation bodies for CAM education often encourage or require community engagement as part of their standards, which promotes best practices and professionalism in CAM services.

RESEARCH AND EVALUATION

Ongoing research and evaluation are vital for assessing the effectiveness of community services, demonstrating their value, and making necessary adjustments. Research can also contribute to the evidence base for CAM practices, enhancing their credibility and integration into broader healthcare systems.

TRAINING AND EDUCATION

Providing continuous education and training for students and practitioners involved in community services ensures that they remain competent and up-to-date with the latest CAM practices and healthcare guidelines. This includes training in cultural competency, communication skills, and interdisciplinary collaboration.

These facilitators play a significant role in enabling CAM schools and departments to offer valuable community services, promoting holistic health and wellness, and integrating CAM practices into wider healthcare and community settings.

COMMUNITY SERVICES BY THE COMPLEMENTARY AND ALTERNATIVE MEDICINE (CAM) SCHOOLS AND DEPARTMENTS: BARRIERS

While CAM schools and departments are committed to providing community services, they face several barriers that can limit their effectiveness and reach. Overcoming these challenges is crucial for maximizing the impact of CAM services on community health and wellness. Here are some of the primary barriers.

FINANCIAL CONSTRAINTS

One of the most significant barriers is financial limitations. Providing community services often requires substantial resources for staffing, equipment, facilities, and supplies. Without adequate funding, CAM schools may struggle to maintain or expand their services. Relying on grants, donations, or institutional budgets that are already stretched thin can limit the scope and sustainability of these programs.

REGULATORY AND LEGAL CHALLENGES

The regulatory landscape for CAM practices varies significantly by region and can be complex. Legal restrictions or the lack of a clear regulatory framework can limit the types of services that CAM practitioners can offer in the community. Additionally, issues related to licensure and insurance reimbursement for CAM therapies can further complicate the provision of services.

LIMITED AWARENESS AND ACCEPTANCE

Despite growing interest in CAM, there is still a lack of awareness and acceptance among the general public and healthcare professionals. Skepticism regarding the efficacy and scientific basis of some CAM practices can hinder collaboration with conventional healthcare providers and limit patient referrals. Overcoming misconceptions and educating the community and healthcare professionals about the benefits and evidence base of CAM practices is an ongoing challenge.

INSUFFICIENT INTEGRATION INTO HEALTHCARE SYSTEMS

The integration of CAM services into mainstream healthcare systems is often limited. This can result in missed opportunities for holistic and integrative care approaches that combine the best of conventional and alternative therapies. Navigating the healthcare system's complexities and establishing CAM as a valued component of comprehensive care requires significant effort and advocacy.

WORKFORCE LIMITATIONS

There can be a shortage of trained CAM practitioners ready to engage in community service activities, especially in underserved areas. Recruiting and retaining skilled professionals who are willing to work in community settings, often for lower compensation than private practice, can be challenging. Additionally, providing adequate training and support for students and practitioners to work effectively in diverse community settings is necessary but resource-intensive.

CULTURAL AND LANGUAGE BARRIERS

Effective CAM community services must be culturally sensitive and accessible to diverse populations. Language barriers, cultural differences, and varying health beliefs can impact the acceptance and utilization of CAM services. Developing culturally competent programs and materials requires a deep understanding of the community's needs and resources, which may not always be readily available.

EVIDENCE AND RESEARCH GAPS

For some CAM practices, there is a need for more robust evidence to support their efficacy and safety. The lack of high-quality research can be a barrier to gaining acceptance from conventional healthcare providers and securing funding for community programs. Investing in research and demonstrating the value and effectiveness of CAM practices are essential for overcoming skepticism and integrating CAM into healthcare.

LOGISTICAL AND OPERATIONAL CHALLENGES

Organizing and managing community service programs involves significant logistical and operational planning. This includes coordinating schedules, securing locations, managing supplies, and ensuring compliance with health and safety regulations. For CAM schools with limited administrative resources, these tasks can be daunting and detract from the primary focus of education and service.

Addressing these barriers requires a multifaceted approach, including advocacy for supportive policies, efforts to increase public and professional awareness of CAM benefits, securing sustainable funding sources, and investing in research and education to bolster the evidence base and workforce for CAM practices.

COMMUNITY SERVICES BY THE COMPLEMENTARY AND ALTERNATIVE MEDICINE (CAM) SCHOOLS AND DEPARTMENTS: EXAMPLES

Community services by CAM schools and departments take many forms, reflecting the diversity of practices within CAM and the varied needs of the communities they serve. Here are some examples of how these institutions contribute to community health and wellness through specific initiatives.

FREE OR LOW-COST CLINICS

Many CAM schools operate clinics that offer free or low-cost services to the community. These clinics often provide a range of treatments, including acupuncture, massage therapy, chiropractic care, and naturopathic medicine. They serve as practical training grounds for students, who work under the supervision of licensed practitioners, offering valuable healthcare services to individuals who might otherwise be unable to afford them.

HEALTH EDUCATION WORKSHOPS AND SEMINARS

CAM schools frequently organize workshops and seminars on various health topics, such as nutrition, herbal medicine, stress management, and the importance of physical activity. These events aim to educate the public about preventive healthcare and holistic wellness strategies, empowering individuals to take an active role in managing their health.

COMMUNITY OUTREACH PROGRAMS

Outreach programs target specific community needs, such as programs designed for seniors focusing on mobility and fall prevention or initiatives aimed at improving mental health through mindfulness and meditation practices. CAM schools may collaborate with community centers, schools, and eldercare facilities to deliver these targeted services.

INTEGRATIVE HEALTHCARE SERVICES

Some CAM schools partner with hospitals and conventional medical clinics to provide integrative healthcare services. These partnerships can offer patients complementary therapies alongside conventional treatments, such as using acupuncture to manage pain or nausea in cancer patients undergoing chemotherapy, enhancing patient care through a holistic approach.

PUBLIC HEALTH INITIATIVES

CAM institutions may engage in public health initiatives that address broader community health issues, such as obesity, diabetes, and substance abuse. These programs often involve interdisciplinary teams that design and implement comprehensive strategies incorporating CAM therapies, lifestyle modifications, and community resources to tackle these challenges.

SUPPORT FOR UNDERSERVED POPULATIONS

Special programs may focus on serving underserved populations, including low-income families, immigrants, and indigenous communities. These services might include culturally tailored health interventions, language translation services, and the integration of traditional healing practices, ensuring that CAM benefits are accessible to all segments of the community.

Environmental Health and Sustainability Programs

Reflecting the holistic philosophy of CAM, some schools initiate programs focused on environmental health and sustainability. These might include community gardens that promote herbal medicine and nutrition, initiatives to reduce environmental toxins, or wellness programs that incorporate nature therapy.

Research and Participatory Health Studies

CAM schools often conduct research and participatory health studies involving the community. These studies can help evaluate the effectiveness of CAM therapies, explore community health needs, and develop new integrative treatment models. Participation in research offers community members the opportunity to contribute to scientific knowledge while potentially benefiting from cutting-edge health interventions.

Mental Health and Wellness Programs

Recognizing the importance of mental health, many CAM schools offer programs specifically designed to address psychological well-being. These might include yoga and meditation classes, art and music therapy sessions, and workshops on coping strategies for stress and anxiety. By making these services accessible to the community, CAM institutions help address the growing need for mental health support.

Herbal Medicine Access and Education

Some CAM schools manage community herbal gardens or apothecaries that provide access to medicinal plants and herbal remedies. Educational programs associated with these gardens teach community members about the cultivation, preparation, and safe use of herbs for health and wellness. This not only promotes natural healing practices but also fosters a deeper connection with the local environment.

Specialized Care Programs

To address specific health conditions or populations, CAM schools may offer specialized care programs. For example, programs geared toward prenatal and postnatal care using CAM therapies such as massage and acupuncture to support women's health during and after pregnancy. Other specialized programs may focus on sports medicine, offering athletes natural and holistic approaches to injury prevention and recovery.

Integrative Pain Management Clinics

With the ongoing opioid crisis, there's a significant focus on finding non-pharmacological approaches to pain management. CAM schools often run pain management clinics that utilize a variety of CAM therapies (like acupuncture, chiropractic adjustments, and massage therapy) to help individuals manage chronic pain without relying on prescription medications, providing a critical service in the face of this public health issue.

Mobile Health Services

To reach populations that may not have easy access to healthcare facilities, some CAM schools operate mobile health services. These services bring CAM practices directly to communities, offering treatments, health screenings, and wellness education at locations like community centers, schools, or even in rural or remote areas. This approach significantly extends the reach of CAM benefits to broader segments of the population.

Health Literacy and Advocacy

CAM schools often take an active role in health literacy and advocacy, providing the community with information on navigating the healthcare system, understanding patient rights, and accessing alternative and complementary health services. These efforts empower individuals to make informed health decisions and advocate for integrative approaches in their healthcare plans.

Disaster Relief and Trauma Recovery

In the aftermath of natural disasters or community traumas, CAM schools and practitioners can provide essential services such as stress relief workshops, trauma-informed yoga sessions, and acupuncture for stress and trauma recovery. These initiatives support community healing processes and demonstrate the role of CAM in addressing acute and post-traumatic conditions.

Volunteer Services Abroad

Some CAM schools extend their community service efforts internationally, offering volunteer services in countries where healthcare resources are limited. These programs can include providing CAM treatments, training local healthcare workers in CAM practices, and engaging in public health projects. This global outreach reflects the universal applicability of CAM principles and practices and their potential to contribute to global health and wellness.

Digital Health and Telehealth Services

With the rise of digital health technologies, some CAM schools have started offering telehealth services, enabling remote consultations and treatments where feasible, such as herbal consultations or guided meditation and stress management sessions. This approach has made CAM more accessible, especially during times when in-person visits are not possible, and has expanded the reach of CAM services to include individuals in remote or underserved areas.

Community Support Groups

CAM schools may facilitate or host support groups for various health conditions or wellness goals, such as groups for chronic illness management, addiction recovery, or healthy living. These support groups often incorporate CAM principles and practices, providing a holistic support system that complements medical treatment and fosters a sense of community and mutual support.

Wellness and Fitness Programs

Recognizing the importance of physical activity in overall health, CAM schools often offer community wellness and fitness programs, including classes in tai chi, qigong, yoga, and other movement-based therapies. These programs not only improve physical health but also promote mental and emotional well-being, demonstrating the interconnectedness of mind and body in health.

NUTRITION AND HEALTHY EATING PROGRAMS

Many CAM disciplines emphasize the role of nutrition in health. CAM schools may offer programs that teach community members about holistic nutrition, including workshops on cooking with medicinal herbs, using food as medicine, and principles of dietary therapy according to various CAM traditions. These programs aim to improve dietary habits and promote a holistic approach to nutrition and wellness.

ENVIRONMENTAL AND COMMUNITY HEALTH PROJECTS

Some CAM schools engage in projects aimed at improving environmental health as a component of community wellness. This can include initiatives to reduce pollution, enhance local green spaces, or promote sustainable living practices. By linking environmental health with human health, these projects embody the holistic principles of CAM and contribute to the well-being of both the community and the planet.

SCHOOL AND YOUTH PROGRAMS

Understanding the importance of early health education, CAM schools may partner with local schools to offer programs for children and adolescents. These can include education on stress management techniques, the importance of nutrition and physical activity, and the introduction to non-invasive CAM practices suitable for young people. By instilling healthy habits early, these programs aim to contribute to the long-term well-being of the community.

CULTURAL PRESERVATION AND INTEGRATION

In communities with rich traditional health practices, CAM schools can play a role in preserving and integrating these traditions into broader health initiatives. This may involve collaborating with traditional healers, documenting indigenous health practices, and offering programs that respect and incorporate local cultural perspectives on health and healing.

POLICY ADVOCACY AND COMMUNITY LEADERSHIP

CAM schools often engage in policy advocacy and leadership roles within the community, working to shape health policies that recognize and integrate CAM practices. By participating in health policy development, CAM institutions can advocate for regulatory changes that support holistic health approaches and ensure that CAM services are included in health planning and decision-making processes.

INTERPROFESSIONAL EDUCATION PROGRAMS

Some CAM schools develop interprofessional education programs that bring together CAM and conventional healthcare students for joint learning experiences. These programs aim to foster mutual understanding and respect among future health professionals, encourage collaborative care models, and ultimately improve patient outcomes by integrating diverse therapeutic approaches.

SPECIALIZED WORKSHOPS FOR SPECIFIC GROUPS

Recognizing the unique health challenges faced by certain populations, CAM schools may offer specialized workshops tailored to the needs of veterans, first responders, or other groups with high levels of occupational stress or trauma. These workshops can include stress reduction techniques, acupuncture for PTSD, and other CAM therapies proven beneficial for these communities.

HEALTH LITERACY CAMPAIGNS

CAM institutions often lead or participate in health literacy campaigns to improve public understanding of health information, including the safe and effective use of CAM therapies. These campaigns can help individuals make informed health decisions, understand the role of CAM in preventive health and chronic disease management, and navigate the healthcare system more effectively.

SUSTAINABLE HEALTHCARE PRACTICES

In line with the holistic philosophy of CAM, some schools advocate for and implement sustainable healthcare practices within their own operations and the wider community. This might involve reducing waste in CAM clinics, promoting the use of sustainable and ethically sourced medicinal products, and educating the community about environmental determinants of health.

COMMUNITY RESILIENCE PROJECTS

CAM schools may engage in projects aimed at building community resilience, particularly in areas prone to natural disasters or economic hardships. These projects can include establishing community wellness centers that offer CAM services as part of disaster preparedness and recovery plans, ensuring that holistic health support is available in times of crisis.

HOLISTIC AGING PROGRAMS

Addressing the health needs of the aging population, CAM schools offer programs focused on holistic aging, promoting wellness and quality of life for seniors. These programs may include CAM therapies for age-related conditions, exercise classes designed for older adults, and workshops on nutrition and healthy aging.

GLOBAL HEALTH INITIATIVES

Beyond local community services, some CAM schools participate in global health initiatives, providing CAM services in international health missions or collaborating with global partners to address health challenges in underserved regions. These initiatives demonstrate the universal applicability of CAM principles and the global commitment of CAM institutions to health and wellness.

TECHNOLOGY AND INNOVATION IN CAM SERVICES

Embracing technological advancements, CAM schools are exploring innovative ways to deliver CAM services, such as using virtual reality for meditation and stress relief, developing apps for health tracking and wellness coaching, and utilizing artificial intelligence to personalize herbal medicine recommendations. These technological initiatives can enhance the accessibility and effectiveness of CAM therapies.

These examples illustrate the diverse ways in which CAM schools and departments contribute to community health and wellness. By providing direct health services, education, and engaging in public health initiatives, CAM institutions play a vital role in promoting holistic health and preventive care within communities.

COMMUNITY SERVICES BY COMPLEMENTARY AND ALTERNATIVE MEDICINE (CAM) SCHOOLS AND DEPARTMENTS: ACHIEVEMENTS

Community services provided by CAM schools and departments have achieved significant impacts in various areas, reflecting their commitment to holistic health, education, and community well-being. These achievements demonstrate the value of CAM in promoting health and wellness, enhancing access to care, and fostering integrative health practices. Here are key areas where CAM community services have made notable contributions:

IMPROVED ACCESS TO HEALTHCARE

CAM schools have greatly expanded access to alternative and complementary healthcare services, particularly for underserved populations. By offering low-cost or free clinics, these institutions have provided critical healthcare options for individuals who might not otherwise afford or access such services. This has been especially important for patients seeking non-pharmacological treatments for pain, chronic conditions, or those preferring holistic approaches to health and wellness.

ENHANCED HEALTH OUTCOMES

Many programs have reported positive health outcomes among their participants, including reduced pain, improved mental health, and better overall well-being. For example, acupuncture clinics run by CAM schools have helped patients manage chronic pain and reduce reliance on opioids, contributing to broader efforts to address the opioid crisis. Similarly, stress reduction programs using meditation and yoga have shown effectiveness in improving mental health and enhancing quality of life.

EDUCATION AND EMPOWERMENT

CAM community services have played a vital role in educating the public about holistic health practices, preventive care, and self-care strategies. Through workshops, seminars, and health fairs, these programs have empowered individuals with knowledge and tools to take charge of their health, make informed decisions, and adopt healthier lifestyles. This educational aspect has also helped to demystify CAM practices, increasing their acceptance and integration into everyday health and wellness routines.

PROMOTION OF INTEGRATIVE HEALTH

CAM schools have been instrumental in promoting the integration of complementary and alternative medicine with conventional healthcare. By collaborating with hospitals, clinics, and other healthcare providers, CAM programs have facilitated a more holistic, patient-centered approach to health care, where the best of conventional and CAM therapies are available to patients. This integrative model has been particularly effective in areas such as pain management, mental health, and chronic disease management.

COMMUNITY ENGAGEMENT AND PARTNERSHIPS

The community services provided by CAM schools have fostered stronger connections and partnerships within communities. Collaborations with community organizations, healthcare providers, and public health initiatives have not only expanded the reach of CAM services but also contributed to a more cohesive and supportive healthcare ecosystem. These partnerships have enabled holistic health initiatives that address social determinants of health and work toward health equity.

RESEARCH AND INNOVATION

Community service programs have also served as platforms for research and innovation in the field of CAM. By collecting data on the effectiveness of various CAM therapies and practices, these programs contribute to the evidence base supporting CAM. This research is crucial for the ongoing development and refinement of CAM practices, ensuring they are safe, effective, and grounded in scientific evidence.

GLOBAL HEALTH CONTRIBUTIONS

Beyond local communities, some CAM schools have extended their services globally, contributing to international health efforts and cross-cultural exchanges in healthcare. These initiatives have not only provided much-needed health services in underserved areas but have also enriched the educational experiences of CAM students and practitioners, exposing them to diverse health challenges and practices around the world.

The achievements of CAM schools and departments in community service reflect a deep commitment to holistic health, education, and social responsibility. Through their efforts, these institutions have made significant contributions to individual and community well-being, advancing the role of complementary and alternative medicine in the broader health landscape.

COMMUNITY SERVICES BY COMPLEMENTARY AND ALTERNATIVE MEDICINE (CAM) SCHOOLS AND DEPARTMENTS: CHALLENGES

Community services provided by CAM schools and departments face several challenges that can impact their effectiveness and sustainability. These challenges often stem from external factors, such as regulatory environments, and internal factors, such as resources and operational capabilities. Addressing these challenges is crucial for maximizing the positive impact of CAM community services. Here's an overview of the primary challenges they face:

FINANCIAL LIMITATIONS

- **Funding**. Securing consistent funding to support the operations, staffing, and resources needed for community services is a major challenge. These programs often rely on donations, grants, or institutional subsidies, which can be unpredictable and insufficient to cover all expenses.
- **Costs**: The costs associated with running community clinics, outreach programs, and educational workshops can be significant. This includes not just the direct costs of supplies and personnel but also indirect costs such as insurance and facility maintenance.

REGULATORY AND LEGAL HURDLES

- **Licensing and Scope of Practice**: CAM practitioners and clinics must navigate complex regulatory environments that vary widely by location. Restrictions on scope of practice can limit the types of services offered, affecting the ability to fully meet community needs.
- **Insurance and Reimbursement**: The lack of consistent insurance coverage for CAM services poses a challenge for both providers and patients. This can limit access for individuals who cannot afford to pay out-of-pocket for CAM therapies.

SKEPTICISM AND ACCEPTANCE

- **Mainstream Medical Community**: Despite growing interest in and evidence for CAM, skepticism remains in parts of the mainstream medical community. This can hinder collaborative efforts and referrals, limiting the reach and impact of CAM community services.
- **Public Perception**: Overcoming misconceptions and lack of awareness about the benefits and evidence base of CAM practices is a constant challenge. Educational efforts are vital but require resources and time to be effective.

OPERATIONAL AND LOGISTICAL CHALLENGES

- **Staffing**: Recruiting and retaining qualified CAM practitioners and volunteers to run community services can be difficult, especially in areas with a limited pool of CAM professionals.
- **Coordination**: Effectively coordinating services, especially in partnership with other organizations or within complex healthcare systems, requires robust operational capabilities and can be logistically challenging.

CULTURAL AND LANGUAGE BARRIERS

- **Diverse Populations**: Serving diverse communities effectively requires cultural competence and often language services. Addressing these needs is essential for effective communication and treatment but can add complexity and cost to service provision.

EVIDENCE AND RESEARCH

- **Efficacy and Outcomes**: There is a continuous need for high-quality research to demonstrate the efficacy of CAM modalities. The lack of robust evidence for some practices can limit funding, acceptance, and integration into mainstream healthcare.
- **Data Collection and Analysis**: Collecting and analyzing data on the impact of community services is crucial for securing funding and support. However, many CAM institutions may lack the resources or expertise to conduct rigorous outcome evaluations.

INTEGRATION WITH HEALTHCARE SYSTEMS

- **Collaboration**: Despite increasing interest in integrative health, achieving meaningful collaboration between CAM and conventional healthcare systems remains a challenge. Differences in approach, language, and values can create barriers to seamless integration.
- **Referrals and Pathways**: Establishing referral pathways between CAM services and conventional healthcare providers requires building trust and understanding across different medical cultures, which can be time-consuming and complex.

Overcoming these challenges requires innovative strategies, advocacy, and continued dialogue between CAM practitioners, healthcare providers, policymakers, and the public. Strengthening the evidence base for CAM, improving regulatory frameworks, and enhancing public and professional education about CAM benefits are key steps toward addressing these obstacles and enhancing the impact of CAM community services.

COMMUNITY SERVICES BY COMPLEMENTARY AND ALTERNATIVE MEDICINE (CAM) SCHOOLS AND DEPARTMENTS: RECOMMENDATIONS FOR THE BEST PRACTICES

To enhance the effectiveness, sustainability, and impact of community services provided by CAM schools and departments, adopting best practices is essential. These recommendations can help navigate challenges, maximize resources, and fulfill the mission of improving community health and wellness through CAM approaches (Al-Worafi, 2020a,b, 2022a,b, 2023a–n, 2024a–d; Hasan et al., 2019). Here are some best practices for CAM community services:

1. **Build Strong Partnerships**
   - *Collaborate with Conventional Healthcare Providers*: Establish relationships with hospitals, clinics, and individual healthcare professionals to integrate CAM services and facilitate patient referrals.
   - *Engage Community Organizations*: Partner with local organizations, schools, and community centers to reach broader audiences and tailor services to meet specific community needs.
2. **Secure Sustainable Funding**
   - *Diversify Funding Sources*: Explore a mix of funding options, including grants, donations, institutional support, and service fees, on a sliding scale to ensure financial sustainability.
   - *Apply for Grants*: Target grants specifically designed for healthcare services, community health initiatives, and educational programs.
3. **Enhance Visibility and Outreach**
   - *Promote Services Effectively*: Use social media, local media, community events, and word-of-mouth to raise awareness about CAM services and their benefits.
   - *Educational Workshops and Seminars*: Offer free or low-cost educational programs to the public to build trust, educate about CAM practices, and demonstrate commitment to community wellness.
4. **Focus on Cultural Competence**
   - *Train Staff and Students*: Ensure that practitioners and students are trained in cultural competence, able to effectively communicate with and treat patients from diverse backgrounds.
   - *Offer Language Services*: Provide translation services or bilingual practitioners to make CAM services accessible to non-English speaking communities.
5. **Implement Evidence-Based Practices**
   - *Stay Informed on Research*: Regularly review and incorporate the latest research findings into practice to ensure that CAM services are evidence-based and effective.
   - *Contribute to Research*: Engage in or support research efforts to contribute to the evidence base for CAM practices, enhancing credibility and support for CAM services.
6. **Ensure Quality and Safety**
   - *Adhere to Professional Standards*: Maintain high standards of professionalism, ethics, and patient care, ensuring that all practitioners are properly trained and credentialed.
   - *Regular Evaluation and Feedback*: Implement mechanisms for regular evaluation of services and feedback from clients to continually improve service quality and responsiveness.
7. **Leverage Technology**
   - *Telehealth Services*: Expand access to CAM services through telehealth, particularly for clients in remote or underserved areas.
   - *Digital Tools for Education and Management*: Utilize digital platforms for client education, appointment scheduling, and practice management to enhance efficiency and reach.

8. **Integrate Services within Healthcare Systems**
   - *Develop Referral Networks*: Create formal referral networks with conventional healthcare providers to facilitate integrative care for patients.
   - *Participate in Healthcare Planning*: Engage in local and regional healthcare planning efforts to ensure that CAM services are included in broader health and wellness initiatives.
9. **Foster Community Engagement**
   - *Community Needs Assessment*: Conduct regular assessments to understand community health needs and preferences, ensuring that CAM services are relevant and responsive.
   - *Volunteer Opportunities*: Encourage student and faculty participation in community service as volunteers, enriching their educational experience and reinforcing a culture of service.
10. **Advocate for CAM**
    - *Policy Advocacy*: Engage in advocacy efforts to support the recognition, integration, and funding of CAM practices within healthcare policies and insurance programs.
    - *Public Education*: Actively participate in public education campaigns to raise awareness about the benefits and potential of CAM approaches to health and wellness.
11. **Implement Interprofessional Education**
    - *Interdisciplinary Collaboration*: Encourage collaboration between CAM and conventional healthcare students through joint educational programs, fostering mutual respect and understanding that can translate into integrated patient care.
    - *Clinical Rotations*: Offer clinical rotations or internships in diverse settings, including hospitals, community clinics, and CAM practices, to expose students to a variety of healthcare models and patient populations.
12. **Focus on Preventive Health Education**
    - *Preventive Workshops and Programs*: Develop and offer programs focused on preventive health, emphasizing lifestyle modifications, nutrition, stress management, and other CAM principles to prevent illness and promote wellness.
    - *Community Health Initiatives*: Engage in community health initiatives that address determinants of health, such as diet, physical activity, and environmental factors, supporting broader public health goals.
13. **Utilize Data and Technology for Service Improvement**
    - *Outcome Measurement*: Systematically measure and analyze outcomes of CAM services to assess effectiveness, guide improvements, and demonstrate value to stakeholders and funders.
    - *Health Information Technology*: Leverage health information technology to manage patient information, track outcomes, and facilitate communication between CAM practitioners and other healthcare providers.
14. **Develop Specialty Programs**
    - *Address Community-Specific Needs*: Design specialty programs that address specific health issues prevalent in the community, such as diabetes management, addiction recovery, or mental health support, utilizing CAM modalities tailored to these conditions.
    - *Special Populations*: Create programs targeted at special populations, including seniors, veterans, and children, to address their unique health needs with appropriate CAM interventions.
15. **Strengthen Community Ties through Service-Learning**
    - *Service-Learning Opportunities*: Integrate service-learning into the curriculum, allowing students to gain hands-on experience in delivering CAM services to the community while learning about social determinants of health and health equity.

- *Community-Based Research*: Involve students and faculty in community-based research projects that address local health concerns, bridging academic research and community health needs.

16. **Advocate for Regulatory Changes**
    - *Engagement in Policy Development*: Actively engage in policy development processes to advocate for regulatory changes that support the integration of CAM into mainstream healthcare, including licensure, scope of practice, and insurance coverage.
    - *Professional Networks*: Participate in professional networks and associations to unify the voice of CAM practitioners in policy discussions and health system planning.

17. **Promote Global Health Engagement**
    - *International Collaborations*: Establish partnerships with CAM institutions and health organizations globally to exchange knowledge, participate in international health initiatives, and contribute to global health education.
    - *Global Health Programs*: Offer global health programs that allow students to experience healthcare delivery in different cultural and socioeconomic settings, expanding their perspectives on health and healing.

18. **Foster a Culture of Continuous Learning**
    - *Professional Development*: Provide ongoing professional development opportunities for faculty and practitioners to stay updated on the latest CAM research, techniques, and best practices.
    - *Lifelong Learning for Graduates*: Encourage alumni and practicing CAM professionals to engage in lifelong learning through workshops, seminars, and continuing education courses, ensuring they remain at the forefront of CAM practice.

19. **Develop a Community Advisory Board**
    - *Engage Local Leaders*: Create a community advisory board that includes local healthcare providers, community leaders, and patient advocates to guide the development and implementation of community services, ensuring they meet local needs.

20. **Enhance Accessibility of Services**
    - *Flexible Scheduling*: Offer services during off-hours or weekends to accommodate those with traditional 9-5 obligations.
    - *Mobile Clinics*: Deploy mobile clinics to reach remote or underserved areas where CAM services are most needed but least available.

21. **Promote Environmental Health**
    - *Sustainable Practices*: Implement and promote environmental sustainability practices within CAM facilities and through community outreach, emphasizing the connection between environmental health and personal well-being.

22. **Offer Mental Health Support**
    - *Specialized Services*: Provide specialized mental health services and support groups, using CAM modalities known to benefit emotional and psychological health.

23. **Foster Youth Engagement**
    - *Youth Programs*: Develop programs specifically designed for children and adolescents, promoting early health education and introducing young people to healthy lifestyle choices.

24. **Encourage Patient Empowerment and Education**
    - *Self-Care Workshops*: Offer workshops that teach self-care practices, empowering patients to take an active role in their health and wellness journey.

25. **Strengthen Alumni Networks**
    - *Engagement in Community Services*: Encourage alumni to participate in community service initiatives, providing a bridge between current students and the broader professional community.

26. **Incorporate Art and Culture into Healing**
- *Cultural Healing Practices*: Integrate local cultural and artistic practices into healing programs, recognizing the role of culture in health and wellness.

27. **Prioritize Health Equity**
- *Address Disparities*: Develop programs specifically aimed at addressing health disparities and promoting health equity within underserved populations.

28. **Innovate in Health Promotion**
- *Health Campaigns*: Launch public health campaigns using innovative approaches to promote healthy living and prevent disease within the community.

29. **Build Resilience Programs**
- *Disaster Response*: Train students and staff in disaster response, preparing them to offer CAM services as part of emergency relief efforts.

30. **Expand Research Initiatives**
- *Community-Based Participatory Research*: Engage the community in participatory research projects to explore the effectiveness of CAM interventions on local health issues.

31. **Integrate Technology in Practice**
- *Digital Health Platforms*: Utilize digital platforms to enhance access to CAM services, including virtual reality for meditation and stress management.

32. **Promote Lifelong Wellness**
- *Wellness Continuum Programs*: Offer programs that address health at all life stages, promoting a continuum of wellness from youth through old age.

33. **Utilize Space Creatively**
- *Community Wellness Spaces*: Create multipurpose spaces within CAM institutions for community use, such as gardens for meditation and classes.

34. **Leverage Local Media**
- *Media Partnerships*: Partner with local media outlets to promote CAM services and educate the public on CAM's benefits.

35. **Foster Global Exchanges**
- *Student Exchange Programs*: Establish student exchange programs with CAM institutions worldwide to share knowledge and practices.

36. **Support the Local Economy**
- *Local Sourcing*: Source supplies and materials from local businesses, supporting the local economy and fostering community ties.

37. **Enhance Language Services**
- *Multilingual Resources*: Expand language services to include more languages, ensuring broader accessibility of CAM services.

38. **Address Chronic Diseases**
- *Chronic Disease Management Programs*: Offer comprehensive programs aimed at managing chronic diseases with CAM modalities, integrating nutritional counseling, physical therapies, and stress reduction techniques.

39. **Foster Innovation and Creativity**
- *Innovation Labs*: Establish innovation labs within CAM schools to develop new CAM technologies, practices, and approaches.

40. **Emphasize Ethical Practice**
- *Ethics Training*: Incorporate ethics training focused on CAM practices, emphasizing the importance of informed consent and cultural sensitivity.

41. **Encourage Volunteering**
- *Volunteer Incentive Programs*: Create incentive programs for students and faculty to volunteer in community service, recognizing their contributions.

42. **Implement Patient Feedback Mechanisms**
    - *Feedback Loops*: Establish mechanisms for patient feedback on CAM services, using insights to continually improve quality and responsiveness.
43. **Strengthen Professional Development**
    - *Continuing Education*: Offer robust continuing education programs for CAM practitioners, focusing on emerging research, clinical skills, and professional ethics.
44. **Promote Healthy Workplaces**
    - *Workplace Wellness Programs*: Offer workplace wellness programs to local businesses, integrating CAM practices to improve employee health and productivity.
45. **Cultivate Community Gardens**
    - *Medicinal Plant Gardens*: Develop community gardens focused on medicinal plants, providing educational and therapeutic resources to the community.
46. **Support Workforce Development**
    - *Career Pathways*: Create programs that support career development in CAM fields, including mentorship, internships, and job placement services.
47. **Enhance Interdisciplinary Learning**
    - *Cross-Disciplinary Courses*: Offer courses that bring together students from CAM and other health disciplines, promoting interdisciplinary learning and collaboration.
48. **Advocate for Community Health**
    - *Public Health Advocacy*: Engage in advocacy efforts to address public health issues at the local and national levels, leveraging CAM's unique perspectives.

Implementing these best practices requires a strategic, collaborative approach that engages all stakeholders, including CAM practitioners, students, patients, healthcare providers, and community leaders. By embracing innovation, prioritizing inclusivity, and fostering partnerships, CAM schools and departments can significantly enhance their community services, contributing to healthier, more resilient communities. By implementing these best practices, CAM schools and departments can enhance the reach, impact, and sustainability of their community services, further integrating CAM into the fabric of community health and wellness initiatives.

CONCLUSION

The role of CAM schools and departments in community services is multifaceted and profoundly impactful, addressing a wide range of health needs and promoting holistic, integrative approaches to wellness. Through the provision of clinical services, educational programs, and public health initiatives, CAM institutions have demonstrated a commitment to improving access to healthcare, enhancing health outcomes, and fostering a culture of health and wellness that transcends traditional healthcare boundaries. The challenges faced by CAM community services, including financial constraints, regulatory hurdles, and the need for greater acceptance within the broader healthcare and public domains, underscore the importance of strategic, innovative approaches to service delivery, funding, and collaboration. By adopting best practices such as developing strong partnerships, leveraging technology, emphasizing cultural competence, and engaging in advocacy and policy development, CAM schools and departments can overcome these obstacles and expand their contributions to community health. Moreover, the recommendations for enhancing CAM community services reflect a comprehensive strategy that encompasses education, research, patient care, and community engagement. These recommendations highlight the importance of integrating CAM into the broader healthcare ecosystem, promoting preventive health, addressing health disparities, and supporting sustainable, community-based health initiatives. In conclusion, the contributions of CAM schools and departments to community services are invaluable, offering unique perspectives and approaches to health and wellness that complement conventional medical practices.

By continuing to innovate, collaborate, and advocate for holistic health, CAM institutions can play a pivotal role in shaping a more inclusive, effective, and resilient healthcare system. The ongoing evolution of CAM community services promises not only to enhance the health and well-being of individual patients but also to contribute to the broader goals of public health and social equity.

REFERENCES

Al-Worafi, Y.M. (Ed.). (2020a). *Drug Safety in Developing Countries: Achievements and Challenges*. Academic Press.

Al-Worafi, Y.M. (2020b). Herbal medicines safety issues. In: Al-Worafi, Y.M. (ed), *Drug Safety in Developing Countries* (pp. 163–178). Academic Press.

Al-Worafi, Y.M. (2022a). *A Guide to Online Pharmacy Education: Teaching Strategies and Assessment Methods*. CRC Press.

Al-Worafi, Y.M. (2022b). Competencies and learning outcomes. In: Al-Worafi, Y.M. (ed), *A Guide to Online Pharmacy Education: Teaching Strategies and Assessment Methods*. CRC Press.

Al-Worafi, Y.M. (2023a). *Patient Safety in Developing Countries: Education, Research, Case Studies*. CRC Press.

Al-Worafi, Y.M. (2023b). *Technology for Drug Safety: Current Status and Future Developments*. Springer Nature.

Al-Worafi, Y.M. (2023c). Patient safety-related issues: Patient care errors and related problems. In: Al-Worafi, Y.M. (ed), *Patient Safety in Developing Countries: Education, Research, Case Studies*. CRC Press.

Al-Worafi, Y.M. (2023d). Patient care errors and related problems: Preventive medicine errors & related problems. In: Al-Worafi, Y.M. (ed), *Patient Safety in Developing Countries: Education, Research, Case Studies*. CRC Press.

Al-Worafi, Y.M. (2023e). Patient care errors and related problems: Patient assessment and diagnostic errors & related problems. In: Al-Worafi, Y.M. (ed), *Patient Safety in Developing Countries: Education, Research, Case Studies*. CRC Press.

Al-Worafi, Y.M. (2023f). Patient care errors and related problems: Non-pharmacological errors & related problems. In: Al-Worafi, Y.M. (ed), *Patient Safety in Developing Countries: Education, Research, Case Studies*. CRC Press.

Al-Worafi, Y.M. (2023g). Patient care errors and related problems: Medical errors & related problems. In: Al-Worafi, Y.M. (ed), *Patient Safety in Developing Countries: Education, Research, Case Studies*. CRC Press.

Al-Worafi, Y.M. (2023h). Patient care errors and related problems: Monitoring errors & related problems. In: Al-Worafi, Y.M. (ed), *Patient Safety in Developing Countries: Education, Research, Case Studies*. CRC Press.

Al-Worafi, Y.M. (2023i). Patient care errors and related problems: Patient education and counselling errors and related problems. In: Al-Worafi, Y.M. (ed), *Patient Safety in Developing Countries: Education, Research, Case Studies*. CRC Press.

Al-Worafi, Y.M. (2023j). Patient safety-related issues: Other medication safety issues. In: Al-Worafi, Y.M. (ed), *Patient Safety in Developing Countries: Education, Research, Case Studies*. CRC Press.

Al-Worafi, Y.M. (2023k). Patient safety culture. In: Al-Worafi, Y.M. (ed), *Patient Safety in Developing Countries: Education, Research, Case Studies*. CRC Press.

Al-Worafi, Y.M. (2023l). Patient safety education: Competencies and learning outcomes. In: Al-Worafi, Y.M. (ed), *Patient Safety in Developing Countries: Education, Research, Case Studies*. CRC Press.

Al-Worafi, Y.M. (Ed.). (2023m). *Clinical Case Studies on Medication Safety*. Academic Press.

Al-Worafi, Y.M. (Ed.). (2023n). *Comprehensive Healthcare Simulation: Pharmacy Education, Practice and Research*. Springer Nature.

Al-Worafi, Y.M. (Ed.). (2024a). *Handbook of Medical and Health Sciences in Developing Countries*. Springer, Cham.

Al-Worafi, Y.M. (2024b). Complementary and alternative medicine (CAM) in developing countries. In: Al-Worafi, Y.M. (ed), *Handbook of Medical and Health Sciences in Developing Countries*. Springer, Cham. https://doi.org/10.1007/978-3-030-74786-2_301-1

Al-Worafi, Y.M. (2024c). Access/equitable access to medical and health sciences in developing countries. In: Al-Worafi, Y.M. (ed), *Handbook of Medical and Health Sciences in Developing Countries*. Springer, Cham. https://doi.org/10.1007/978-3-030-74786-2_147-1

Al-Worafi, Y.M. (2024d). Community services by medical and health sciences schools in developing countries: Overview. In: Al-Worafi, Y.M. (ed), *Handbook of Medical and Health Sciences in Developing Countries*. Springer, Cham. https://doi.org/10.1007/978-3-030-74786-2_159-1

Hasan, S., Al-Omar, M.J., AlZubaidy, H., and Al-Worafi, Y.M. (2019). Use of medications in Arab countries. In: Laher, I. (ed), *Handbook of Healthcare in the Arab World* (p. 42). Springer, Cham.

16 Complementary and Alternative Medicine (CAM) Education
Library and Education Resources

INTRODUCTION

Complementary and alternative medicine (CAM) encompasses a broad range of practices, products, and therapies that are not generally considered part of conventional medicine. CAM education focuses on integrating these alternative approaches with traditional medicine to provide holistic care. This interdisciplinary field of study aims to equip healthcare professionals with knowledge, skills, and competencies to apply CAM strategies effectively and safely in patient care.

Library and education resources for CAM are vital for both students and practitioners. These resources include textbooks, journals, online databases, and multimedia materials covering various CAM modalities such as herbal medicine, acupuncture, homeopathy, and mind-body practices. Libraries dedicated to CAM offer access to scientific research, case studies, and clinical trials that support evidence-based practice in the field. Online platforms and digital libraries have become increasingly important, providing remote access to a wealth of information.

Education in CAM also involves experiential learning opportunities, such as workshops, seminars, and clinical rotations, allowing students to apply theoretical knowledge in real-world settings. Accredited programs in CAM, offered by colleges and universities, often include courses on ethics, legal issues, and the integration of CAM into conventional healthcare systems. Continuing education programs and certifications are available for healthcare professionals seeking to expand their expertise in specific CAM modalities.

Overall, the growth of CAM education and the availability of diverse library and educational resources reflect the increasing acceptance and integration of alternative therapies in healthcare. These resources support the ongoing research, development, and application of CAM practices, contributing to more comprehensive and personalized patient care.

RATIONALITY AND IMPORTANCE OF TO THE LIBRARIES IN THE COMPLEMENTARY AND ALTERNATIVE MEDICINE (CAM) SCHOOLS/DEPARTMENTS

The presence of well-equipped libraries in CAM schools and departments is critical for several reasons, underpinning both the rationality and importance of these resources in the field of CAM education and practice.

RATIONALITY

1. **Evidence-Based Practice**: The foundation of modern healthcare, including CAM, is evidence-based practice. Libraries provide access to a wealth of scientific literature, including journals, books, and databases that are essential for students, educators, and practitioners to find evidence supporting the efficacy and safety of CAM therapies.

This evidence is crucial for integrating CAM practices into conventional healthcare in an informed and ethical manner.

2. **Research Support**: CAM libraries support research by providing resources necessary for the investigation of alternative therapies, understanding their mechanisms of action, and evaluating their effectiveness and safety. These libraries facilitate access to both conventional medical literature and specialized CAM resources, which is essential for conducting comprehensive literature reviews and staying abreast of the latest findings in the field.

3. **Educational Excellence**: Libraries are central to the educational mission of CAM schools and departments. They offer a range of materials that support curriculum development, teaching, and learning. This includes textbooks, multimedia materials, and online courses that cover various CAM modalities. Libraries also provide spaces for study and collaboration, enhancing the learning environment for students.

4. **Professional Development**: For practitioners, CAM libraries are a resource for continuous learning and professional development. They offer access to continuing education materials, professional guidelines, and the latest research, helping practitioners to enhance their skills, update their knowledge, and provide the highest quality care to patients.

IMPORTANCE

1. **Quality Patient Care**: By supporting evidence-based practice and professional development, CAM libraries contribute directly to the quality of patient care. Practitioners who have access to the latest research and resources are better equipped to make informed decisions about integrating CAM therapies into patient treatment plans.

2. **Innovation and Advancement**: Libraries play a crucial role in fostering innovation and the advancement of the CAM field. By facilitating research and providing access to a broad range of resources, libraries help to stimulate new ideas, promote the development of new therapies, and contribute to the body of knowledge in CAM.

3. **Interdisciplinary Collaboration**: CAM libraries serve as a bridge between conventional medicine and alternative therapies, promoting interdisciplinary collaboration. Access to diverse resources enables healthcare professionals from different backgrounds to learn about each other's practices, fostering mutual respect and integrated approaches to patient care.

4. **Public Health and Wellness**: Finally, CAM libraries contribute to public health and wellness by supporting the education of practitioners who are knowledgeable about a wide range of therapeutic options. This diversity in healthcare practices can lead to more personalized and holistic approaches to health and wellness, benefiting individuals and communities.

In summary, libraries in CAM schools and departments are indispensable for fostering an evidence-based, research-oriented, and comprehensive educational environment. They are fundamental to the ongoing development of CAM practices and their integration into mainstream healthcare, ultimately contributing to improved patient outcomes and the advancement of healthcare as a whole.

EDUCATIONAL RESOURCES FOR COMPLEMENTARY AND ALTERNATIVE MEDICINE (CAM) STUDENTS

Educational resources for students pursuing studies in CAM are diverse and tailored to support a wide range of learning objectives, from foundational knowledge to advanced clinical skills. These resources play a crucial role in preparing students for careers in healthcare that integrate traditional and alternative medicine practices. Here are some key educational resources that are instrumental for CAM students.

TEXTBOOKS AND REFERENCE BOOKS

- **Core Textbooks**: Cover fundamental concepts, theories, and practices across various CAM modalities, including herbal medicine, acupuncture, chiropractic, naturopathy, and mind-body therapies.
- **Specialized Reference Books**: Provide in-depth information on specific areas within CAM, such as botanical pharmacognosy, nutritional supplements, and energy medicine.

ONLINE DATABASES AND JOURNALS

- **Scientific Databases**: Platforms like PubMed, ScienceDirect, and the Cochrane Library offer access to peer-reviewed articles, clinical trial reports, and systematic reviews relevant to CAM research and practice.
- **CAM-Specific Journals**: Publications such as the *Journal of Alternative and Complementary Medicine* and *Evidence-Based Complementary and Alternative Medicine* publish the latest research findings, reviews, and clinical studies focused on CAM.

MULTIMEDIA RESOURCES

- **Instructional Videos and Webinars**: Visual aids that demonstrate techniques, procedures, and case studies, making complex concepts more accessible.
- **Online Courses and MOOCs**: Massive Open Online Courses (MOOCs) and other online platforms offer courses on various CAM topics, providing flexibility and access to leading experts in the field.

CLINICAL PRACTICE GUIDELINES AND PROTOCOLS

- **Professional Guidelines**: Documents and manuals provided by professional CAM organizations outline standards of practice, ethical considerations, and safety protocols.
- **Clinical Toolkits**: Collections of assessment tools, patient handouts, and treatment planning resources to support clinical decision-making.

WORKSHOPS, SEMINARS, AND CONFERENCES

- **Hands-On Workshops**: Offer practical experience in specific CAM techniques, such as massage therapy, acupuncture needle placement, or herbal medicine compounding.
- **Professional Seminars and Conferences**: Provide opportunities for networking, continuing education, and exposure to cutting-edge research and clinical practices.

SIMULATION AND VIRTUAL REALITY (VR) TOOLS

- **Clinical Simulations**: Software and VR platforms simulate real-life patient encounters and clinical scenarios, allowing students to practice and refine their diagnostic and treatment skills in a risk-free environment.

COMMUNITY AND CLINICAL INTERNSHIPS

- **Internship Programs**: Placement in healthcare settings that offer CAM services, providing invaluable real-world experience and mentorship from experienced practitioners.

PATIENT EDUCATION MATERIALS

- **Brochures and Handouts**: Resources designed to educate patients on various CAM therapies, their benefits, and what to expect during treatment.

PROFESSIONAL ASSOCIATIONS AND NETWORKS

- **Membership in CAM Organizations**: Offers access to a community of professionals, along with resources such as newsletters, policy updates, and professional development opportunities.

By leveraging these diverse educational resources, CAM students can gain a comprehensive understanding of alternative medicine practices, evidence-based research skills, and the clinical competencies needed to provide holistic and patient-centered care. These resources are pivotal not only for academic success but also for the professional development of students as they transition into CAM practitioners.

EDUCATIONAL RESOURCES FOR COMPLEMENTARY AND ALTERNATIVE MEDICINE (CAM) EDUCATORS

For educators in the field of CAM, having access to a wide range of educational resources is crucial for designing effective curricula, staying updated on the latest research, and providing the highest quality education to students. Here's an overview of key resources that can support CAM educators in their roles.

CURRICULUM DEVELOPMENT AND PEDAGOGICAL RESOURCES

- **Curriculum Guides and Standards**: Documents from professional CAM organizations and accrediting bodies that outline competencies, learning objectives, and educational standards for various CAM disciplines.
- **Pedagogical Tools**: Resources on innovative teaching methods, including problem-based learning, flipped classrooms, and experiential learning, tailored for CAM education.

RESEARCH DATABASES AND LIBRARIES

- **Specialized CAM Databases**: Access to databases like the National Center for Complementary and Integrative Health (NCCIH) and others that offer comprehensive research, systematic reviews, and clinical guidelines specific to CAM.
- **University and Medical School Libraries**: Institutional subscriptions to medical and scientific journals, eBooks, and online resources that provide a wealth of information relevant to CAM teaching and research.

PROFESSIONAL DEVELOPMENT AND CONTINUING EDUCATION

- **Workshops and Seminars**: Opportunities for educators to learn new teaching strategies, assessment methods, and updates in CAM practice and research.
- **Online Courses and Webinars**: Platforms offering courses on the latest CAM research, educational technology, and pedagogy.

Networking and Collaborative Platforms

- **Professional Associations**: Membership in CAM and integrative health organizations offers networking opportunities, access to conferences, and collaborative research possibilities.
- **Online Forums and Social Media Groups**: Spaces for CAM educators to share experiences, teaching materials, and to discuss challenges and solutions in CAM education.

Multimedia and Digital Resources

- **Instructional Videos and Demonstrations**: Visual aids that can be integrated into lectures or online learning platforms to demonstrate CAM techniques and procedures.
- **Virtual Reality (VR) and Simulation Tools**: Advanced tools for creating immersive learning experiences, especially useful for clinical skills training in areas like acupuncture, massage therapy, and physical assessments.

Clinical Practice and Case Studies

- **Case Study Libraries**: Collections of real-world cases that can be used for teaching clinical decision-making, diagnostic skills, and treatment planning in CAM.
- **Clinical Guidelines and Protocols**: Access to up-to-date clinical guidelines that inform best practices in CAM treatments and patient care.

Patient Education Materials

- **Health Literacy Resources**: Materials designed to enhance patient education and communication skills, crucial for preparing students to effectively engage with patients about CAM therapies.

Ethical and Legal Resources

- **Ethics Guidelines**: Resources on the ethical considerations in CAM practice and research, including informed consent, cultural competence, and professional boundaries.
- **Regulatory Updates**: Information on licensure, certification, and legal regulations affecting CAM practices in various jurisdictions.

By leveraging these resources, CAM educators can enhance their teaching, contribute to the evidence base of CAM practices, and prepare students for successful careers in healthcare. These resources not only support the academic and professional development of educators but also ensure that CAM education remains dynamic, evidence-based, and aligned with the evolving landscape of healthcare.

EDUCATIONAL RESOURCES FOR COMPLEMENTARY AND ALTERNATIVE MEDICINE (CAM) RESEARCHERS

CAM researchers play a pivotal role in exploring the efficacy, safety, and mechanisms of CAM therapies, contributing to the body of evidence that supports or refutes their use in healthcare. For these researchers, a robust array of educational and research resources is essential to facilitate high-quality, innovative studies. Below is an overview of key resources beneficial for CAM researchers.

SCIENTIFIC DATABASES AND LIBRARIES

- **PubMed**: Offers access to a vast collection of biomedical literature, including CAM-related research articles, reviews, and clinical trials.
- **Cochrane Library**: A resource for systematic reviews and meta-analyses on healthcare interventions, including CAM practices, providing evidence of their effectiveness.
- **ScienceDirect and SpringerLink**: These platforms provide access to a wide range of journals and books covering various aspects of CAM research.
- **CAM-Specific Research Databases**: Databases such as CAM on PubMed and the National Center for Complementary and Integrative Health (NCCIH) database offer specialized resources focused on CAM research.

RESEARCH NETWORKS AND CONSORTIA

- **International Research Networks**: Joining networks like the International Society for Complementary Medicine Research (ISCMR) can provide opportunities for collaboration, access to exclusive research findings, and partnerships.
- **National and Regional CAM Research Groups**: These groups offer platforms for sharing research methodologies, funding opportunities, and insights into specific CAM modalities.

FUNDING SOURCES AND GRANT OPPORTUNITIES

- **Government Health Agencies**: Agencies such as the NCCIH in the United States provide funding for CAM research projects, including exploratory studies, clinical trials, and interdisciplinary research.
- **Private Foundations and Non-Profit Organizations**: Foundations focused on health and wellness, as well as those with a specific interest in CAM, often offer grants and scholarships for CAM research.

STATISTICAL ANALYSIS AND RESEARCH METHODOLOGY RESOURCES

- **Statistical Software Tutorials**: Learning resources for software like SPSS, SAS, and R, which are crucial for analyzing research data.
- **Methodology Workshops and Online Courses**: These resources can enhance skills in research design, data collection, and analysis specific to CAM studies.

PROFESSIONAL DEVELOPMENT AND CONTINUING EDUCATION

- **Workshops and Conferences**: Attending CAM research conferences and workshops can provide updates on the latest research methods, findings, and trends in CAM.
- **Online Learning Platforms**: Platforms like Coursera, edX, and others offer courses on research methodology, biostatistics, and specific CAM modalities.

ETHICAL AND REGULATORY GUIDANCE

- **Institutional Review Board (IRB) Guidelines**: Resources on ethical considerations in research, particularly for clinical trials involving human participants.
- **Regulatory Agencies**: Information from agencies like the FDA or EMA on the approval process for new CAM therapies and products.

Collaborative Tools and Software

- **Research Collaboration Platforms**: Tools like ResearchGate and Academia.edu allow researchers to share their work, collaborate with others, and find partners for joint studies.
- **Data Sharing and Management Platforms**: Services such as Figshare and Dryad provide platforms for storing, sharing, and managing research data in accordance with open science principles.

Academic Writing and Publication Resources

- **Guidelines for Authors**: Resources from leading CAM journals on preparing and submitting manuscripts for publication.
- **Writing Workshops and Software**: Tools and courses to assist in academic writing, literature review, and reference management, such as EndNote or Zotero.

By utilizing these resources, CAM researchers can advance their investigations, contribute to the evidence base of CAM practices, and engage in meaningful scientific discourse. These resources not only support the research process but also foster professional growth and development in the CAM research community.

THE ROLE OF TECHNOLOGY IN COMPLEMENTARY AND ALTERNATIVE MEDICINE (CAM) LIBRARIES AND EDUCATIONAL RESOURCES

The integration of technology in CAM libraries and educational resources has significantly transformed how information is accessed, learned, and applied in the field of CAM. Technology plays a multifaceted role, enhancing the accessibility, quality, and effectiveness of educational and informational resources for students, educators, practitioners, and researchers alike. Here's an overview of the pivotal role technology plays in CAM libraries and educational resources.

Digital Libraries and Databases

Technology has enabled the creation of digital libraries and databases that offer instant access to a vast array of CAM resources, including journals, books, research articles, and clinical guidelines. These resources are crucial for evidence-based practice and research in CAM, allowing users to easily find and retrieve up-to-date information on various therapies, their effectiveness, and safety profiles.

E-Learning Platforms and Online Courses

The advent of e-learning platforms has revolutionized CAM education, providing flexible learning opportunities that can reach a wider audience. Online courses, webinars, and MOOCs (Massive Open Online Courses) make it possible for students and practitioners to learn about different CAM modalities, research methodologies, and clinical practices from anywhere in the world, often with the opportunity to earn certification or continuing education credits.

Virtual Reality (VR) and Augmented Reality (AR)

VR and AR technologies are being increasingly used in CAM education to provide immersive learning experiences. These technologies can simulate clinical scenarios, allowing students to practice skills such as acupuncture, massage techniques, or yoga poses in a virtual environment. This hands-on approach enhances learning by enabling students to visualize and practice procedures in a safe and controlled setting.

MOBILE APPLICATIONS

Mobile apps related to CAM provide convenient access to information, tools for practice, and patient education resources. From herbal medicine databases and yoga tutorials to meditation and mindfulness apps, these tools support both self-directed learning and the integration of CAM practices into daily life and clinical care.

ONLINE FORUMS AND SOCIAL MEDIA

Technology facilitates community building and professional networking through online forums, social media groups, and platforms dedicated to CAM. These spaces allow students, educators, practitioners, and researchers to share insights, ask questions, discuss the latest research findings, and collaborate on projects or ideas.

TELEHEALTH AND REMOTE CONSULTATIONS

The use of telehealth technology has expanded the reach of CAM practitioners, enabling them to offer consultations and certain types of therapy remotely. This is particularly beneficial for patients in remote or underserved areas, improving access to CAM services and supporting patient education and self-care practices.

RESEARCH AND DATA ANALYSIS TOOLS

Advanced software and analytical tools support CAM research by enabling sophisticated data collection, analysis, and interpretation. Technology facilitates the handling of large datasets, complex statistical analyses, and the visualization of research findings, contributing to the rigor and quality of CAM research.

INTERACTIVE PATIENT EDUCATION TOOLS

Interactive websites, apps, and digital platforms provide patients with accessible information on CAM therapies, helping them to make informed decisions about their health care. These tools often include features like symptom checkers, therapy comparison charts, and personalized wellness plans.

PERSONALIZED LEARNING ENVIRONMENTS

Adaptive learning technologies and AI-driven platforms can tailor educational content to meet the individual needs of students. By analyzing user responses and progress, these platforms adjust the difficulty level and types of resources presented, ensuring that each student receives a personalized learning experience. This approach can be particularly beneficial in CAM education, where the breadth of modalities and therapies requires a nuanced understanding tailored to each learner's background and interests.

BIG DATA AND PREDICTIVE ANALYTICS

The utilization of big data and predictive analytics in CAM research is growing, offering insights into patterns, trends, and outcomes associated with various therapies. By analyzing large datasets from electronic health records, social media, and other digital platforms, researchers can uncover valuable information about the effectiveness of CAM practices, patient satisfaction, and potential areas for integration into conventional care.

Digital Consent and Ethical Considerations

Technology also addresses ethical considerations in CAM practice and research, particularly regarding patient consent and data privacy. Digital platforms can facilitate the process of obtaining informed consent for treatments or participation in research studies, ensuring transparency and ease of access to information for patients. Moreover, blockchain and other secure data management technologies can protect patient data, addressing privacy concerns associated with electronic health records and telehealth consultations.

Wearable Technologies and Self-Monitoring Tools

Wearable devices and self-monitoring tools empower patients to take an active role in their health and wellness. These technologies can track physical activity, sleep patterns, heart rate, and other health indicators, providing data that can inform CAM practices such as fitness, nutrition, and stress management. Integrating this data into CAM education and practice supports a more holistic view of health, emphasizing the importance of lifestyle factors in overall wellbeing.

Simulation Software for Research and Education

Advanced simulation software enables researchers and educators to model complex biological systems and health conditions, facilitating a deeper understanding of how CAM therapies can impact health. These simulations can explore the mechanisms of action of various therapies at the cellular or systemic level, offering insights that are difficult to obtain through traditional research methods.

Online Patient Communities

Technology fosters the creation of online patient communities where individuals can share their experiences with CAM therapies, seek advice, and provide support to one another. These communities can be invaluable resources for both CAM practitioners and researchers, offering real-world insights into patient needs, experiences, and outcomes associated with CAM therapies.

Continuing Professional Development

E-learning platforms and digital libraries provide ongoing professional development opportunities for CAM practitioners. Through online courses, workshops, and webinars, practitioners can stay updated on the latest research, clinical practices, and regulatory changes in the field of CAM, ensuring that their knowledge and skills remain current.

As technology continues to evolve, its integration into CAM libraries and educational resources will likely unveil new possibilities for enhancing education, research, and practice in complementary and alternative medicine. The ongoing challenge and opportunity lie in harnessing these technological advancements to improve health outcomes, patient satisfaction, and the integration of CAM into holistic healthcare models.

In conclusion, technology plays an essential role in advancing CAM education and practice by enhancing access to information, supporting innovative teaching and learning methods, facilitating research, and improving patient care. As technology continues to evolve, its integration into CAM libraries and educational resources will undoubtedly deepen, offering new opportunities for growth and development in the field.

COMPLEMENTARY AND ALTERNATIVE MEDICINE (CAM) EDUCATION LIBRARIES AND EDUCATIONAL RESOURCES: FACILITATORS

In the realm of CAM education, libraries and educational resources serve as crucial facilitators for a range of activities that support learning, teaching, research, and clinical practice. These resources not only provide access to information but also enable the CAM community to engage with the material in innovative and effective ways. Below are several key roles these resources play as facilitators in CAM education:

1. **Enhancing Access to Information**
   - *Digital Libraries and Databases*: Provide comprehensive access to a wide array of resources, including peer-reviewed articles, books, and clinical guidelines, facilitating research and evidence-based practice in CAM.
   - *Open Access Resources*: Enable free access to scholarly articles and textbooks, making it easier for students and practitioners in locations with limited resources to obtain valuable information.
2. **Supporting Curriculum Development and Instruction**
   - *Curricular Materials*: Specialized textbooks and multimedia resources aid in the development of CAM curricula that are both comprehensive and aligned with current industry standards.
   - *Online Learning Platforms*: Offer flexible teaching and learning modalities, including online courses, webinars, and virtual labs, accommodating diverse learning styles and schedules.
3. **Fostering Research and Evidence-Based Practice**
   - *Research Databases*: Facilitate access to current and historical research, supporting the integration of evidence-based practices into CAM education and clinical care.
   - *Statistical Analysis Tools*: Provide researchers with the means to analyze data effectively, contributing to the development of a robust evidence base for CAM therapies.
4. **Promoting Professional Development and Lifelong Learning**
   - *Continuing Education Resources*: Online courses, workshops, and seminars offer opportunities for professionals to update their knowledge and skills in response to evolving practices and new research findings.
   - *Certification and Training Programs*: Enable practitioners to specialize in specific CAM modalities, enhancing their expertise and career opportunities.
5. **Facilitating Clinical Skills Development**
   - *Simulation Tools and Virtual Reality*: Allow students to practice and refine their clinical skills in a safe, controlled environment, preparing them for real-world patient interactions.
   - *Clinical Case Studies*: Provide practical examples and scenarios that help students and practitioners understand the application of CAM therapies in various clinical contexts.
6. **Encouraging Interdisciplinary Collaboration**
   - *Collaborative Platforms and Forums*: Online forums, social media groups, and professional networks promote dialogue and collaboration among CAM practitioners, educators, and researchers from diverse disciplines.
   - *Conferences and Seminars*: Serve as venues for sharing research, teaching strategies, and clinical practices, fostering an interdisciplinary approach to health care.
7. **Supporting Patient Education and Engagement**
   - *Patient Education Materials*: Libraries and online resources offer pamphlets, videos, and interactive tools that practitioners can use to educate patients about CAM therapies and their potential benefits and risks.

- *Health Apps and Wearables*: Provide patients with tools to actively participate in their health and wellness, encouraging engagement and self-management.

8. **Advancing Global and Cultural Perspectives**
   - *Cultural Competence Resources*: Materials that offer insights into how CAM practices vary across cultures, enhancing the cultural competence of practitioners and the inclusivity of care.
   - *International Research and Collaboration Tools*: Facilitate access to global research findings and collaborative opportunities, broadening the scope and impact of CAM education and practice.

CAM education libraries and educational resources play a multifaceted role as facilitators in the ecosystem of CAM learning and practice. By providing access to a wealth of information, supporting innovative educational practices, and fostering a community of collaboration and continuous learning, these resources are indispensable in advancing the field of complementary and alternative medicine.

COMPLEMENTARY AND ALTERNATIVE MEDICINE (CAM) EDUCATION LIBRARIES AND EDUCATIONAL RESOURCES: BARRIERS

While CAM education libraries and educational resources serve as vital facilitators in the learning and practice of CAM, they also face several barriers that can hinder their effectiveness and accessibility. These challenges can affect students, educators, practitioners, and researchers alike, impacting the overall quality and integration of CAM into healthcare. Understanding these barriers is crucial for developing strategies to overcome them and enhance the availability and utility of CAM resources.

1. **Limited Funding and Resources**
   - *Financial Constraints*: Many CAM institutions and libraries may face financial constraints, limiting their ability to subscribe to journals, purchase books, and invest in advanced technological tools for education and research.
   - *Resource Allocation*: Limited funding can also affect the development and maintenance of digital resources and platforms, restricting access to up-to-date and comprehensive CAM information.

2. **Variability in Quality and Accreditation**
   - *Inconsistent Quality*: There is a wide variability in the quality of CAM educational materials and resources, making it challenging for users to discern reliable and evidence-based information from anecdotal or non-scientific content.
   - *Accreditation Standards*: The lack of uniform accreditation standards for CAM educational programs and resources can contribute to inconsistencies in the quality of education and training provided.

3. **Access and Equity Issues**
   - *Digital Divide*: Not all students and practitioners have equal access to digital resources due to geographical, socioeconomic, or institutional barriers, leading to disparities in education and practice opportunities.
   - *Cultural and Language Barriers*: Limited availability of CAM resources in languages other than English or that are culturally insensitive can hinder learning and practice in diverse communities.

4. **Regulatory and Legal Restrictions**
   - *Regulatory Hurdles*: Legal and regulatory constraints can limit the scope of CAM research and the dissemination of information, particularly regarding new or emerging therapies.

- *Intellectual Property Issues*: Copyright and intellectual property laws may restrict the sharing and use of educational materials, complicating the development of open-access resources.

5. **Challenges in Research and Evidence Base**
   - *Research Funding*: CAM research often struggles to secure funding from traditional sources, impacting the ability to conduct high-quality studies and expand the evidence base.
   - *Skepticism toward CAM*: Skepticism from within the conventional medical community can create barriers to the integration of CAM resources and practices in mainstream healthcare education and practice.

6. **Technological Limitations**
   - *Technology Adoption*: The rapid pace of technological change can outstrip the ability of institutions to adopt and integrate new tools, leading to outdated resources and methods.
   - *User Skills and Training*: There may be a gap in digital literacy and technical skills among students, educators, and practitioners, limiting the effective use of digital CAM resources.

7. **Interdisciplinary Collaboration Challenges**
   - *Cultural Differences*: Differences in terminology, practices, and philosophies between CAM and conventional medicine can hinder effective interdisciplinary collaboration and integration.
   - *Institutional Silos*: Academic and healthcare institutions often operate in silos, impeding the sharing of resources and knowledge between CAM and other healthcare disciplines.

Overcoming these barriers requires a multifaceted approach, including increased investment in CAM libraries and resources, the development of clear accreditation standards, efforts to enhance digital access and literacy, and fostering a culture of openness and collaboration between CAM and conventional medicine. Addressing these challenges is essential for ensuring that CAM education and resources are accessible, reliable, and integrated into the broader healthcare landscape.

COMPLEMENTARY AND ALTERNATIVE MEDICINE (CAM) EDUCATION LIBRARIES AND EDUCATIONAL RESOURCES: ACHIEVEMENTS

The field of CAM has seen significant achievements in the development and expansion of educational libraries and resources over the years. These successes have contributed to the growth and integration of CAM into mainstream healthcare education and practice. The achievements in CAM education libraries and resources reflect a broader acceptance and recognition of the value of alternative and integrative health practices. Below are key achievements in this area:

1. **Expansion of Digital Libraries and Databases**
   The creation and expansion of digital libraries and specialized databases have greatly improved access to CAM research and educational materials. Platforms such as the National Center for Complementary and Integrative Health (NCCIH) and PubMed Central have curated vast collections of CAM-related content, making it easier for students, educators, and practitioners to find reliable information and research on various CAM modalities.

2. **Growth of Online Education and Training**
   There has been a significant increase in online CAM education and training programs, including degree programs, certifications, and continuing education courses. These programs offer flexibility and accessibility, enabling learners from diverse backgrounds and geographical locations to pursue CAM education and professional development.

3. **Advancements in Multimedia and Interactive Learning Tools**

   The use of multimedia resources, virtual reality (VR), and augmented reality (AR) in CAM education has enhanced the learning experience by providing interactive and immersive learning tools. These technologies have been particularly effective in teaching anatomy, herbal medicine compounding, and clinical skills such as acupuncture and massage therapy.

4. **Increased Interdisciplinary Collaboration**

   There has been a notable increase in interdisciplinary collaboration between CAM and conventional medical fields. This has led to the development of integrated curricula and joint research projects, fostering a holistic approach to healthcare education and practice. Such collaborations have also contributed to the growing acceptance and inclusion of CAM in mainstream healthcare systems.

5. **Enhanced Access to CAM Research Funding**

   Although CAM research has historically faced funding challenges, there has been progress in securing research grants from both governmental and private sources. This increased funding support has enabled more rigorous research into CAM therapies, contributing to the evidence base and informing clinical practice.

6. **Development of Professional Standards and Accreditation**

   The establishment of professional standards and accreditation processes for CAM educational programs and institutions has been a significant achievement. These standards ensure the quality and credibility of CAM education, promoting excellence in training and practice.

7. **Globalization of CAM Education**

   CAM education and resources have become increasingly globalized, with educational materials and courses available in multiple languages and cultural contexts. This globalization has facilitated the exchange of knowledge and practices across borders, enriching the field of CAM with diverse perspectives and approaches.

8. **Patient Education and Empowerment**

   The development of patient-centered CAM resources, including educational websites, apps, and online communities, has empowered patients to take an active role in their healthcare. These resources provide accessible information on CAM therapies, enabling patients to make informed decisions in collaboration with their healthcare providers.

9. **Recognition of CAM in Healthcare Policies**

   There has been a growing recognition of CAM within healthcare policies and guidelines, reflecting its integration into healthcare systems. This recognition is partly due to the efforts of educators and practitioners to demonstrate the efficacy and safety of CAM practices through education and research.

These achievements highlight the dynamic growth and evolving landscape of CAM education and resources. By continuing to build on these successes, the field of CAM can further its integration into healthcare education, research, and practice, ultimately enhancing patient care and promoting a more holistic approach to health and wellness.

COMPLEMENTARY AND ALTERNATIVE MEDICINE (CAM) EDUCATION LIBRARIES AND EDUCATIONAL RESOURCES: CHALLENGES

While the field of CAM has seen significant growth and integration into mainstream healthcare, CAM education libraries and educational resources face a number of challenges. These challenges can impact the quality, accessibility, and effectiveness of CAM education and practice. Addressing these issues is crucial for the continued development and acceptance of CAM within the broader healthcare landscape. Key challenges include:

1. **Lack of Standardization across CAM Educational Content**
   - *Variability in Quality*: There's a wide range of quality in CAM educational materials, with some resources lacking evidence-based backing or rigorous scientific validation. This variability can lead to misinformation and inconsistency in CAM education and practice.
   - *Diverse Modalities*: The vast array of CAM practices, each with its own historical, cultural, and theoretical background, complicates efforts to standardize educational content and outcomes.

2. **Limited Research and Evidence Base**
   - *Funding for Research*: CAM research often receives less funding than conventional medicine, limiting the ability to conduct large-scale, high-quality studies.
   - *Evidence Quality*: Many CAM modalities have been under-researched, leading to a scarcity of high-quality evidence to support their efficacy and inform educational content.

3. **Integration with Conventional Healthcare Education**
   - *Curricular Inclusion*: Integrating CAM education into conventional healthcare curricula can be challenging due to scheduling constraints, varying levels of acceptance among faculty, and the already crowded curriculum of medical and health sciences education.
   - *Interdisciplinary Understanding*: There is often a lack of understanding and acceptance between practitioners of conventional medicine and CAM, which can hinder collaborative learning and integrative care practices.

4. **Regulatory and Accreditation Challenges**
   - *Licensing and Regulation*: The regulation of CAM practices and education varies widely between and within countries, affecting the standardization of training and practice.
   - *Accreditation*: There's a need for more comprehensive accreditation standards for CAM educational programs to ensure consistency in the quality of education across institutions.

5. **Access to Quality Resources**
   - *Financial Constraints*: CAM libraries and educational institutions may face financial limitations that restrict access to journals, databases, and other resources necessary for a comprehensive education.
   - *Digital Divide*: Not all students and practitioners have equal access to digital educational resources, especially in low-resource settings or among populations with limited technological literacy.

6. **Cultural and Ethical Considerations**
   - *Cultural Sensitivity*: CAM practices often have cultural origins and significance. Educational resources must navigate these aspects sensitively to avoid cultural appropriation or misinterpretation.
   - *Ethical Use*: Ensuring that CAM is practiced and taught ethically, particularly in areas involving endangered species or traditional knowledge, is a challenge that requires careful consideration and guidance.

7. **Keeping Pace with Technological Advancements**
   - *Rapid Technological Change*: The fast pace of technological advancement presents a challenge for CAM education libraries to stay updated and incorporate the latest digital tools and resources effectively.
   - *Training and Skills*: There is a continuous need for training students, educators, and practitioners in the use of new technologies and digital resources for CAM education and practice.

8. **Public Perception and Legitimacy**
   - *Skepticism*: Despite growing acceptance, CAM still faces skepticism from parts of the medical community and the public, affecting the resources allocated for CAM education and research.
   - *Misinformation*: The proliferation of misinformation about CAM on the internet and social media complicates efforts to provide accurate and reliable CAM education.

9. **Globalization vs. Localization**
   - *Balancing Global Standards and Local Practices*: As CAM education seeks to establish global standards, there's a challenge in preserving the localized, traditional knowledge that forms the basis of many CAM practices. Ensuring that educational resources are both internationally relevant and locally applicable requires careful curation and respect for cultural nuances.
   - *Cross-Cultural Exchange*: While globalization offers opportunities for cross-cultural exchange and learning, it also poses challenges in terms of ensuring the accuracy and cultural sensitivity of translated materials and practices.

10. **Professional Development and Career Pathways**
    - *Clear Career Pathways*: There's often a lack of clear, structured career pathways for CAM practitioners post-graduation. This uncertainty can affect students' decisions to enter CAM fields and may impact the development of professional standards.
    - *Continuing Education*: Keeping pace with evolving practices and new research findings requires ongoing professional development opportunities, which may not be readily available or accessible to all CAM practitioners.

11. **Interprofessional Education and Practice**
    - *Collaborative Skills*: The challenge of fostering effective interprofessional education and collaboration skills among CAM and conventional healthcare students is significant. These skills are crucial for integrated care models but can be difficult to develop in siloed educational systems.
    - *Recognition and Respect*: Achieving mutual recognition and respect between CAM and conventional healthcare professionals is essential for integrated patient care but remains a challenge due to differing paradigms and historical prejudices.

12. **Digital Health Literacy**
    - *Navigating Digital Resources*: As more CAM educational resources become digital, there's a growing need to ensure that both students and practitioners have the digital literacy skills required to effectively navigate these resources.
    - *Critical Evaluation of Online Information*: With the vast amount of CAM information available online, teaching critical evaluation skills to discern reliable from unreliable information is increasingly important.

13. **Sustainability and Environmental Considerations**
    - *Sustainable Practices*: Many CAM modalities rely on natural resources, which raises the challenge of ensuring sustainable practices in both the sourcing of materials and the teaching of these practices.
    - *Environmental Impact*: Addressing the environmental impact of CAM practices, including the carbon footprint of global herb sourcing and the use of animal products, is a growing concern that needs to be integrated into CAM education and practice.

14. **Ethics and Patient Safety**
    - *Informed Consent*: Ensuring that CAM practitioners are trained to obtain informed consent, particularly in modalities that are less understood by the general public, is a challenge.
    - *Safety Standards*: Developing and teaching clear safety standards across the diverse range of CAM modalities, many of which have varying degrees of regulation and oversight, is crucial for patient safety.

15. **Innovative Educational Methods**
   - *Adopting New Pedagogies*: There's a challenge in adopting and integrating innovative educational methods, such as problem-based learning and simulation-based training, into CAM education due to traditional teaching approaches and resource constraints.
   - *Measuring Educational Outcomes*: Developing effective tools and methods for measuring educational outcomes in CAM is challenging but necessary for continuous improvement and accreditation.

Addressing these challenges requires collaborative efforts from educators, practitioners, researchers, policymakers, and other stakeholders in the CAM and conventional healthcare communities. Through targeted initiatives, increased funding for research, and the development of standardized curricula and accreditation processes, the field of CAM can continue to evolve and contribute positively to global healthcare.

COMPLEMENTARY AND ALTERNATIVE MEDICINE (CAM) EDUCATION LIBRARIES AND EDUCATIONAL RESOURCES: RECOMMENDATIONS FOR THE BEST PRACTICES

To enhance the effectiveness, accessibility, and quality of CAM education, libraries, and resources, adopting best practices is crucial. These recommendations aim to address existing challenges and optimize the contribution of CAM to healthcare education and practice (Al-Worafi, 2020a,b, 2022a,b, 2023a–n, 2024a–c; Hasan et al., 2019). Here are key best practices for CAM education libraries and educational resources:

1. **Ensure Access to High-Quality and Evidence-Based Resources**
   - *Curate Collections*: Rigorously select and curate library collections to include peer-reviewed, evidence-based resources on CAM practices, ensuring they are up-to-date and scientifically valid.
   - *Open Access*: Advocate for and contribute to open access publishing in the CAM field to increase the availability of high-quality research and educational materials.
2. **Foster Interdisciplinary Collaboration and Integration**
   - *Interprofessional Education*: Develop and implement interprofessional education programs that include both CAM and conventional medicine, promoting mutual understanding and collaboration.
   - *Cross-Disciplinary Research*: Encourage and facilitate cross-disciplinary research initiatives that explore the integration of CAM with conventional healthcare practices.
3. **Adopt Innovative Educational Technologies**
   - *Utilize Digital Tools*: Incorporate digital tools and technologies, such as e-learning platforms, virtual reality (VR), and simulation-based learning, to enhance the interactivity and accessibility of CAM education.
   - *Mobile Learning*: Develop mobile apps and resources for flexible learning and reference to support students and practitioners in accessing information on-the-go.
4. **Promote Cultural Competence and Global Perspectives**
   - *Cultural Sensitivity Training*: Include cultural competence training in CAM curricula to ensure respectful and appropriate use of traditional practices from various cultures.
   - *Global Exchange Programs*: Establish global exchange and collaboration programs to share knowledge, practices, and research across different cultural and healthcare contexts.

5. **Enhance Research Capacity and Evidence Base**
   - *Research Support*: Provide support for CAM research through funding opportunities, mentorship programs, and access to research databases and tools.
   - *Critical Appraisal Skills*: Teach critical appraisal skills to enable students and practitioners to evaluate the quality of research and the reliability of various CAM resources.
6. **Support Professional Development and Lifelong Learning**
   - *Continuing Education*: Offer continuing education programs and resources for CAM practitioners to keep their knowledge and skills current with the latest research and practices.
   - *Career Guidance*: Provide career counseling and resources to help students navigate the diverse opportunities in CAM fields and develop clear professional pathways.
7. **Prioritize Ethics and Patient Safety**
   - *Ethical Guidelines*: Develop and disseminate clear ethical guidelines for CAM practice and research, emphasizing patient safety, informed consent, and professional integrity.
   - *Safety Education*: Include comprehensive safety education in CAM curricula, covering potential interactions, contraindications, and best practices for patient care.
8. **Implement Quality Standards and Accreditation**
   - *Accreditation Processes*: Support the development and implementation of accreditation processes for CAM educational programs to ensure consistency and quality in education.
   - *Quality Assurance*: Establish quality assurance mechanisms for CAM resources and libraries, including regular reviews and updates of materials.
9. **Engage with the Community and Stakeholders**
   - *Public Education*: Engage in public education initiatives to improve awareness and understanding of CAM practices and their potential benefits and risks.
   - *Stakeholder Collaboration*: Collaborate with healthcare providers, policymakers, and professional organizations to advocate for the integration of CAM into healthcare systems and policies.
10. **Leverage Data and Analytics for Continuous Improvement**
    - *Feedback Mechanisms*: Implement systems for collecting feedback from users of CAM libraries and educational resources to inform continuous improvement.
    - *Analytics*: Use data analytics to monitor the usage and impact of resources, adjusting strategies to meet the evolving needs of students and practitioners.
11. **Implement Sustainable Practices**
    - *Environmental Sustainability*: Integrate principles of sustainability into the teaching and practice of CAM, emphasizing the ethical sourcing of natural products and minimizing environmental impact.
    - *Sustainable Development Goals (SDGs)*: Align CAM education and practices with the United Nations Sustainable Development Goals to promote health and wellness in a way that supports ecological balance and global health equity.
12. **Enhance Digital Health Literacy**
    - *Digital Literacy Training*: Include digital health literacy as a core component of CAM education, preparing students and practitioners to effectively utilize digital resources, engage with telehealth technologies, and navigate the digital healthcare landscape.
    - *Critical Evaluation of Digital Content*: Teach strategies for critically evaluating the credibility and quality of online CAM resources, helping users discern evidence-based information from misinformation.
13. **Support Language Diversity and Accessibility**
    - *Multilingual Resources*: Develop and provide access to CAM educational materials in multiple languages, enhancing accessibility for non-English speakers and supporting cultural diversity.

- *Accessibility Features*: Ensure that digital CAM resources and technologies are designed with accessibility in mind, accommodating users with disabilities through adaptive technologies and inclusive design principles.

14. **Promote Ethical Use of Traditional Knowledge**
    - *Respect for Indigenous Knowledge*: Incorporate ethical guidelines that respect and protect indigenous and traditional knowledge in CAM practices, preventing exploitation and ensuring that such knowledge is used in a way that benefits the original communities.
    - *Collaborative Partnerships*: Foster partnerships with indigenous and local communities to co-create educational materials that accurately and respectfully represent traditional practices, ensuring mutual benefits and knowledge exchange.

15. **Expand Community-Based Learning Opportunities**
    - *Service-Learning*: Incorporate service-learning opportunities into CAM education, allowing students to apply their knowledge in real-world settings, engage with diverse communities, and understand the social determinants of health.
    - *Community Engagement Projects*: Encourage students and practitioners to participate in community engagement projects that address health disparities and promote wellness through CAM practices, fostering a sense of social responsibility.

16. **Leverage Artificial Intelligence and Big Data**
    - *AI in CAM Research*: Utilize artificial intelligence (AI) and machine learning tools to analyze large datasets in CAM research, uncovering patterns and insights that can inform evidence-based practices and personalized medicine approaches.
    - *Predictive Analytics*: Apply predictive analytics to CAM practices to enhance patient care, such as personalizing treatment plans based on patient data and predicting outcomes of CAM interventions.

17. **Encourage Patient Participation in Learning**
    - *Patient-Centered Resources*: Develop patient-centered educational materials that empower individuals to participate actively in their healthcare decisions, fostering a collaborative approach to CAM practices.
    - *Feedback Loops*: Create mechanisms for patient feedback on CAM educational materials and practices, ensuring that resources are responsive to patient needs and experiences.

18. **Strengthen Global Networks and Partnerships**
    - *International Collaboration*: Strengthen international networks and partnerships among CAM educational institutions, facilitating the exchange of knowledge, research collaborations, and shared resources across borders.
    - *Global Health Initiatives*: Engage in global health initiatives that incorporate CAM practices, contributing to international efforts to improve health outcomes and reduce health inequities.

19. **Promote Data Sharing and Open Science**
    - *Open Access Initiatives*: Encourage and participate in open access initiatives to increase the availability of CAM research findings, making them accessible to a wider audience without paywall restrictions.
    - *Data Repositories*: Support the establishment of or contribute to data repositories for CAM research, facilitating data sharing among researchers and enabling meta-analyses and systematic reviews to build a stronger evidence base.

20. **Utilize Mixed-Methods Research**
    - *Embrace Qualitative Research*: Incorporate qualitative research methods to capture the experiential and subjective aspects of CAM therapies, providing depth and context to quantitative findings.

- *Integrated Research Approaches*: Promote integrated research approaches that combine qualitative and quantitative methods, offering a more holistic understanding of CAM practices and their outcomes.

21. **Innovate in Clinical Training and Simulation**
    - *Advanced Simulation Tools*: Invest in and utilize advanced simulation tools that mimic real-life patient interactions and CAM therapy administrations, enhancing clinical training without compromising patient safety.
    - *Telehealth Training*: Include telehealth training in CAM education programs, preparing practitioners for remote delivery of certain CAM services, a practice that has expanded significantly in the healthcare sector.

22. **Strengthen Alumni Networks and Lifelong Learning**
    - *Alumni Engagement*: Develop strong alumni networks that provide ongoing professional support, opportunities for continuing education, and platforms for sharing clinical experiences and insights.
    - *Lifelong Learning Platforms*: Offer accessible lifelong learning platforms for CAM practitioners, featuring updates on research, emerging practices, and professional development opportunities.

23. **Focus on Wellness and Preventative Care Education**
    - *Integrate Wellness Principles*: Embed principles of wellness and preventive care within CAM education, emphasizing the role of CAM in maintaining health, preventing illness, and enhancing quality of life.
    - *Community Wellness Programs*: Engage students and practitioners in community wellness programs that demonstrate the practical application of CAM strategies in promoting public health.

24. **Adopt Adaptive Learning Technologies**
    - *Personalized Learning Paths*: Implement adaptive learning technologies that adjust the educational content based on the learner's progress and understanding, offering personalized learning paths for CAM students.
    - *Feedback-Driven Learning*: Use technology platforms that provide immediate feedback to learners, facilitating a more dynamic and responsive educational experience.

25. **Engage in Policy Advocacy and Healthcare Integration**
    - *Policy Engagement*: Encourage active engagement in healthcare policy discussions and advocacy efforts to integrate CAM practices into mainstream healthcare systems and insurance coverage.
    - *Intersectoral Collaboration*: Foster collaborations between the CAM community, healthcare providers, policymakers, and insurers to develop integrated care models that include CAM as a complementary approach to health and wellness.

26. **Prioritize Ethical Considerations in CAM Practices**
    - *Ethics in Education*: Incorporate ethics as a core component of CAM education, covering topics such as informed consent, patient autonomy, and the ethical use of natural resources.
    - *Ethical Practice Guidelines*: Develop and disseminate ethical practice guidelines for CAM practitioners, ensuring responsible and patient-centered care.

27. **Enhance Community and Patient Involvement**
    - *Community-Based Participatory Research (CBPR)*: Adopt CBPR approaches in CAM research to involve community members in the research process, ensuring that studies are responsive to community needs and interests.
    - *Patient Advisory Boards*: Establish patient advisory boards for CAM programs and clinics, involving patients in decision-making processes and program development to ensure services meet patient needs and preferences.

Adopting these best practices can significantly enhance the role of CAM education libraries and resources in supporting evidence-based, culturally competent, and safe CAM practices. By doing so, the CAM community can better contribute to holistic, patient-centered healthcare and the ongoing integration of CAM into mainstream health systems.

CONCLUSION

The exploration of best practices for CAM education, libraries, and educational resources reveals a multifaceted approach aimed at enhancing the quality, accessibility, and integration of CAM within the broader healthcare landscape. These practices underscore the importance of evidence-based resources, interdisciplinary collaboration, innovative educational technologies, and a commitment to ethical and culturally competent care. Key to advancing CAM education and practice is the emphasis on high-quality, accessible, and evidence-based resources that support the diverse needs of students, practitioners, researchers, and patients. This involves not only the curation of digital libraries and databases but also the adoption of open access policies to widen the dissemination of CAM research findings. Interdisciplinary collaboration stands out as a critical component, bridging the gap between CAM and conventional medicine. By fostering mutual understanding and respect, these efforts can lead to integrated curricula, joint research projects, and ultimately, holistic patient care models that leverage the strengths of both CAM and conventional therapies. The role of technology, particularly through e-learning platforms, virtual and augmented reality, and mobile applications, is pivotal in creating interactive, flexible, and engaging learning environments. These technologies not only enhance the educational experience but also prepare practitioners for the evolving landscape of healthcare delivery, including telehealth and digital health applications. Cultural competence and a global perspective are essential in ensuring that CAM practices are taught and applied in a manner that respects the diversity of traditions and patients they serve. This includes integrating principles of sustainability, ethical use of traditional knowledge, and engagement with indigenous and local communities. The challenges facing CAM education, such as standardization, research funding, regulatory hurdles, and integration with conventional healthcare education, require concerted efforts from all stakeholders. Addressing these challenges will involve enhancing research capacity, adopting mixed-methods research, innovating in clinical training, and engaging in policy advocacy. Ultimately, the recommendations for best practices in CAM education libraries and resources are geared toward fostering a healthcare environment that values holistic, patient-centered care. By implementing these strategies, the CAM community can further its contribution to health and wellness, ensuring that CAM practices are accessible, reliable, and integrated into the broader healthcare system. This holistic approach not only benefits patients by providing diverse treatment options but also enriches the healthcare field with a broad spectrum of knowledge and therapeutic techniques, promoting overall health and wellbeing in society.

REFERENCES

Al-Worafi, Y.M. (Ed.). (2020a). *Drug Safety in Developing Countries: Achievements and Challenges*. Academic Press.

Al-Worafi, Y.M. (2020b). Herbal medicines safety issues. In: Al-Worafi, Y.M. (ed), *Drug Safety in Developing Countries* (pp. 163–178). Academic Press.

Al-Worafi, Y.M. (2022a). *A Guide to Online Pharmacy Education: Teaching Strategies and Assessment Methods*. CRC Press.

Al-Worafi, Y.M. (2022b). Competencies and learning outcomes. In: Al-Worafi, Y.M. (ed), *A Guide to Online Pharmacy Education: Teaching Strategies and Assessment Methods*. CRC Press.

Al-Worafi, Y.M. (2023a). *Patient Safety in Developing Countries: Education, Research, Case Studies*. CRC Press.

Al-Worafi, Y.M. (2023b). *Technology for Drug Safety: Current Status and Future Developments*. Springer Nature.

Al-Worafi, Y.M. (2023c). Patient safety-related issues: Patient care errors and related problems. In: Al-Worafi, Y.M. (ed), *Patient Safety in Developing Countries: Education, Research, Case Studies*. CRC Press.

Al-Worafi, Y.M. (2023d). Patient care errors and related problems: Preventive medicine errors & related problems. In: Al-Worafi, Y.M. (ed), *Patient Safety in Developing Countries: Education, Research, Case Studies*. CRC Press.

Al-Worafi, Y.M. (2023e). Patient care errors and related problems: Patient assessment and diagnostic errors & related problems. In: Al-Worafi, Y.M. (ed), *Patient Safety in Developing Countries: Education, Research, Case Studies*. CRC Press.

Al-Worafi, Y.M. (2023f). Patient care errors and related problems: Non-pharmacological errors & related problems. In: Al-Worafi, Y.M. (ed), *Patient Safety in Developing Countries: Education, Research, Case Studies*. CRC Press.

Al-Worafi, Y.M. (2023g). Patient care errors and related problems: Medical errors & related problems. In: Al-Worafi, Y.M. (ed), *Patient Safety in Developing Countries: Education, Research, Case Studies*. CRC Press.

Al-Worafi, Y.M. (2023h). Patient care errors and related problems: Monitoring errors & related problems. In: Al-Worafi, Y.M. (ed), *Patient Safety in Developing Countries: Education, Research, Case Studies*. CRC Press.

Al-Worafi, Y.M. (2023i). Patient care errors and related problems: Patient education and counselling errors and related problems. In: Al-Worafi, Y.M. (ed), *Patient Safety in Developing Countries: Education, Research, Case Studies*. CRC Press.

Al-Worafi, Y.M. (2023j). Patient safety-related issues: Other medication safety issues. In: Al-Worafi, Y.M. (ed), *Patient Safety in Developing Countries: Education, Research, Case Studies*. CRC Press.

Al-Worafi, Y.M. (2023k). Patient safety culture. In: Al-Worafi, Y.M. (ed), *Patient Safety in Developing Countries: Education, Research, Case Studies*. CRC Press.

Al-Worafi, Y.M. (2023l). Patient safety education: Competencies and learning outcomes. In: Al-Worafi, Y.M. (ed), *Patient Safety in Developing Countries: Education, Research, Case Studies*. CRC Press.

Al-Worafi, Y.M. (Ed.). (2023m). *Clinical Case Studies on Medication Safety*. Academic Press.

Al-Worafi, Y.M. (Ed.). (2023n). *Comprehensive Healthcare Simulation: Pharmacy Education, Practice and Research*. Springer Nature.

Al-Worafi, Y.M. (Ed.). (2024a). *Handbook of Medical and Health Sciences in Developing Countries*. Springer, Cham.

Al-Worafi, Y.M. (2024b). Complementary and alternative medicine (CAM) in developing countries. In: Al-Worafi, Y.M. (ed), *Handbook of Medical and Health Sciences in Developing Countries*. Springer, Cham. https://doi.org/10.1007/978-3-030-74786-2_301-1

Al-Worafi, Y.M. (2024c). Access/equitable access to medical and health sciences in developing countries. In: Al-Worafi, Y.M. (ed), *Handbook of Medical and Health Sciences in Developing Countries*. Springer, Cham. https://doi.org/10.1007/978-3-030-74786-2_147-1

Hasan, S., Al-Omar, M.J., AlZubaidy, H., and Al-Worafi, Y.M. (2019). Use of medications in Arab countries. In: Laher, I. (ed), *Handbook of Healthcare in the Arab World* (p. 42). Springer, Cham.

17 Complementary and Alternative Medicine (CAM) Education
Interprofessional Education

INTRODUCTION

Interprofessional Healthcare Education is an approach to teaching healthcare that emphasizes the importance of collaboration among various healthcare professionals to improve patient outcomes. This educational strategy recognizes that the complex health needs of populations can be more effectively addressed by teams of healthcare providers working together, rather than by professionals working in isolation. It focuses on preparing students from different healthcare disciplines—such as medicine, nursing, pharmacy, and others—to work in coordinated efforts, understanding each other's roles, communicating effectively, and providing care that is holistic and patient-centered. The goal is to foster a healthcare environment where teamwork and collaboration lead to safer, more effective, and more efficient patient care. CAM education integrates the study of non-conventional medical practices with traditional medical curricula. CAM refers to a wide range of practices, products, and therapies that are not typically part of conventional medical care or that may not have been fully integrated into the mainstream healthcare system. Examples include herbal medicine, acupuncture, massage therapy, and mindfulness techniques, among others. CAM education aims to provide healthcare professionals with knowledge about these practices, enabling them to understand the potential benefits, limitations, and the scientific evidence supporting various CAM therapies. The inclusion of CAM in healthcare education reflects a growing recognition of the value of holistic and patient-centered care approaches, acknowledging that patients often use these therapies alongside conventional medical treatments. The integration of Interprofessional Healthcare and CAM Education represents a holistic approach to healthcare education that aims to prepare future healthcare professionals to work collaboratively across disciplines and to incorporate a wide range of therapeutic practices into patient care. This combined approach promotes a more inclusive, patient-centered model of care that acknowledges the importance of diverse medical traditions and practices in achieving optimal health outcomes. It encourages healthcare providers to consider all factors that influence health, wellness, and disease, including physical, psychological, social, and spiritual dimensions, and to work together to provide care that is respectful of and responsive to individual patient preferences, needs, and values.

RATIONALITY AND IMPORTANCE OF INTERPROFESSIONAL HEALTHCARE AND COMPLEMENTARY AND ALTERNATIVE MEDICINE (CAM) EDUCATION

The rationale and importance of integrating Interprofessional Healthcare and CAM education into the training of healthcare professionals are grounded in the evolving needs of healthcare delivery and patient care. This integration is driven by several key factors.

DOI: 10.1201/9781003327202-17

RATIONALITY

1. **Complex Healthcare Needs**: The complexity of patient healthcare needs in the 21st century requires a multidisciplinary approach. Chronic diseases, mental health issues, and the aging population demand collaborative efforts from a variety of healthcare professionals, including those trained in CAM practices, to offer comprehensive care.
2. **Patient-Centered Care**: There is a growing emphasis on patient-centered care, which focuses on respecting and responding to individual patient preferences, needs, and values. Integrating CAM into healthcare education acknowledges that many patients are already using these therapies and seeks to inform healthcare providers about their uses and evidence base, ensuring that patients receive informed comprehensive care.
3. **Evidence-Based CAM Practices**: As research into CAM practices grows, there is an increasing body of evidence supporting the efficacy of certain CAM therapies for specific conditions. Healthcare professionals must be educated about these therapies to make informed decisions regarding patient care.
4. **Cultural Competency**: The global diversity of patients and their healthcare practices necessitates a culturally competent healthcare system. Understanding and integrating CAM practices into healthcare education can improve the cultural competency of healthcare providers, as CAM often includes traditional and indigenous practices.

IMPORTANCE

1. **Improved Patient Outcomes**: Interprofessional collaboration has been shown to improve patient outcomes by reducing medical errors, enhancing patient safety, and increasing patient satisfaction. A healthcare team that understands and utilizes a variety of approaches, including CAM, can offer more personalized and effective treatment plans.
2. **Holistic Healthcare**: Integrating CAM into healthcare education promotes a more holistic approach to patient care that considers not just the physical aspects of health but also the psychological, social, and spiritual dimensions. This comprehensive approach can lead to better health outcomes and an improved quality of life for patients.
3. **Healthcare System Efficiency**: Interprofessional healthcare teams can improve the efficiency of the healthcare system by ensuring that resources are used effectively, reducing duplication of services, and enhancing coordination of care among different providers.
4. **Informed Decision-Making**: Educating healthcare professionals about CAM enables them to guide their patients in making informed decisions about their use of CAM therapies. This is particularly important given the vast amount of unverified information available to patients on the internet and through other sources.
5. **Addressing Healthcare Disparities**: By embracing a range of conventional and alternative approaches, healthcare education can help address disparities in healthcare access and outcomes. It acknowledges the value of diverse healing traditions and practices, potentially leading to more equitable healthcare delivery.

In summary, the integration of Interprofessional Healthcare and CAM Education is a response to the need for a more collaborative, holistic, and patient-centered approach to healthcare. It aims to prepare healthcare professionals to work effectively in multidisciplinary teams and to incorporate a wide range of evidence-based practices into patient care, ultimately improving health outcomes and patient satisfaction.

INTERPROFESSIONAL HEALTHCARE AND COMPLEMENTARY AND ALTERNATIVE MEDICINE (CAM) EDUCATION: FACILITATORS

Facilitating the integration of Interprofessional Healthcare and CAM education into the healthcare curriculum involves overcoming various barriers and leveraging multiple facilitators. These facilitators are essential for creating an educational environment that supports the development of healthcare professionals capable of delivering comprehensive, patient-centered care. Here are some key facilitators.

INSTITUTIONAL SUPPORT AND LEADERSHIP

- **Commitment from Educational Institutions**: Universities, colleges, and healthcare training institutions must prioritize the integration of interprofessional and CAM education into their curricula. This includes providing resources, time, and space for the development and implementation of these programs.
- **Leadership Advocacy**: Strong leadership and advocacy from within healthcare and educational institutions can drive the change needed for the integration of these approaches. Leaders can facilitate collaboration between different departments and professionals, fostering an environment that supports interprofessional learning and the inclusion of CAM.

CURRICULUM DEVELOPMENT AND INTEGRATION

- **Innovative Curriculum Design**: Developing curricula that naturally integrate interprofessional education (IPE) and CAM within existing healthcare training programs. This might involve case-based learning, simulation exercises, and clinical rotations that expose students to collaborative practice and CAM therapies.
- **Accreditation Standards**: Accrediting bodies for healthcare education programs can facilitate integration by incorporating requirements for IPE and CAM education into their standards, ensuring that institutions adopt these approaches to meet accreditation criteria.

INTERDISCIPLINARY COLLABORATION

- **Cross-Disciplinary Partnerships**: Building partnerships between different healthcare disciplines, as well as between conventional and CAM practitioners, can enhance educational programs. These partnerships can provide students with exposure to a range of perspectives and practices.
- **Professional Development Opportunities**: Offering professional development and continuing education in IPE and CAM for faculty and healthcare providers can help create a cadre of educators and practitioners who are well-versed in these areas.

RESEARCH AND EVIDENCE BASE

- **Research on IPE and CAM**: Conducting and disseminating research on the effectiveness of interprofessional education and CAM practices can provide an evidence base that supports their integration into healthcare education. This includes studies on patient outcomes, healthcare efficiency, and educational methodologies.
- **Inclusion of Evidence-Based CAM**: Educating healthcare students and professionals about the evidence base for various CAM practices can facilitate their acceptance and integration into patient care. This requires a critical evaluation of CAM therapies to distinguish between those supported by evidence and those lacking scientific validation.

CULTURAL AND ORGANIZATIONAL CHANGE

- **Promoting Cultural Competence**: Training that emphasizes cultural competence and the value of diverse healthcare traditions can facilitate the inclusion of CAM in healthcare education. This approach respects and integrates various cultural perspectives on health and healing.
- **Organizational Change Management**: Implementing change management strategies within healthcare and educational institutions to support the shift toward interprofessional and CAM education. This involves addressing resistance to change and fostering an organizational culture that values collaboration and holistic care.

TECHNOLOGY AND RESOURCES

- **Leveraging Technology**: Using technology, such as online platforms, simulations, and virtual reality, can facilitate the teaching of IPE and CAM concepts. These tools can help overcome logistical challenges associated with bringing students from different disciplines together and can provide access to CAM learning resources.
- **Access to CAM Resources and Experts**: Providing access to CAM practitioners, clinics, and resources as part of the educational experience allows students to gain direct exposure to CAM practices and the role they can play in integrated healthcare.

Facilitating the integration of Interprofessional Healthcare and CAM Education requires a multifaceted approach that addresses curriculum development, institutional support, interdisciplinary collaboration, and the promotion of a culture that values holistic, patient-centered care.

INTERPROFESSIONAL HEALTHCARE AND COMPLEMENTARY AND ALTERNATIVE MEDICINE (CAM) EDUCATION: BARRIERS

While the integration of Interprofessional Healthcare and Complementary and Alternative Medicine (CAM) Education into healthcare training programs holds significant promise for enhancing patient care, several barriers can impede its implementation. Overcoming these obstacles is crucial for the successful incorporation of these educational components. Here are some of the key barriers.

INSTITUTIONAL AND CULTURAL BARRIERS

1. **Resistance to Change**: Traditional healthcare education and practice often operate in silos, with each discipline working independently. There can be significant resistance among faculty and practitioners to adopting interprofessional and integrative approaches, partly due to deeply ingrained professional identities and historical prejudices against CAM practices.
2. **Lack of Institutional Support**: Adequate funding, resources, and administrative support are essential for developing and implementing interprofessional and CAM curricula. Without institutional commitment, these programs may struggle to get off the ground.
3. **Curricular Overload**: Healthcare education programs are already densely packed with required courses and competencies. Finding space in the curriculum for additional content on interprofessional education and CAM can be challenging.

EDUCATIONAL AND TRAINING BARRIERS

1. **Lack of Qualified Educators**: There is a shortage of educators who are adequately prepared to teach both interprofessional competencies and CAM therapies. This gap can be due to a lack of training opportunities or incentives for faculty development in these areas.

2. **Inconsistent Standards and Accreditation**: The variability in standards for CAM education and practice, both within and across countries, complicates its integration into mainstream healthcare education. Similarly, accreditation standards for interprofessional education can be inconsistent or lacking, making it difficult to ensure quality and uniformity in educational outcomes.

3. **Diverse Backgrounds and Languages**: Interprofessional education requires bringing together students from diverse healthcare disciplines, who often have different levels of education, terminologies, and professional cultures. This diversity can create communication barriers and hinder effective teamwork.

CLINICAL AND PRACTICAL BARRIERS

1. **Limited Access to Interprofessional Clinical Sites**: Providing students with real-world experiences in interprofessional and integrative healthcare settings is critical. However, there may be a limited number of clinical sites that offer such experiences, particularly for CAM practices that are not widely integrated into mainstream healthcare.

2. **Professional Scope of Practice Issues**: Legal and regulatory frameworks defining the scope of practice for various healthcare professionals can limit their ability to practice interprofessional care or integrate CAM therapies into patient care plans.

RESEARCH AND EVIDENCE-BASED PRACTICE BARRIERS

1. **Lack of Robust Evidence for CAM**: Although research on CAM is growing, the evidence base for many CAM therapies remains limited or contested. This can lead to skepticism among healthcare professionals and educators about incorporating CAM into the curriculum.

2. **Challenges in Researching Interprofessional Education Outcomes**: Measuring the impact of interprofessional education on healthcare outcomes can be complex and challenging, making it difficult to build a strong evidence base to support its implementation.

SOCIETAL AND PATIENT-CENTERED BARRIERS

1. **Patient Awareness and Acceptance**: Patients' lack of awareness or misconceptions about CAM therapies and interprofessional care models can influence their acceptance and utilization of these approaches.

2. **Cultural and Societal Attitudes**: Cultural and societal attitudes toward CAM and interprofessional care can vary widely, affecting the willingness of both patients and providers to embrace these approaches.

Overcoming these barriers requires concerted efforts from educational institutions, healthcare organizations, accrediting bodies, and policymakers. Strategies might include curriculum reform, faculty development, research initiatives, and policy changes to support the integration of Interprofessional Healthcare and CAM Education into healthcare training and practice.

INTERPROFESSIONAL HEALTHCARE AND COMPLEMENTARY AND ALTERNATIVE MEDICINE (CAM) EDUCATION: ACHIEVEMENTS

The integration of interprofessional healthcare and CAM education into healthcare training and practice has seen various achievements despite the challenges and barriers. These successes demonstrate progress in moving toward a more holistic, patient-centered approach to healthcare. Here are some notable achievements in this area.

Enhanced Interprofessional Collaboration

- **Improved Teamwork and Communication**: Programs focusing on interprofessional education have successfully fostered better teamwork and communication among healthcare professionals. This improvement is crucial for patient safety and effective care coordination, leading to enhanced patient outcomes.
- **Development of Interprofessional Education Frameworks**: The creation and implementation of frameworks and models for interprofessional education, such as the Interprofessional Education Collaborative (IPEC) competencies in the United States, have provided a structured approach to integrating these principles into healthcare education.

Increased Recognition and Integration of CAM

- **Inclusion in National Healthcare Policies**: Several countries have recognized the importance of CAM by incorporating it into national healthcare policies and insurance schemes, acknowledging its role in maintaining health and treating illness.
- **Growth in Research and Evidence Base**: There has been a significant increase in research focused on CAM, leading to a stronger evidence base for certain therapies. This research has facilitated the integration of evidence-based CAM practices into conventional healthcare settings.

Curriculum Development and Reform

- **Incorporation into Healthcare Curricula**: Many medical, nursing, pharmacy, and other health profession schools have incorporated CAM and interprofessional education components into their curricula. This inclusion prepares future healthcare professionals to work collaboratively and to understand and respect a wide range of therapeutic approaches.
- **Interdisciplinary Programs and Degrees**: The development of interdisciplinary degree programs and certificates in integrative health and wellness is an achievement that highlights the growing recognition of the importance of both CAM and interprofessional education in healthcare.

Professional Development and Continuing Education

- **Expansion of Continuing Education Opportunities**: There has been an expansion of continuing education opportunities in both interprofessional collaboration and CAM for practicing healthcare professionals. These programs help update the knowledge and skills of the current workforce to include these important areas.

Patient-Centered Care Initiatives

- **Promotion of Holistic, Patient-Centered Care**: The integration of interprofessional and CAM education has played a role in the broader movement toward holistic, patient-centered care. This approach considers the physical, emotional, and spiritual well-being of patients, aligning with the principles of both interprofessional care and CAM.

Policy and Accreditation Changes

- **Accreditation Requirements**: Accreditation bodies for healthcare education in some regions have begun to require interprofessional education components, ensuring that new healthcare professionals are prepared to work effectively in team-based care settings.

- **Support for Integrative Medicine Practices**: Policies and accreditation standards have increasingly recognized and supported integrative medicine practices, which combine conventional medical treatments with CAM approaches in a coordinated way.

PUBLIC AWARENESS AND ACCEPTANCE

- **Increased Public Interest and Utilization**: There has been a notable increase in public interest in and utilization of CAM therapies, partly due to greater awareness and understanding of these options through education and advocacy. This shift reflects a broader acceptance of diverse approaches to health and wellness.

These achievements represent significant steps forward in the integration of Interprofessional Healthcare and CAM Education into the healthcare system. However, continued efforts are needed to address ongoing challenges, expand these initiatives, and fully realize the potential benefits for patient care and the healthcare system as a whole.

INTERPROFESSIONAL HEALTHCARE AND COMPLEMENTARY AND ALTERNATIVE MEDICINE (CAM) EDUCATION: CHALLENGES

The integration of interprofessional healthcare and CAM education into healthcare training and practice faces several challenges. These challenges stem from systemic, cultural, educational, and logistical issues that can impede the successful adoption and implementation of these innovative approaches to healthcare education. Addressing these challenges is essential for realizing the full potential of interprofessional and CAM education in improving healthcare outcomes and patient care. Key challenges include.

SYSTEMIC AND INSTITUTIONAL CHALLENGES

- **Resource Allocation**: Limited financial and human resources can hinder the development and implementation of interprofessional and CAM education programs. Institutions may struggle to allocate the necessary funds and faculty time for these initiatives.
- **Curricular Space**: Integrating new content into already crowded healthcare curricula is challenging. Finding space for interprofessional and CAM education requires reevaluating and possibly restructuring existing courses and content, which can meet resistance.
- **Institutional Resistance**: There can be institutional inertia and resistance to change, especially when it involves adopting educational approaches that significantly differ from traditional models. Overcoming skepticism from faculty and administrators who are unfamiliar with or skeptical of CAM practices is a significant hurdle.

EDUCATIONAL AND PEDAGOGICAL CHALLENGES

- **Standardization and Accreditation**: Establishing standardized competencies and accreditation standards for interprofessional and CAM education is complex. The diversity of practices within CAM and the varied educational objectives across health professions complicate efforts to create cohesive, universally accepted standards.
- **Faculty Development**: There is a need for more faculty who are adequately trained in both interprofessional education principles and CAM practices. Educating and training faculty members to competently teach these subjects and to model interprofessional collaboration is a significant challenge.

- **Evaluation and Assessment**: Developing effective methods for evaluating student competencies in interprofessional collaboration and CAM understanding is challenging. Assessment tools must accurately reflect students' abilities to work in teams and their knowledge of CAM practices, which can be difficult to quantify.

CULTURAL AND PROFESSIONAL CHALLENGES

- **Professional Silos and Hierarchies**: Healthcare education and practice are often compartmentalized into distinct professional silos with deeply ingrained hierarchies. This segmentation can create barriers to interprofessional collaboration and mutual respect among different healthcare disciplines, including those that practice CAM.
- **Evidence Base for CAM**: The variability in the quality and quantity of evidence supporting different CAM practices can be a barrier to their acceptance and integration into mainstream healthcare education and practice. Skepticism toward CAM among healthcare professionals educated in conventional medicine can be a significant obstacle.
- **Cultural Competence and Sensitivity**: Integrating CAM into healthcare education requires cultural competence and sensitivity to the diverse origins and traditions of CAM practices. Addressing cultural and ethical considerations respectfully and accurately is a challenge in curriculum development.

LOGISTICAL AND PRACTICAL CHALLENGES

- **Clinical Training Opportunities**: Providing students with clinical training experiences that reflect interprofessional and CAM principles is logistically challenging. There are limited training sites where students can observe and practice these approaches in a real-world setting.
- **Interprofessional Coordination**: Organizing schedules and coordinating training activities across different healthcare disciplines is a logistical challenge. It requires significant coordination to align the schedules and educational objectives of various programs.

STRENGTHENING COLLABORATION ACROSS DISCIPLINES

- **Interdisciplinary Research Initiatives**: Promoting research projects that involve multiple healthcare disciplines, including CAM, can foster collaboration and mutual respect. These initiatives can also enhance the evidence base for interprofessional practices and CAM therapies, addressing skepticism and enhancing curriculum development.
- **Joint Educational Programs**: Developing joint degrees or certificates that bridge traditional healthcare disciplines with CAM can encourage collaboration from the outset of a healthcare professional's education. This approach can help dismantle professional silos and build a foundation for interdisciplinary respect and teamwork.

ENHANCING POLICY SUPPORT AND ADVOCACY

- **Policy Reforms**: Advocacy for policy reforms at the national and institutional levels can support the integration of interprofessional and CAM education. Policies that incentivize the adoption of these educational models, such as through funding or accreditation incentives, can drive systemic change.
- **Professional Associations and Networks**: Leveraging professional associations and networks to promote interprofessional and CAM education can raise awareness and support among healthcare professionals. These organizations can play a pivotal role in advocating for curriculum changes, providing resources, and facilitating professional development opportunities.

LEVERAGING TECHNOLOGY AND INNOVATION

- **Digital and Simulation-Based Learning Tools**: Exploiting digital platforms and simulation-based learning can overcome logistical challenges related to interprofessional and CAM education. Virtual reality, online collaborative platforms, and simulation scenarios can provide students with interactive, multidisciplinary learning experiences that are difficult to replicate in traditional settings.
- **Open Educational Resources (OER)**: Developing and sharing open educational resources related to interprofessional healthcare and CAM can help disseminate knowledge and best practices widely, facilitating curriculum development and faculty training across institutions.

FOSTERING A CULTURE OF LIFELONG LEARNING

- **Continuing Education and Professional Development**: Encouraging a culture of lifelong learning among healthcare professionals can help ensure that interprofessional collaboration and CAM integration continue to evolve and improve. Continuing education programs that focus on these areas can update healthcare professionals on the latest practices, research, and collaborative techniques.
- **Reflective Practice and Feedback**: Incorporating reflective practices and feedback mechanisms into healthcare education and professional development can help learners and practitioners continuously improve their interprofessional collaboration skills and CAM knowledge. This approach encourages self-assessment and adaptation to new information and experiences.

EMPHASIZING PATIENT AND COMMUNITY ENGAGEMENT

- **Patient-Centered Education Models**: Integrating patient and community perspectives into healthcare education can highlight the importance of interprofessional collaboration and CAM in meeting diverse health needs. Involving patients and community members in educational activities can provide valuable insights into the real-world impact of these approaches on care delivery.
- **Community-Based Learning Experiences**: Offering community-based learning experiences can expose students to the practical applications of interprofessional healthcare and CAM in diverse settings. These experiences can enhance students' understanding of the social determinants of health, cultural competency, and the value of holistic care approaches.

FUTURE DIRECTIONS AND SOLUTIONS

To address these challenges, innovative solutions and dedicated efforts from educational institutions, healthcare organizations, accrediting bodies, and policymakers are needed. Solutions might include developing new funding models, revising curricula to prioritize interprofessional and CAM education, enhancing faculty development programs, and fostering partnerships with healthcare organizations that practice interprofessional and integrative care. Additionally, ongoing research to strengthen the evidence base for CAM and innovative pedagogical approaches to interprofessional education can support the further integration of these essential components into healthcare training and practice.

INTERPROFESSIONAL HEALTHCARE AND COMPLEMENTARY AND ALTERNATIVE MEDICINE (CAM) EDUCATION: RECOMMENDATIONS FOR THE BEST PRACTICES

Implementing best practices in the integration of interprofessional healthcare and CAM education within healthcare training and practice is vital for enhancing patient care, promoting holistic health, and fostering a collaborative healthcare environment (Al-Worafi, 2020a,b, 2022a,b, 2023a–n, 2024a–c; Hasan et al., 2019). The following recommendations outline best practices for achieving these goals effectively.

CURRICULUM DEVELOPMENT AND INTEGRATION

1. **Incorporate Evidence-Based CAM Content**: Ensure that the CAM content integrated into the curriculum is evidence-based, distinguishing between practices supported by research and those not yet substantiated by scientific evidence.
2. **Develop Interprofessional Core Competencies**: Establish core competencies for interprofessional education that all healthcare students must achieve, focusing on teamwork, communication, ethics, and understanding of roles and responsibilities.
3. **Use Innovative Teaching Methods**: Employ innovative and interactive teaching methods, such as case-based learning, simulations, and problem-based learning, to engage students in real-world scenarios requiring interprofessional collaboration and the application of CAM knowledge.

FACULTY DEVELOPMENT AND SUPPORT

1. **Provide Interprofessional Teaching Training**: Offer training and professional development opportunities for faculty in interprofessional education principles and CAM, ensuring they are equipped to teach these subjects effectively.
2. **Foster Faculty Collaboration**: Encourage collaboration among faculty from different healthcare disciplines, including CAM practitioners, to develop and deliver the curriculum. This approach models interprofessional collaboration for students.
3. **Recognize and Reward Interprofessional Teaching**: Implement policies to recognize and reward faculty efforts and achievements in interprofessional education and CAM integration, promoting continued engagement and innovation.

CLINICAL AND PRACTICAL EXPERIENCE

1. **Offer Interprofessional Clinical Rotations**: Provide clinical rotations and practical experiences that are explicitly designed for interprofessional teams, allowing students to practice collaborative care in real-world settings.
2. **Integrate CAM Practitioners into Clinical Training**: Involve CAM practitioners in clinical education, either through direct participation in clinical rotations or through guest lectures and workshops, to expose students to diverse perspectives and practices.
3. **Utilize Community-Based Educational Opportunities**: Engage with community health organizations and CAM providers to offer students experiential learning opportunities outside of traditional healthcare settings, enhancing their understanding of holistic and community health.

RESEARCH AND CONTINUOUS IMPROVEMENT

1. **Encourage Research on Interprofessional and CAM Education**: Support research initiatives aimed at evaluating the outcomes of interprofessional and CAM education on student learning, healthcare practice, and patient care outcomes.
2. **Use Feedback for Continuous Curriculum Improvement**: Implement mechanisms for continuous feedback from students, faculty, healthcare professionals, and patients to inform the ongoing development and improvement of the interprofessional and CAM curriculum.

POLICY AND ADVOCACY

1. **Advocate for Supportive Policies**: Engage with healthcare policy makers, accreditation bodies, and educational institutions to advocate for policies that support the integration of interprofessional and CAM education, including funding, curriculum standards, and accreditation requirements.
2. **Promote Public Awareness**: Work toward increasing public awareness and understanding of the benefits of interprofessional care and CAM practices through outreach, education, and community engagement efforts.

TECHNOLOGY AND RESOURCES

1. **Leverage Technology for Interprofessional Learning**: Utilize online platforms, virtual simulations, and other digital tools to facilitate interprofessional learning experiences, especially in situations where the physical co-location of students from different disciplines is challenging.
2. **Ensure Access to Quality CAM Resources**: Provide students and faculty with access to high-quality resources on CAM, including databases, journals, and professional networks, to support their learning and practice.

ENGAGEMENT AND PARTNERSHIP

1. **Build Partnerships with Healthcare Organizations**: Develop strategic partnerships with healthcare organizations that practice interprofessional teamwork and integrate CAM into patient care. These partnerships can provide practical training opportunities for students and facilitate research collaborations.
2. **Engage Community and Patient Advocates**: Involve patients, community health advocates, and representatives from diverse cultural backgrounds in curriculum development and educational activities. Their perspectives can enrich the learning experience and ensure that education is aligned with community needs and patient-centered care principles.

CULTURAL COMPETENCE AND DIVERSITY

1. **Incorporate Cultural Competence Training**: Embed cultural competence training within interprofessional and CAM education to prepare students to provide respectful, informed care to patients from diverse cultural and ethnic backgrounds.
2. **Diversify Educational Content and Perspectives**: Ensure that the curriculum reflects a broad spectrum of cultural perspectives, including indigenous and global healing practices, to broaden students' understanding and appreciation of diverse approaches to health and wellness.

Assessment and Evaluation

1. **Develop Multidisciplinary Assessment Tools**: Create assessment tools that are designed to evaluate students' competencies in interprofessional collaboration and CAM knowledge. These should measure not only individual knowledge but also the ability to work effectively within a team.
2. **Implement Reflective Practice**: Encourage students to engage in reflective practice throughout their education, fostering a habit of self-assessment and continuous learning in interprofessional settings and CAM.

Leadership and Advocacy

1. **Foster Leadership Skills**: Incorporate leadership training within interprofessional and CAM education to empower students to become advocates for collaborative practice and integrative healthcare models in their future careers.
2. **Promote Advocacy for Integrative Healthcare**: Encourage students and faculty to engage in advocacy efforts for the recognition and integration of CAM and interprofessional care in healthcare policies, practices, and research funding.

Technological Innovation

1. **Explore Emerging Technologies**: Utilize emerging technologies such as artificial intelligence (AI), machine learning, and telehealth platforms to facilitate interprofessional education and practice, as well as to provide innovative ways to integrate CAM into healthcare delivery.
2. **Develop Online Collaborative Learning Platforms**: Use online platforms to foster collaboration among students from different health disciplines, including forums for discussion, shared learning resources, and virtual teamwork exercises.

Sustainability and Scalability

1. **Ensure Program Sustainability**: Develop strategies for the long-term sustainability of interprofessional and CAM education programs, including securing ongoing funding, embedding programs within institutional structures, and building a pipeline of trained educators.
2. **Scalability and Adaptation**: Create adaptable interprofessional and CAM education models that can be scaled and customized to fit different educational settings, healthcare systems, and community needs, facilitating broader implementation.

Continuous Feedback and Improvement

1. **Establish Continuous Improvement Processes**: Set up mechanisms for ongoing evaluation and refinement of interprofessional and CAM education programs based on feedback from all stakeholders, including students, faculty, healthcare professionals, and patients.
2. **Benchmarking and Best Practices Sharing**: Engage in benchmarking against best practices in interprofessional and CAM education and share findings and innovations with the wider educational and healthcare community to foster continuous improvement and innovation.

ADAPTIVE LEARNING ENVIRONMENTS

1. **Create Flexible Learning Modules**: Design learning modules that can be adapted to different learners' needs, including varying levels of prior knowledge about CAM and interprofessional practice. This approach accommodates diverse learning styles and speeds.
2. **Utilize Adaptive Technology**: Implement adaptive learning technologies that personalize the educational experience, allowing students to focus on areas where they need improvement and advance more quickly through areas where they have demonstrated competence.

INCLUSIVITY IN EDUCATION AND PRACTICE

1. **Promote Inclusivity in Healthcare Teams**: Educate students about the importance of inclusivity in healthcare teams, emphasizing the value of diverse perspectives, including those of patients, CAM practitioners, and professionals from various healthcare disciplines.
2. **Address Health Disparities**: Incorporate discussions and learning objectives related to health disparities and social determinants of health into interprofessional and CAM education, preparing students to address these issues in their practice.

DYNAMIC CURRICULUM UPDATES

1. **Regularly Update Curriculum Content**: Establish a process for regularly updating curriculum content to reflect the latest research findings, clinical guidelines, and best practices in both interprofessional healthcare and CAM, ensuring that students receive the most current information.
2. **Incorporate Emerging Health Trends**: Stay abreast of emerging health trends and public health challenges, integrating relevant content into the curriculum to prepare students for the evolving healthcare landscape.

COMMUNITY AND GLOBAL HEALTH PERSPECTIVES

1. **Engage with Global Health Initiatives**: Integrate global health perspectives into interprofessional and CAM education, exposing students to healthcare challenges and practices from around the world and preparing them for work in diverse settings.
2. **Community Engagement Projects**: Encourage students to participate in community engagement projects that involve interprofessional teams and CAM practices, enhancing their understanding of community health needs and the practical application of their skills.

PROFESSIONAL DEVELOPMENT AND LIFELONG LEARNING

1. **Encourage Lifelong Learning**: Instill the importance of lifelong learning in students, highlighting the need for continuous professional development in interprofessional practice and CAM to adapt to changes in healthcare practices and patient needs.
2. **Develop Mentorship Programs**: Establish mentorship programs that connect students with experienced professionals in interprofessional and CAM fields, facilitating knowledge transfer, professional development, and career guidance.

ETHICAL CONSIDERATIONS AND PATIENT SAFETY

1. **Emphasize Ethical Considerations**: Ensure that ethical considerations in interprofessional collaboration and CAM practices are a core component of the curriculum, including informed consent, patient autonomy, and cultural sensitivity.

2. **Focus on Patient Safety**: Train students to prioritize patient safety in interprofessional settings and when integrating CAM into patient care, including understanding the interactions between CAM therapies and conventional treatments.

RESEARCH AND INNOVATION

1. **Encourage Student Research**: Provide opportunities for students to engage in research on interprofessional collaboration and CAM, fostering a culture of inquiry and innovation.
2. **Innovate Healthcare Solutions**: Encourage interdisciplinary teams of students to work on innovative healthcare solutions that combine interprofessional collaboration and CAM, potentially addressing unmet healthcare needs.

EMBRACING DIGITAL HEALTH INNOVATIONS

1. **Integrate Digital Health into Curricula**: Educate students about digital health technologies, including telehealth, health informatics, and mobile health applications, and their role in facilitating interprofessional collaboration and CAM practices.
2. **Prepare for the Digital Transformation**: Equip students with the skills to navigate the digital transformation in healthcare, ensuring they can effectively use electronic health records, data analytics, and digital communication tools in an interprofessional setting.

FOSTERING RESILIENCE AND WELL-BEING

1. **Incorporate Resilience Training**: Include training on resilience and well-being for healthcare professionals within the curriculum, addressing the high-stress nature of healthcare work and teaching strategies for managing stress and preventing burnout.
2. **Promote Self-Care Practices**: Educate students on the importance of self-care, including CAM practices that support personal well-being, and encourage their integration into personal and professional life to maintain a healthy work-life balance.

EXPANDING INTERDISCIPLINARY RESEARCH

1. **Foster Interdisciplinary Research Collaborations**: Encourage collaborations between healthcare professionals, CAM practitioners, and researchers from various disciplines to explore innovative healthcare solutions and enhance the evidence base for interprofessional practice and CAM.
2. **Support Translational Research**: Promote translational research that bridges the gap between scientific discoveries and clinical application, ensuring that advancements in interprofessional healthcare and CAM benefit patients directly.

ENHANCING GLOBAL HEALTH COMPETENCIES

1. **Develop Global Health Partnerships**: Establish partnerships with institutions and organizations around the world to provide students with international learning experiences and perspectives on interprofessional healthcare and CAM.
2. **Incorporate Global Health Challenges**: Integrate content on global health challenges, such as infectious diseases, climate change impacts on health, and global health equity, into the curriculum, preparing students to think broadly about health solutions.

LEADERSHIP AND CHANGE MANAGEMENT

1. **Cultivate Leadership Skills**: Beyond clinical skills, cultivate leadership and change management skills among students, preparing them to lead and advocate for integrative, collaborative healthcare models in their future careers.
2. **Prepare for Healthcare System Changes**: Educate students about healthcare system dynamics, policy, and management, enabling them to navigate and influence the system for the integration of interprofessional practices and CAM.

ETHICS AND CULTURAL COMPETENCY

1. **Strengthen Ethical Reasoning**: Deepen students' understanding of ethical reasoning and decision-making in complex interprofessional and CAM scenarios, emphasizing respect for diverse beliefs and practices.
2. **Enhance Cultural Competency**: Further enhance cultural competency education to prepare students for effectively communicating and collaborating with patients and colleagues from diverse cultural and ethnic backgrounds.

CONTINUOUS QUALITY IMPROVEMENT

1. **Implement Quality Improvement Projects**: Encourage students to participate in quality improvement projects that address challenges in interprofessional collaboration and CAM integration, applying their knowledge to real-world healthcare improvements.
2. **Use Data for Decision-Making**: Teach students to use data and evidence for continuous quality improvement in healthcare practices, including the evaluation of interprofessional teamwork and CAM interventions.

ADVANCING TECHNOLOGICAL COMPETENCY

1. **Incorporate AI and Machine Learning**: Educate healthcare professionals about the potential of artificial intelligence (AI) and machine learning in healthcare, including applications in diagnosis, treatment planning, and personalized medicine, and their implications for interprofessional collaboration and CAM.
2. **Enhance Telehealth Training**: As telehealth becomes a mainstay in healthcare delivery, ensure that healthcare professionals are adept at delivering compassionate, effective care remotely, including the use of CAM therapies that can be facilitated through telehealth.

PROMOTING ENVIRONMENTAL HEALTH AND SUSTAINABILITY

1. **Integrate Environmental Health**: Educate on the intersection of environmental health and healthcare, emphasizing how interprofessional teams can address environmental factors affecting patient health and how CAM practices can support environmental sustainability.
2. **Sustainability in Healthcare Practices**: Encourage sustainable healthcare practices within interprofessional teams, including the use of sustainable materials and resources in CAM therapies, to promote environmental health.

STRENGTHENING COMMUNITY AND PUBLIC HEALTH INTEGRATION

1. **Community Health Engagement**: Foster strong ties with communities by involving healthcare students in community health assessments and projects, highlighting the role of interprofessional teams and CAM in addressing community health needs.

2. **Public Health Preparedness**: Include training on public health preparedness and response, emphasizing how interprofessional healthcare teams and CAM practitioners can collaborate in public health crises, such as pandemics or natural disasters.

ENHANCING COMMUNICATION AND INTERPERSONAL SKILLS

1. **Develop Advanced Communication Skills**: Beyond basic communication training, provide advanced coursework in emotional intelligence, conflict resolution, and cross-cultural communication to enhance the effectiveness of interprofessional teams and the therapeutic alliance in CAM practices.
2. **Patient Advocacy Training**: Train healthcare professionals in patient advocacy, empowering them to support patients in navigating the healthcare system, accessing interprofessional care, and making informed decisions about CAM therapies.

FOSTERING INNOVATION AND ENTREPRENEURSHIP

1. **Encourage Healthcare Innovation**: Stimulate innovation in healthcare delivery by encouraging students to develop novel solutions that integrate interprofessional collaboration and CAM, supporting entrepreneurship in healthcare.
2. **Innovation Labs and Hackathons**: Organize innovation labs or hackathons that bring together students from healthcare, engineering, business, and other disciplines to create innovative solutions for integrating interprofessional healthcare and CAM.

BUILDING RESILIENT HEALTHCARE SYSTEMS

1. **Systems Thinking**: Incorporate systems thinking into healthcare education, enabling future healthcare professionals to understand and improve the complex systems within which interprofessional teams and CAM practices operate.
2. **Resilience in Healthcare Delivery**: Teach strategies for building resilience in healthcare delivery systems, ensuring that they can maintain high levels of care, including interprofessional collaboration and CAM, during times of stress or change.

Implementing these recommendations requires a coordinated effort among educational institutions, healthcare organizations, policymakers, and the broader healthcare community. By adopting these best practices, healthcare education can more effectively prepare future healthcare professionals to work collaboratively across disciplines and incorporate a wide range of therapeutic approaches into patient care, ultimately leading to improved health outcomes and patient satisfaction.

CONCLUSION

The integration of interprofessional healthcare and CAM education into the healthcare system is a complex but increasingly necessary endeavor. As healthcare continues to evolve toward a more holistic, patient-centered model, the need for healthcare professionals who are adept in interprofessional collaboration and knowledgeable about CAM therapies becomes ever more critical. The recommendations provided aim to address the multifaceted challenges and leverage the opportunities inherent in this integration, focusing on curriculum development, faculty support, clinical experience, technological competency, environmental health, community engagement, communication skills, innovation, and system resilience. Key takeaways include the importance of evidence-based CAM education, the development of core competencies in interprofessional collaboration, the use of innovative teaching and learning technologies, and the emphasis on cultural competence and ethical practice. Moreover, fostering a culture of lifelong learning, resilience, and adaptability among healthcare professionals will ensure that the workforce remains responsive to the changing healthcare landscape and patient needs. Achieving these goals requires concerted efforts from educational institutions, healthcare organizations, accrediting bodies, policymakers, and the healthcare community at large. It also necessitates a commitment to continuous quality improvement, research, and innovation to ensure that interprofessional and CAM education remains relevant, effective, and aligned with the goal of improving health outcomes. In conclusion, the integration of Interprofessional Healthcare and CAM Education represents a forward-thinking approach to healthcare education and practice. By preparing healthcare professionals to work collaboratively across disciplines and to incorporate a wide range of therapeutic approaches into patient care, we can advance toward a more inclusive, effective, and holistic healthcare system. This endeavor, though challenging, holds the promise of transforming healthcare delivery to better meet the complex health needs of populations, enhance patient satisfaction, and improve the overall health and well-being of communities around the world.

REFERENCES

Al-Worafi, Y.M. (Ed.). (2020a). *Drug Safety in Developing Countries: Achievements and Challenges.* Academic Press.

Al-Worafi, Y.M. (2020b). Herbal medicines safety issues. In: Al-Worafi, Y.M. (ed), *Drug Safety in Developing Countries* (pp. 163–178). Academic Press.

Al-Worafi, Y.M. (2022a). *A Guide to Online Pharmacy Education: Teaching Strategies and Assessment Methods.* CRC Press.

Al-Worafi, Y.M. (2022b). Competencies and learning outcomes. In: Al-Worafi, Y.M. (ed), *A Guide to Online Pharmacy Education: Teaching Strategies and Assessment Methods.* CRC Press.

Al-Worafi, Y.M. (2023a). *Patient Safety in Developing Countries: Education, Research, Case Studies.* CRC Press.

Al-Worafi, Y.M. (2023b). *Technology for Drug Safety: Current Status and Future Developments.* Springer Nature.

Al-Worafi, Y.M. (2023c). Patient safety-related issues: Patient care errors and related problems. In: Al-Worafi, Y.M. (ed), *Patient Safety in Developing Countries: Education, Research, Case Studies.* CRC Press.

Al-Worafi, Y.M. (2023d). Patient care errors and related problems: Preventive medicine errors & related problems. In: Al-Worafi, Y.M. (ed), *Patient Safety in Developing Countries: Education, Research, Case Studies.* CRC Press.

Al-Worafi, Y.M. (2023e). Patient care errors and related problems: Patient assessment and diagnostic errors & related problems. In: Al-Worafi, Y.M. (ed), *Patient Safety in Developing Countries: Education, Research, Case Studies.* CRC Press.

Al-Worafi, Y.M. (2023f). Patient care errors and related problems: Non-pharmacological errors & related problems. In: Al-Worafi, Y.M. (ed), *Patient Safety in Developing Countries: Education, Research, Case Studies.* CRC Press.

Al-Worafi, Y.M. (2023g). Patient care errors and related problems: Medical errors & related problems. In: *Patient Safety in Developing Countries: Education, Research, Case Studies.* CRC Press.

Al-Worafi, Y.M. (2023h). Patient care errors and related problems: Monitoring errors & related problems. In: Al-Worafi, Y.M. (ed), *Patient Safety in Developing Countries: Education, Research, Case Studies.* CRC Press.

Al-Worafi, Y.M. (2023i). Patient care errors and related problems: Patient education and counselling errors and related problems. In: Al-Worafi, Y.M. (ed), *Patient Safety in Developing Countries: Education, Research, Case Studies.* CRC Press.

Al-Worafi, Y.M. (2023j). Patient safety-related issues: Other medication safety issues. In: Al-Worafi, Y.M. (ed), *Patient Safety in Developing Countries: Education, Research, Case Studies.* CRC Press.

Al-Worafi, Y.M. (2023k). Patient safety culture. In: Al-Worafi, Y.M. (ed), *Patient Safety in Developing Countries: Education, Research, Case Studies.* CRC Press.

Al-Worafi, Y.M. (2023l). Patient safety education: Competencies and learning outcomes. In: Al-Worafi, Y.M. (ed), *Patient Safety in Developing Countries: Education, Research, Case Studies.* CRC Press.

Al-Worafi, Y.M. (Ed.). (2023m). *Clinical Case Studies on Medication Safety.* Academic Press.

Al-Worafi, Y.M. (Ed.). (2023n). *Comprehensive Healthcare Simulation: Pharmacy Education, Practice and Research.* Springer Nature.

Al-Worafi, Y.M. (Ed.). (2024a). *Handbook of Medical and Health Sciences in Developing Countries.* Springer, Cham.

Al-Worafi, Y.M. (2024b). Complementary and alternative medicine (CAM) in developing countries. In: Al-Worafi, Y.M. (ed), *Handbook of Medical and Health Sciences in Developing Countries.* Springer, Cham. https://doi.org/10.1007/978-3-030-74786-2_301-1

Al-Worafi, Y.M. (2024c). Access/equitable access to medical and health sciences in developing countries. In: Al-Worafi, Y.M. (ed), *Handbook of Medical and Health Sciences in Developing Countries.* Springer, Cham. https://doi.org/10.1007/978-3-030-74786-2_147-1

Hasan, S., Al-Omar, M.J., AlZubaidy, H., and Al-Worafi, Y.M. (2019). Use of medications in Arab countries. In: Laher, I. (ed), *Handbook of Healthcare in the Arab World* (p. 42). Springer, Cham.

18 Healthcare Professionals, Students, and Patients/Public Perspectives on Complementary and Alternative Medicine (CAM) Education and Training

INTRODUCTION

The perspectives of healthcare professionals on complementary and alternative medicine (CAM) education and training encompass a variety of viewpoints, which reflect the diverse backgrounds, training, and professional cultures within the healthcare industry. CAM includes a broad range of practices, products, and therapies outside the realm of conventional medicine, such as herbal medicine, acupuncture, massage therapy, chiropractic, and mindfulness. The increasing interest and usage of CAM among the general public have prompted a discussion regarding its integration into healthcare professionals' education and training.

RECOGNITION OF CAM'S IMPORTANCE

Many healthcare professionals recognize the importance of understanding CAM practices due to the high prevalence of CAM use among patients. They acknowledge that having knowledge about CAM can improve patient-provider communication, enable them to offer more comprehensive care, and assist in advising patients on safe and effective CAM use. This recognition often leads to support for incorporating CAM education and training into medical, nursing, and pharmacy curricula.

VARIABILITY IN CAM EDUCATION AND TRAINING

Despite the recognized importance, there is considerable variability in how CAM is incorporated into the education and training of healthcare professionals. Some institutions offer extensive CAM courses and electives, while others provide minimal coverage. This variability can be attributed to factors such as differing perceptions of CAM's legitimacy within the medical community, lack of standardized CAM educational curricula, and the challenge of integrating CAM into already crowded professional training programs.

CHALLENGES IN CAM EDUCATION

A significant challenge in CAM education and training is the wide range of practices that CAM encompasses, many of which have varying levels of evidence supporting their efficacy. This leads to difficulties in deciding which modalities should be taught and how they should be integrated

DOI: 10.1201/9781003327202-18

with conventional medical education. Additionally, there is often a lack of high-quality research on certain CAM practices, making it challenging to teach evidence-based use of CAM.

Advocacy for Evidence-Based CAM Education

There is a growing advocacy among healthcare professionals for evidence-based CAM education. This approach emphasizes the importance of integrating CAM into healthcare education in a way that is informed by scientific evidence and clinical efficacy. The goal is to equip healthcare professionals with the knowledge to critically evaluate CAM practices and guide their patients in making informed decisions about their use.

Interprofessional Education

The concept of interprofessional education, where students from various healthcare disciplines learn together and from each other, is also being applied to CAM education. This approach can foster a more collaborative and holistic view of patient care, recognizing the potential benefits of integrating CAM with conventional treatments. Interprofessional education in CAM encourages healthcare professionals to understand and respect the roles of various CAM practitioners, promoting team work and improving patient outcomes. The perspectives of healthcare professionals on CAM education and training reflect a balance between recognizing the value of CAM in addressing patients' needs and the challenges of integrating CAM into conventional medical education. As the demand for CAM continues to grow, there is a clear need for further development of evidence-based CAM education and training programs that can prepare healthcare professionals to provide informed, safe, and comprehensive care to their patients.

HEALTHCARE PROFESSIONALS PERSPECTIVES ON COMPLEMENTARY AND ALTERNATIVE MEDICINE (CAM) EDUCATION AND TRAINING

The perspectives of healthcare professionals on CAM education and training are varied and complex, reflecting broader debates within the medical community about the role and efficacy of CAM practices. These perspectives can be influenced by a professional's background, clinical experiences, and the prevailing attitudes within their specialty toward CAM. The integration of CAM into healthcare education and training programs is a key area of focus, given the growing interest among patients in these therapies. Here are several key themes that characterize healthcare professionals' perspectives on CAM education and training:

Support for CAM Education

Many healthcare professionals support the inclusion of CAM education within medical, nursing, and pharmacy curricula. They argue that understanding CAM practices is essential for comprehensive patient care, given the high usage rates of these therapies among the general population. Professionals advocating for CAM education often emphasize the importance of being able to discuss CAM therapies knowledgeably with patients, guide them on safe use, and integrate CAM practices into treatment plans when appropriate.

Concerns About Evidence and Efficacy

Skepticism among healthcare professionals regarding CAM education often centers on concerns about the evidence base and efficacy of certain CAM practices. Critics argue that integrating CAM into healthcare education could lend undue legitimacy to practices that are not supported by robust scientific evidence. These professionals stress the importance of evidence-based medicine in clinical training and are wary of including CAM content that does not meet these rigorous standards.

THE CHALLENGE OF INTEGRATION

Integrating CAM education into existing healthcare training programs presents logistical and curricular challenges. Medical, nursing, and pharmacy curricula are already densely packed with content, and finding space for CAM education can be difficult. Additionally, there is the challenge of determining what content should be included, given the wide range of practices encompassed by CAM, from well-evidenced practices like acupuncture and certain herbal medicines to more controversial and less evidenced therapies.

THE NEED FOR STANDARDIZATION

There is a recognized need for standardization in CAM education for healthcare professionals. Without standardized curricula, the content and quality of CAM education can vary significantly between institutions, leading to disparities in knowledge and competency among practitioners. Standardizing CAM education could help ensure that all healthcare professionals receive a baseline level of training on the most important and evidence-supported CAM practices.

ADVOCACY FOR INTERPROFESSIONAL LEARNING

Interprofessional learning is seen as a valuable approach to CAM education, bringing together students and practitioners from various healthcare disciplines to learn about, from, and with each other. This approach can foster a more holistic and collaborative approach to patient care, acknowledging the potential benefits of integrating conventional and CAM therapies. It encourages mutual respect among professionals from different backgrounds and can enhance communication and teamwork skills.

PATIENT-CENTERED CARE

Ultimately, many healthcare professionals view CAM education and training through the lens of patient-centered care. They recognize that patients are increasingly interested in and using CAM therapies, and that being informed about these practices is essential for engaging patients in discussions about their healthcare choices. Education in CAM is thus seen as part of providing holistic, patient-centered care that respects patients' values, preferences, and needs.

In conclusion, healthcare professionals' perspectives on CAM education and training reflect a broader debate about the role of CAM in healthcare, balancing skepticism about evidence and efficacy with a desire to meet patient needs and provide holistic care. The discussion points toward a need for evidence-based, standardized, and patient-centered approaches to CAM education within healthcare training programs.

HEALTHCARE STUDENTS PERSPECTIVES ON COMPLEMENTARY AND ALTERNATIVE MEDICINE (CAM) EDUCATION AND TRAINING

The perspectives of healthcare students on CAM education and training are crucial as these future professionals navigate the evolving landscape of healthcare, where CAM plays an increasingly significant role. These perspectives are shaped by a variety of factors including personal experiences, cultural backgrounds, and the level of exposure to CAM practices within their educational programs. Here are several key themes that illustrate healthcare students' views on CAM education and training:

INTEREST AND OPENNESS

Many healthcare students express a strong interest and openness toward learning about CAM. This interest is often driven by a recognition of the growing demand for CAM among patients and a desire to offer holistic care that incorporates a wide range of therapeutic options. Students frequently advocate for more comprehensive CAM education within their curricula to better meet the needs of a diverse patient population.

PERCEIVED GAPS IN EDUCATION

A common concern among healthcare students is the perceived gap in their education regarding CAM. Students often report that their training provides limited exposure to CAM, focusing predominantly on conventional medicine. They express a desire for more balanced education that includes evidence-based CAM practices, enabling them to make informed recommendations to patients and integrate CAM into their future practice where appropriate.

DEMAND FOR EVIDENCE-BASED CAM EDUCATION

Healthcare students typically emphasize the importance of evidence-based medicine in their training and express a desire for CAM education to adhere to the same standards. They seek reliable, scientifically backed information on the efficacy and safety of CAM modalities to make informed decisions in their professional practice. Students often call for increased research into CAM practices to be integrated into their educational programs.

CHALLENGES IN INTEGRATING CAM INTO CURRICULA

Students acknowledge several challenges in integrating CAM education into existing healthcare curricula. These include limited time and space within packed schedules, determining the appropriate depth and breadth of CAM content, and finding qualified instructors with expertise in both CAM and conventional medicine. There is a recognition that addressing these challenges requires thoughtful curriculum design and prioritization of the most relevant CAM topics.

THE ROLE OF CAM IN HOLISTIC CARE

Healthcare students frequently highlight the role of CAM in providing holistic care that addresses physical, emotional, mental, and spiritual aspects of health. They view education in CAM as essential for developing a more comprehensive approach to patient care, one that recognizes the value of alternative therapies in promoting well-being and preventive health.

INTERPROFESSIONAL LEARNING OPPORTUNITIES

There is a growing appreciation among healthcare students for interprofessional learning opportunities that include CAM. Such opportunities allow students from different health disciplines to learn from and with each other, fostering a collaborative approach to patient care that includes conventional and CAM therapies. Students value the chance to understand the roles of various health professionals, including CAM practitioners, in providing integrated care.

PATIENT-CENTERED APPROACH

Ultimately, healthcare students view CAM education as part of a broader patient-centered approach to healthcare. They recognize that understanding and respecting patients' healthcare choices, including the use of CAM, is crucial for building trust and effectively managing patient care. Students advocate for education that prepares them to discuss CAM openly with patients, respecting their values and preferences while providing evidence-based guidance.

In summary, healthcare students' perspectives on CAM education and training reflect a blend of interest, recognition of educational gaps, and a strong desire for evidence-based, comprehensive training that prepares them for the realities of a diverse healthcare environment. Their views underscore the importance of integrating CAM into healthcare education to foster holistic, patient-centered care that aligns with the preferences and needs of the population they will serve.

PATIENTS AND THE PUBLICS PERSPECTIVES ON COMPLEMENTARY AND ALTERNATIVE MEDICINE (CAM)

The perspectives of patients and the general public on CAM are diverse and have evolved significantly over recent years. The growing interest in and use of CAM reflects broader societal shifts toward holistic, personalized, and preventive healthcare approaches. Here are several key themes that illustrate the perspectives of patients and the public on CAM.

INCREASING ACCEPTANCE AND UTILIZATION

There has been a noticeable increase in the acceptance and utilization of CAM among patients and the general public. This trend is driven by a variety of factors, including dissatisfaction with conventional medicine, a desire for more natural or holistic treatment options, and a growing emphasis on wellness and preventive care. Many individuals turn to CAM as a way to take more control over their health and to seek treatments that align with their personal beliefs and preferences.

DESIRE FOR INTEGRATION

A significant portion of patients and the public express a desire for the integration of CAM with conventional healthcare. They advocate for a more comprehensive approach to health that combines the best of conventional medicine and CAM to address the full spectrum of health needs. This perspective is often rooted in the belief that such integration can lead to more personalized, effective, and holistic care.

SAFETY AND EFFICACY CONCERNS

While there is considerable interest in CAM, concerns about the safety and efficacy of certain CAM practices persist among patients and the general public. The variability in regulatory standards and the lack of scientific evidence supporting some CAM therapies contribute to these concerns. Many individuals emphasize the importance of reliable, accessible information on the safety and effectiveness of CAM modalities to make informed healthcare decisions.

SEEKING INFORMATION AND GUIDANCE

Patients and the public often seek information and guidance on CAM from a variety of sources, including healthcare providers, the Internet, and social networks. There is a strong demand for trustworthy, evidence-based information that can help individuals navigate the complex landscape of CAM. Many express the need for healthcare providers to be knowledgeable about CAM and open to discussing it as part of the care process.

PERSONAL EMPOWERMENT

For many, CAM represents a pathway to personal empowerment in managing their health. It allows individuals to explore treatments that resonate with their personal health philosophies, lifestyle choices, and cultural beliefs. This aspect of CAM is particularly valued in the context of chronic conditions, where conventional medicine may not offer complete solutions.

CULTURAL AND SOCIAL INFLUENCES

Cultural and social influences play a significant role in shaping attitudes toward CAM. In some cultures, CAM practices are deeply embedded in traditional healthcare systems and are widely accepted and utilized. In others, CAM is seen as a complementary approach to enhance well-being or fill gaps in conventional healthcare. The social environment, including trends and endorsements by influencers or celebrities, can also impact perceptions and acceptance of CAM.

ECONOMIC CONSIDERATIONS

Economic factors, including the cost of CAM therapies and the extent to which they are covered by health insurance, influence perspectives and utilization. Some individuals view CAM as a cost-effective alternative to conventional treatments, particularly for managing chronic conditions or promoting wellness. However, the out-of-pocket costs for CAM therapies can be a barrier for others, highlighting the need for broader insurance coverage and affordability.

In summary, the perspectives of patients and the public on CAM are characterized by a blend of enthusiasm and caution, reflecting a desire for holistic, integrated, and personalized healthcare options alongside concerns about safety, efficacy, and access to reliable information. These perspectives underscore the importance of addressing the needs and preferences of individuals in the evolving healthcare landscape, including the integration of evidence-based CAM into mainstream healthcare practices.

RECOMMENDATIONS

Given the varied perspectives of healthcare professionals, students, and the public on CAM, and the evolving landscape of healthcare that increasingly incorporates CAM, several recommendations can be made to address the diverse needs, concerns, and interests of these groups (Al-Worafi, 2020a,b, 2022a,b, 2023a–n, 2024a–c; Hasan et al., 2019). These recommendations aim to enhance the integration, education, and regulation of CAM in a way that promotes safe, effective, and patient-centered care.

FOR HEALTHCARE SYSTEMS AND REGULATORY BODIES

1. **Standardize CAM Education**: Develop standardized curricula for CAM education across medical, nursing, and other healthcare professional training programs. This should include evidence-based practices, safety protocols, and ethical considerations.
2. **Enhance CAM Research**: Increase funding and support for research into CAM practices to build a stronger evidence base for their safety and efficacy. Encourage collaboration between conventional and CAM researchers.
3. **Regulate CAM Practices**: Implement and enforce regulations for CAM practices to ensure practitioner competency, treatment safety, and consumer protection. This includes licensing of CAM practitioners and regulation of CAM products.

4. **Improve Insurance Coverage**: Work with insurance companies to expand coverage of evidence-based CAM therapies, making them more accessible and affordable for patients.
5. **Promote Interprofessional Collaboration**: Encourage collaboration between conventional healthcare providers and CAM practitioners to facilitate integrated care that meets the diverse needs of patients.

For Healthcare Providers

1. **Seek CAM Education**: Pursue continuing education opportunities in CAM to enhance understanding and competency in discussing and integrating CAM therapies into patient care where appropriate.
2. **Practice Open Communication**: Engage in open, non-judgmental conversations with patients about CAM, encouraging them to share their experiences and preferences. Provide evidence-based advice on CAM use.
3. **Adopt a Holistic Approach**: Incorporate a holistic view of patient care that considers physical, emotional, mental, and spiritual well-being, recognizing the potential role of CAM in addressing these aspects.

For CAM Practitioners

1. **Engage in Professional Development**: Continue to engage in professional development and education to stay current with the latest research and best practices in CAM.
2. **Collaborate with Healthcare Providers**: Seek opportunities for collaboration with conventional healthcare providers to offer integrated care and ensure patients receive coordinated, comprehensive treatment.
3. **Adhere to Ethical Standards**: Maintain high ethical standards in practice, including transparency about the evidence supporting CAM therapies and clear communication about treatment goals and expectations.

For Patients and the Public

1. **Seek Reliable Information**: Look for information on CAM from reputable sources, including healthcare providers and scientifically valid websites, to make informed decisions about CAM use.
2. **Discuss CAM with Healthcare Providers**: Communicate openly with healthcare providers about any CAM practices being considered or used to ensure safe and coordinated care.
3. **Consider Evidence and Safety**: Evaluate the evidence supporting CAM therapies and consider safety implications, especially in relation to existing health conditions and conventional treatments.

For Educational Institutions

1. **Incorporate CAM into Health Curricula**: Include CAM education within healthcare curricula to prepare future healthcare professionals to meet patient needs and preferences for integrated care.
2. **Promote Critical Thinking**: Teach students to critically evaluate the evidence for and against CAM practices, fostering an evidence-based approach to all aspects of healthcare.
3. **Foster Interprofessional Education**: Facilitate interprofessional education that includes students from conventional and CAM disciplines, promoting mutual respect and collaboration.

Implementing these recommendations requires concerted efforts from all stakeholders involved in healthcare delivery and education. By addressing the educational, regulatory, and collaborative needs surrounding CAM, the healthcare system can better accommodate the growing interest in CAM and ensure that patients receive safe, effective, and holistic care.

CONCLUSION

In conclusion, the integration of CAM into the broader healthcare landscape presents a multifaceted challenge that mirrors the complexity and diversity of healthcare needs, preferences, and practices today. The perspectives of healthcare professionals, students, and the public on CAM underscore a collective movement toward a more holistic, patient-centered approach to health and wellness. This shift calls for a balanced embrace of CAM, grounded in evidence-based practice, ethical standards, and a commitment to patient safety and quality care. The recommendations provided aim to foster a healthcare environment where CAM is not seen as an alternative to conventional medicine but as a complementary component of a comprehensive care model. This entails rigorous education and training for healthcare professionals, enhanced research into CAM practices, robust regulatory frameworks, and open, informed dialogue between providers and patients. Such efforts can demystify CAM, bolster its evidence base, and ensure its safe, informed, and effective use. As we move forward, the integration of CAM into healthcare will likely continue to evolve, shaped by ongoing research, technological advances, and changing societal attitudes toward health and wellness. The goal should not be to blur the lines between conventional and alternative medicine indiscriminately but to create a seamless continuum of care that leverages the best of both worlds to meet the diverse needs of patients. By prioritizing patient welfare, evidence-based practice, and interdisciplinary collaboration, the healthcare community can navigate the challenges and opportunities presented by CAM, ensuring that all patients receive the holistic, respectful, and effective care they deserve.

REFERENCES

Al-Worafi, Y.M. (Ed.). (2020a). *Drug Safety in Developing Countries: Achievements and Challenges*. Academic Press.

Al-Worafi, Y.M. (2020b). Herbal medicines safety issues. In: Al-Worafi, Y.M. (ed), *Drug Safety in Developing Countries* (pp. 163–178). Academic Press.

Al-Worafi, Y.M. (2022a). *A Guide to Online Pharmacy Education: Teaching Strategies and Assessment Methods*. CRC Press.

Al-Worafi, Y.M. (2022b). Competencies and learning outcomes. In: Al-Worafi, Y.M. (ed), *A Guide to Online Pharmacy Education: Teaching Strategies and Assessment Methods*. CRC Press.

Al-Worafi, Y.M. (2023a). *Patient Safety in Developing Countries: Education, Research, Case Studies*. CRC Press.

Al-Worafi, Y.M. (2023b). *Technology for Drug Safety: Current Status and Future Developments*. Springer Nature.

Al-Worafi, Y.M. (2023c). Patient safety-related issues: Patient care errors and related problems. In: Al-Worafi, Y.M. (ed), *Patient Safety in Developing Countries: Education, Research, Case Studies*. CRC Press.

Al-Worafi, Y.M. (2023d). Patient care errors and related problems: Preventive medicine errors & related problems. In: Al-Worafi, Y.M. (ed), *Patient Safety in Developing Countries: Education, Research, Case Studies*. CRC Press.

Al-Worafi, Y.M. (2023e). Patient care errors and related problems: Patient assessment and diagnostic errors & related problems. In: Al-Worafi, Y.M. (ed), *Patient Safety in Developing Countries: Education, Research, Case Studies*. CRC Press.

Al-Worafi, Y.M. (2023f). Patient care errors and related problems: Non-pharmacological errors & related problems. In: Al-Worafi, Y.M. (ed), *Patient Safety in Developing Countries: Education, Research, Case Studies*. CRC Press.

Al-Worafi, Y.M. (2023g). Patient care errors and related problems: Medical errors & related problems. In: Al-Worafi, Y.M. (ed), *Patient Safety in Developing Countries: Education, Research, Case Studies*. CRC Press.

Al-Worafi, Y.M. (2023h). Patient care errors and related problems: Monitoring errors & related problems. In: Al-Worafi, Y.M. (ed), *Patient Safety in Developing Countries: Education, Research, Case Studies*. CRC Press.

Al-Worafi, Y.M. (2023i). Patient care errors and related problems: Patient education and counselling errors and related problems. In: Al-Worafi, Y.M. (ed), *Patient Safety in Developing Countries: Education, Research, Case Studies*. CRC Press.

Al-Worafi, Y.M. (2023j). Patient safety-related issues: Other medication safety issues. In: Al-Worafi, Y.M. (ed), *Patient Safety in Developing Countries: Education, Research, Case Studies*. CRC Press.

Al-Worafi, Y.M. (2023k). Patient safety culture. In: Al-Worafi, Y.M. (ed), *Patient Safety in Developing Countries: Education, Research, Case Studies*. CRC Press.

Al-Worafi, Y.M. (2023l). Patient safety education: Competencies and learning outcomes. In: Al-Worafi, Y.M. (ed), *Patient Safety in Developing Countries: Education, Research, Case Studies*. CRC Press.

Al-Worafi, Y.M. (Ed.). (2023m). *Clinical Case Studies on Medication Safety*. Academic Press.

Al-Worafi, Y.M. (Ed.). (2023n). *Comprehensive Healthcare Simulation: Pharmacy Education, Practice and Research*. Springer Nature.

Al-Worafi, Y.M. (Ed.). (2024a). *Handbook of Medical and Health Sciences in Developing Countries*. Springer, Cham.

Al-Worafi, Y.M. (2024b). Complementary and alternative medicine (CAM) in developing countries. In: Al-Worafi, Y.M. (ed), *Handbook of Medical and Health Sciences in Developing Countries*. Springer, Cham. https://doi.org/10.1007/978-3-030-74786-2_301-1

Al-Worafi, Y.M. (2024c). Access/equitable access to medical and health sciences in developing countries. In: Al-Worafi, Y.M. (ed), *Handbook of Medical and Health Sciences in Developing Countries*. Springer, Cham. https://doi.org/10.1007/978-3-030-74786-2_147-1

Hasan, S., Al-Omar, M.J., AlZubaidy, H., and Al-Worafi, Y.M. (2019). Use of medications in Arab countries. In: Laher, I. (ed), *Handbook of Healthcare in the Arab World* (p. 42). Springer, Cham.

19 Complementary and Alternative Medicine (CAM) Education
Achievements, Challenges

INTRODUCTION

Complementary and alternative medicine (CAM) encompasses a broad range of practices, products, and therapies that are not generally considered part of conventional medicine. CAM education, therefore, involves training and learning about these various approaches to health and healing that fall outside the realm of mainstream medical practices. The education in CAM is designed to provide healthcare professionals, as well as individuals interested in personal health and wellness, with knowledge and skills in alternative approaches to prevention, diagnosis, and treatment of various health conditions. CAM education covers a wide array of topics, including but not limited to herbal medicine, acupuncture, mind-body therapies (such as meditation, yoga, and tai chi), manual therapies (like massage and chiropractic care), and energy therapies. Programs may range from introductory courses to more advanced studies, including certification and degree programs for those wishing to specialize in a particular area of CAM. Educational opportunities are offered through various institutions, including universities, community colleges, specialized schools, and online platforms, catering to a diverse audience from medical professionals to laypersons with an interest in alternative health practices. The curriculum in CAM education typically emphasizes a holistic approach to health, focusing on the integration of physical, psychological, social, and spiritual aspects of well-being. It also stresses the importance of the patient-practitioner relationship, the use of natural and less invasive interventions, and the body's inherent ability to heal itself. Ethics, cultural competence, and evidence-based practice are also crucial components of CAM education, ensuring that practitioners understand the importance of scientific research in validating CAM therapies and are aware of the ethical considerations in recommending these alternatives to conventional treatments. Moreover, CAM education often includes training on how to critically evaluate the literature on CAM practices, understand the mechanisms behind these therapies, and integrate CAM approaches into conventional medical settings. This is particularly important as the demand for CAM continues to grow, with more patients seeking complementary approaches alongside their standard medical care. Healthcare professionals trained in CAM can provide more comprehensive care by advising patients on safe and effective CAM options and incorporating these practices into treatment plans when appropriate. Overall, CAM education aims to broaden the healthcare landscape by equipping individuals with knowledge about a variety of healing practices. It promotes a more inclusive, patient-centered approach to healthcare, recognizing the value of diverse medical traditions and the potential benefits of integrating conventional and alternative therapies for optimal health outcomes.

DOI: 10.1201/9781003327202-19

COMPLEMENTARY AND ALTERNATIVE MEDICINE (CAM) EDUCATION: ACHIEVEMENTS

The field of CAM has seen significant achievements in education, reflecting a growing acceptance and integration of CAM practices within the broader healthcare landscape. These achievements have contributed to a more holistic and patient-centered approach to health and wellness, both in clinical settings and in public health strategies. Here are some notable advancements in CAM education:

1. **Incorporation into Medical School Curricula**: A substantial achievement in CAM education is its incorporation into the curricula of many medical schools around the world. This integration ensures that future healthcare providers are aware of CAM modalities, understand their uses and limitations, and can communicate effectively about CAM with their patients. It fosters a more comprehensive approach to patient care, recognizing the value of combining conventional medicine with CAM practices where evidence supports their safety and efficacy.

2. **Accreditation of CAM Programs**: The establishment and recognition of accreditation standards for CAM educational programs, including those for chiropractic, acupuncture, naturopathy, and herbal medicine, represent significant progress. Accreditation ensures that CAM practitioners receive a high standard of education and training, which is crucial for patient safety and the effectiveness of CAM treatments.

3. **Research and Evidence-Based Practice**: There has been a notable increase in research aimed at assessing the efficacy, safety, and mechanisms of CAM modalities. Many CAM educational programs now emphasize evidence-based practice, teaching students how to critically evaluate the scientific literature on CAM therapies. This focus helps integrate CAM into conventional healthcare settings based on solid evidence, improving outcomes and patient satisfaction.

4. **Expansion of Continuing Education Opportunities**: The growth of continuing education courses in CAM for healthcare professionals is another achievement. These opportunities allow practitioners to stay informed about the latest CAM research, technologies, and practices, ensuring that they can offer the most current and effective care to their patients. Continuing education in CAM also facilitates interdisciplinary learning and collaboration, bridging the gap between conventional and alternative medicine practitioners.

5. **Public Health Initiatives and Community Education**: CAM education has extended beyond healthcare professionals to include public health initiatives and community education programs. These efforts aim to empower individuals with knowledge about CAM practices, enabling them to make informed decisions about their health and wellness. Such initiatives also promote awareness and understanding of CAM among the general public, increasing its accessibility and acceptance.

6. **International Collaboration and Standardization**: The global CAM community has made strides in fostering international collaboration and standardization in CAM education. This includes the development of shared educational resources, competencies, and guidelines, facilitating a more unified and rigorous approach to CAM training worldwide. International collaboration also supports the exchange of knowledge and best practices, enhancing the quality and diversity of CAM education.

7. **Professional Recognition and Credentialing**: The development and recognition of professional credentials for CAM practitioners is a significant achievement. Credentialing processes, including certification and licensure, have been established for various CAM professions, such as acupuncture, massage therapy, and naturopathy. These credentials not only ensure that practitioners meet specific educational and professional standards but also enhance the credibility of CAM professions in the eyes of both the medical community and the public. This recognition is crucial for integrating CAM practitioners into the

healthcare system, facilitating collaboration with conventional healthcare providers, and ensuring that patients have access to qualified CAM practitioners.

8. **Digital and Online Learning Platforms**: The expansion of digital and online learning platforms dedicated to CAM education has dramatically increased access to CAM knowledge. These platforms offer a range of educational resources, from introductory courses for the general public to advanced professional development programs for healthcare providers. The availability of online learning has made CAM education more flexible and accessible, allowing learners from diverse backgrounds and geographic locations to explore the field of CAM, thus democratizing education in this field.

9. **Interprofessional Education (IPE) Programs**: The integration of CAM into Interprofessional Education (IPE) programs is another noteworthy achievement. IPE programs bring together students from various healthcare disciplines, including CAM and conventional medicine, to learn from and with each other. This approach fosters mutual respect and understanding among future healthcare providers, encouraging a collaborative, team-based approach to patient care. By including CAM in IPE, these programs prepare healthcare professionals to work effectively in integrated healthcare settings, where multiple modalities are used to achieve the best patient outcomes.

10. **Public and Private Funding for CAM Education**: Increased funding from both public and private sources for CAM education and research has significantly contributed to the field's achievements. This funding has supported the development of new educational programs, scholarships for students, and research projects aimed at expanding the evidence base for CAM practices. Financial support for CAM education has been essential for advancing the field, improving educational quality, and fostering innovation in CAM teaching and practice.

11. **Global Health Initiatives**: CAM education has played a role in global health initiatives, particularly in areas where conventional medical resources are limited or where traditional medicine is an integral part of the culture. Educational programs that incorporate CAM can offer culturally relevant, sustainable healthcare solutions. These initiatives often focus on local health needs and priorities, leveraging CAM practices alongside conventional medicine to address public health challenges, such as chronic disease management, mental health, and maternal-child health.

12. **Patient Advocacy and Empowerment**: Finally, CAM education has contributed to greater patient advocacy and empowerment. By educating patients about their health and the range of available treatment options, including CAM, individuals are better equipped to make informed healthcare decisions. This empowerment is crucial in a patient-centered healthcare model, where the values, preferences, and needs of the patient guide all aspects of care. CAM education for patients and the public fosters a more engaged and informed healthcare consumer, which is essential for improving health outcomes and satisfaction.

13. **Standardization of CAM Curricula**: The effort to standardize CAM curricula across educational institutions is a significant advancement. This standardization ensures that all students receive a consistent and comprehensive education in CAM practices, regardless of where they study. It sets a benchmark for the knowledge and skills that CAM practitioners should possess, fostering a high level of professionalism and competence in the field. This effort also facilitates the transferability of credits and recognition of qualifications across different regions and countries, promoting greater mobility among CAM practitioners.

14. **Collaboration with Healthcare Systems**: An important achievement in CAM education is the growing collaboration between CAM educational institutions and mainstream healthcare systems. This collaboration has led to the development of integrative medicine departments and clinics where CAM practitioners work alongside conventional healthcare providers. Such integrative settings offer practical training opportunities for students and serve as models for how CAM can be incorporated into conventional healthcare in a way

that benefits patients. These partnerships also facilitate research and clinical trials that contribute to the evidence base for CAM practices.

15. **Development of Specialized Journals and Publications**: The establishment of specialized journals and publications focusing on CAM education and research represents another key achievement. These platforms provide a venue for disseminating research findings, sharing best practices, and discussing challenges and innovations in CAM education. They contribute to the ongoing development of the field by fostering an academic and professional dialogue among educators, researchers, practitioners, and students.

16. **Advancements in Regulatory Frameworks**: Progress in developing regulatory frameworks for CAM professions in various countries has significantly impacted CAM education. Regulatory frameworks establish legal recognition of CAM practices and set standards for education, training, and practice. This legal recognition not only legitimizes CAM professions but also ensures that education programs are designed to meet regulatory requirements, thus safeguarding the quality of CAM education and the safety of patients receiving CAM treatments.

17. **Community-Based Learning and Service**: CAM education has increasingly incorporated community-based learning and service components. These initiatives involve students in providing CAM services to underserved populations or participating in community health projects. This hands-on experience not only enriches students' education but also instills a sense of social responsibility and a commitment to serving diverse communities. It provides students with a broader understanding of health disparities and the potential role of CAM in addressing these challenges.

18. **Focus on Sustainability and Global Health**: CAM education has increasingly emphasized sustainability and its relevance to global health challenges. Many CAM practices advocate for natural, less invasive interventions, and a holistic view of health that includes environmental factors. Educating students about the principles of sustainability within the context of healthcare encourages future practitioners to consider the environmental impact of their practices and to advocate for sustainable health solutions.

19. **Emphasis on Personal and Professional Development**: Lastly, CAM education often places a strong emphasis on the personal and professional development of students. This includes fostering skills such as empathy, mindfulness, and self-care, which are crucial for effective patient care. By prioritizing the development of the whole person, CAM education contributes to the cultivation of compassionate, reflective practitioners who are well-equipped to support the holistic well-being of their patients.

20. **Global Standards for CAM Education**: Efforts to establish global standards for CAM education ensure consistency in training and practice worldwide, fostering international recognition of CAM qualifications and facilitating cross-border practice.

21. **Innovative Teaching Methods**: The adoption of innovative teaching methods, such as simulation-based learning, virtual reality (VR), and augmented reality (AR), enhances the learning experience for CAM students, providing immersive, hands-on training in a safe and controlled environment.

22. **Patient Simulation Labs**: The use of patient simulation laboratories in CAM education allows students to practice clinical skills in a realistic, interactive setting, enhancing their preparedness for real-world patient care.

23. **E-Learning Platforms and MOOCs**: The proliferation of e-learning platforms and massive open online courses (MOOCs) in CAM education democratizes access to CAM knowledge, allowing a global audience to learn about CAM practices.

24. **Integration of Genomics and Personalized Medicine**: CAM education increasingly integrates concepts of genomics and personalized medicine, preparing practitioners to offer personalized CAM interventions based on genetic profiles and individual health needs.

25. **Focus on Interdisciplinary Research**: There's an increasing focus on interdisciplinary research involving CAM and conventional medicine, leading to a deeper understanding of how CAM practices can complement traditional treatments, enhancing patient outcomes.
26. **Expansion into Veterinary CAM**: The expansion of CAM education into veterinary medicine, including practices such as acupuncture and herbal medicine for animals, meets the growing demand for holistic pet care.
27. **Public Health Campaigns and Awareness**: CAM educational institutions often participate in public health campaigns and awareness programs, promoting the safe and informed use of CAM practices among the general population.
28. **Cultural Competence Training**: CAM education emphasizes cultural competence, preparing practitioners to understand and respect the diverse cultural backgrounds and health beliefs of their patients, ensuring culturally sensitive care.
29. **Entrepreneurship and Business Skills**: Some CAM programs incorporate entrepreneurship and business skills training, equipping graduates to open and successfully run their own CAM practices or wellness centers.
30. **Regenerative Medicine and CAM**: CAM education is exploring the intersection with regenerative medicine, such as the use of stem cells and platelet-rich plasma (PRP) in conjunction with traditional CAM practices for enhanced healing.
31. **Collaborative International Exchange Programs**: The development of international exchange programs allows CAM students and practitioners to study and work abroad, promoting cross-cultural exchange and learning.
32. **Advanced Degrees and PhD Programs**: The establishment of advanced degrees and PhD programs in CAM fields supports the development of a scholarly foundation for CAM practices, encouraging research and academic excellence.
33. **Mental Health Focus**: CAM education increasingly addresses mental health, training practitioners in CAM modalities that support mental and emotional well-being, such as mindfulness and yoga.
34. **Sustainability in CAM Practices**: CAM education includes a focus on sustainability, teaching students about sustainable harvesting of medicinal plants and the ecological impact of health interventions.
35. **Ethics and Legal Aspects**: Enhanced training on the ethics and legal aspects of CAM practice prepares students to navigate the complex regulatory landscape and make ethical decisions in their practice.
36. **Telehealth Training**: With the rise of telehealth, CAM education now includes training on how to effectively deliver CAM services remotely, ensuring practitioners can offer their services to a broader audience.
37. **Community Engagement and Service Learning**: CAM programs are increasingly incorporating service learning and community engagement projects, encouraging students to apply their skills in real-world settings and underserved communities.
38. **Wellness and Prevention**: CAM education places a strong emphasis on wellness and prevention, training practitioners to work with patients on lifestyle interventions that promote long-term health and prevent disease.
39. **Integrative Health Centers as Educational Sites**: The use of integrative health centers as educational sites provides CAM students with clinical experience in settings where CAM is practiced alongside conventional medicine.
40. **Advocacy and Policy Involvement**: CAM education encourages involvement in advocacy and policy, preparing practitioners to engage in the political process and advocate for the integration of CAM into healthcare systems.

These achievements in CAM education reflect a broader shift toward a more inclusive understanding of health and wellness, recognizing the value of integrating various healing traditions. As CAM continues to evolve, ongoing research, education, and policy development will be critical to further-ing its integration into mainstream healthcare and ensuring that CAM practices are used safely, effectively, and ethically.

COMPLEMENTARY AND ALTERNATIVE MEDICINE (CAM) EDUCATION: CHALLENGES IN THE HIGH-INCOME COUNTRIES

CAM education faces several challenges in high-income countries, despite the growing acceptance and integration of CAM practices within the broader healthcare landscape. These challenges can impact the development, delivery, and perception of CAM education, as well as its integration into mainstream healthcare systems. Here are some of the key challenges:

1. **Regulatory and Standardization Issues**: One of the primary challenges is the lack of uniform regulatory and accreditation standards for CAM education and practice. This can lead to variability in the quality of CAM education programs and make it difficult for stu-dents to discern which programs offer rigorous and comprehensive training. Furthermore, without standardized credentials, it can be challenging for CAM practitioners to gain rec-ognition and integrate into mainstream healthcare settings.

2. **Integration into Mainstream Healthcare**: Despite some progress, CAM still faces hur-dles in being fully integrated into mainstream healthcare systems. This is partly due to skepticism from within the conventional medical community, differences in healthcare philosophies, and a lack of understanding of CAM modalities among some healthcare professionals. These factors can hinder collaboration between CAM practitioners and con-ventional healthcare providers, limiting opportunities for CAM graduates.

3. **Research and Evidence Base**: Although there has been an increase in research on CAM, there is still a need for more high-quality, evidence-based research to support the efficacy and safety of various CAM practices. The lack of robust evidence can affect the cred-ibility of CAM education and its practices, making it difficult to incorporate CAM into evidence-based healthcare.

4. **Financial and Insurance Challenges**: In many high-income countries, CAM treatments are not always covered by insurance plans, which can limit patients' access to CAM ser-vices and reduce the demand for CAM practitioners. This lack of insurance coverage can also affect the viability of CAM practices, making it a less attractive career option for potential students.

5. **Curricular Integration Challenges**: Integrating CAM education into the curricula of established medical and healthcare professional schools can be challenging. Issues include limited curriculum time, competing educational priorities, and resistance from faculty who may prioritize conventional medical education. This can result in insufficient exposure to CAM principles and practices for healthcare students.

6. **Cultural and Perceptual Barriers**: Cultural and perceptual barriers can also pose chal-lenges to CAM education. Skepticism about the legitimacy and efficacy of CAM practices among both healthcare professionals and the public can impact the acceptance and uti-lization of CAM. Overcoming these barriers requires significant education and aware-ness-raising efforts.

7. **Ethical and Safety Considerations**: Ensuring that CAM education adequately addresses ethical and safety considerations is crucial. Students must be taught to understand the limi-tations of CAM practices, recognize when it is appropriate to refer patients to conventional medical professionals, and navigate the ethical implications of offering alternative treat-ments, especially when evidence is limited.

8. **Resource Limitations**: Developing comprehensive CAM educational programs can be resource-intensive, requiring access to qualified faculty, clinical training sites, and

educational materials that accurately reflect the current state of CAM research and practice. In high-income countries, where education costs are often high, securing the necessary resources to develop and maintain high-quality CAM programs can be challenging.

9. **Public Education and Awareness**: Educating the public about the value and potential benefits of CAM, as well as how to access safe and effective CAM services, remains a challenge. Misinformation and misconceptions about CAM can persist, requiring ongoing public education efforts by CAM educational institutions and practitioners.

10. **Professional Isolation**: CAM practitioners may experience professional isolation, especially in regions where CAM is not widely accepted or integrated into the healthcare system. This can limit professional development opportunities and reduce the potential for interdisciplinary collaboration.

11. **Professional Recognition and Legitimacy**: CAM disciplines struggle with varying levels of professional recognition across high-income countries. The absence of uniform licensure and regulatory standards for many CAM practices complicates professional legitimacy and public trust. Achieving parity with conventional healthcare professions requires navigating diverse regulatory landscapes, which can be a formidable barrier to CAM education and practice.

12. **Interprofessional Education and Collaboration**: While there is a push toward IPE that includes CAM, challenges remain in creating truly integrative IPE frameworks. Differences in terminology, diagnostic approaches, and treatment philosophies between CAM and conventional medicine can impede effective collaboration and learning. Overcoming these differences to foster mutual respect and understanding between future CAM practitioners and their conventional counterparts is crucial but challenging.

13. **Funding for CAM Education and Research**: Securing funding for CAM education programs and research is significantly challenging in environments where conventional medicine predominates. Limited funding affects the ability to conduct rigorous scientific research needed to build the evidence base for CAM practices, develop high-quality educational materials, and support students financially.

14. **Access to Clinical Training Sites**: CAM students often face difficulties in securing clinical training opportunities that are as structured and comprehensive as those available to conventional medical students. This is partly due to the smaller number of established CAM clinics and practitioners who can provide such training, as well as hesitancy from conventional medical facilities to host CAM students.

15. **Quality Control and Assurance**: Ensuring the quality of CAM education across various institutions poses a challenge. Without universally accepted accreditation standards and quality benchmarks, the educational experience and competence of CAM graduates can vary widely, impacting the overall perception and acceptance of CAM professions.

16. **Technological Integration**: Integrating technology into CAM education, such as telehealth, electronic health records, and advanced diagnostic tools, requires resources and training that may not be readily available in all CAM educational institutions. Keeping pace with technological advancements in healthcare is essential for CAM practitioners to remain relevant and integrated within the broader healthcare system.

17. **Cultural Sensitivity and Adaptation**: CAM practices often originate from diverse cultural traditions. Educating students about the cultural roots and significance of these practices while ensuring they are adapted and applied respectfully and appropriately in a multicultural society is a nuanced challenge. This includes navigating issues of cultural appropriation and ensuring culturally competent care.

18. **Balancing Tradition with Modernity**: CAM education must navigate the delicate balance between preserving traditional knowledge and embracing scientific inquiry and evidence-based practice. This tension can manifest in curricular design, teaching methodologies, and the integration of CAM into conventional healthcare settings.

19. **Evolving Healthcare Policies**: CAM education and practice must adapt to an ever-changing policy landscape that affects healthcare delivery, professional licensing, and insurance reimbursement. Keeping abreast of and influencing policy developments requires active engagement and advocacy from the CAM community, which can be resource-intensive and challenging.

20. **Societal Attitudes and Expectations**: Changing societal attitudes and expectations toward CAM can be slow. While there is growing interest in holistic and preventive approaches to health, skepticism and misinformation still exist. CAM education institutions have a role in public engagement and outreach to shift perceptions and educate the public on the value, potential benefits, and limitations of CAM.

21. **Environmental and Sustainability Concerns**: With an emphasis on natural products and therapies, CAM education must also address sustainability concerns related to the sourcing and use of medicinal plants and other natural resources. Teaching sustainable practices and advocating for environmental stewardship within the CAM community presents additional challenges but is essential for the long-term viability of many CAM modalities.

22. **Digital Literacy and Online Presence**: As CAM education and practice increasingly move online, ensuring digital literacy among CAM educators and practitioners becomes critical. Developing a strong online presence, including the use of social media for education and outreach, presents challenges in terms of resources, skills, and maintaining professional integrity in digital spaces.

23. **Adapting to Rapid Changes in Healthcare**: The healthcare sector is rapidly evolving, with advances in biotechnology, genomics, and personalized medicine reshaping the landscape. CAM education must remain agile, updating curricula to reflect these advances and exploring how CAM can complement emerging healthcare technologies and approaches.

24. **Ethical Sourcing and Fair Trade Practices**: With CAM's reliance on herbal medicines and natural products, ethical sourcing and adherence to fair trade practices become significant concerns. Educating students about the ethical complexities of global supply chains and the importance of sustainability is essential but challenging, requiring a deep understanding of global health and economics.

25. **Navigating the Digital Health Revolution**: The rise of digital health technologies, including wearable devices and health apps, intersects with many CAM practices that emphasize self-monitoring and wellness. CAM education needs to incorporate these technologies, teaching students how to evaluate and integrate digital health tools into their practice responsibly.

26. **Addressing Health Disparities**: CAM education must also tackle the challenge of health disparities, ensuring that CAM practices are accessible and relevant to diverse populations, including marginalized and underserved communities. This involves not only cultural competence but also advocacy for health equity and accessibility.

27. **Intersecting with Pharmaceutical Sciences**: As interest grows in natural products and their pharmacological applications, CAM education faces the challenge of intersecting more closely with pharmaceutical sciences. This requires developing curricula that cover pharmacognosy, drug interactions, and the evidence-based use of herbal medicines, necessitating a bridging of disciplines that have traditionally been separate.

28. **Mental Health Integration**: Integrating mental health more thoroughly into CAM education reflects an understanding of the mind-body connection central to many CAM practices. Addressing the stigma around mental health, teaching holistic approaches to mental wellness, and understanding the interplay between mental and physical health are crucial areas for development.

29. **Challenges in Patient Education**: CAM practitioners often take on the role of educators for their patients, requiring skills in communication and patient education that may not be adequately emphasized in all CAM programs. Training in these areas is essential for

practitioners to effectively convey the benefits, limitations, and appropriate use of CAM therapies to their patients.

30. **Legal and Insurance Navigation**: The legal landscape and insurance systems in high-income countries present ongoing challenges for CAM education and practice. Understanding and navigating these systems is crucial for CAM practitioners, requiring education on legal rights, professional responsibilities, and how to advocate for greater insurance coverage of CAM therapies.

31. **Building Interdisciplinary Research Capacities**: Enhancing capacities for interdisciplinary research within CAM education is critical for advancing the evidence base of CAM practices. This involves fostering collaborations across disciplines, training in research methodologies that are appropriate for CAM studies, and securing funding for CAM research projects.

32. **Enhancing Public Health Integration**: CAM's integration into public health initiatives and policies is an area for growth. CAM education can play a role by training practitioners in public health principles, preventive care, and the role of CAM in addressing public health challenges, such as chronic disease management and pandemic responses.

33. **Professional Networking and Mentorship**: Building professional networks and mentorship opportunities for CAM students and practitioners can be challenging in environments where CAM is less established. Developing these networks is crucial for professional growth, knowledge exchange, and the advancement of CAM practices.

34. **Evolving Patient-Practitioner Dynamics**: CAM education must adapt to changing patient-practitioner dynamics, where patients are increasingly informed and active participants in their healthcare decisions. This shift requires training CAM practitioners in shared decision-making, patient-centered care, and navigating informed consent in the context of CAM therapies.

35. **Sustaining Traditional Knowledge**: As CAM practices are often rooted in traditional knowledge, there's a challenge in sustaining and respectfully integrating this knowledge into modern CAM education. This involves ethical considerations, respecting intellectual property rights, and ensuring that traditional knowledge holders are recognized and involved in the educational process.

Addressing these challenges requires concerted efforts from educators, practitioners, policymakers, and researchers. Strategies include advocating for regulatory reforms, enhancing the quality and quantity of CAM research, improving CAM education standards, and fostering collaboration between CAM and conventional healthcare providers. Overcoming these hurdles is essential for fully realizing the potential of CAM to contribute to comprehensive, holistic, and patient-centered care.

COMPLEMENTARY AND ALTERNATIVE MEDICINE (CAM) EDUCATION: CHALLENGES IN THE MIDDLE-INCOME COUNTRIES

CAM education in middle-income countries faces a unique set of challenges. While there is often a rich cultural tradition of using CAM practices in these regions, integrating these practices into formal education and healthcare systems can be complicated by economic, structural, and social factors. Here are some of the key challenges:

1. **Limited Financial Resources**: One of the primary challenges is the limited financial resources available for education and healthcare in many middle-income countries. This scarcity can affect the development and maintenance of CAM educational programs, research into CAM practices, and access to CAM treatments. Funding constraints can also

limit the availability of scholarships for students interested in pursuing CAM education, thereby affecting the diversity of the student body and the accessibility of CAM education for underprivileged groups.

2. **Regulatory and Legislative Frameworks**: In many middle-income countries, there is a lack of clear regulatory and legislative frameworks governing CAM practices and education. This can lead to issues with standardization, quality control, and recognition of CAM qualifications both within and across countries. Without proper regulation, there is also a risk of unethical practices and the exploitation of vulnerable populations.

3. **Integration into Mainstream Healthcare Systems**: Integrating CAM into mainstream healthcare systems poses significant challenges. These include resistance from conventional medical practitioners, lack of understanding or acceptance of CAM practices among healthcare professionals, and insufficient infrastructure to support an integrated approach to healthcare. This situation can lead to CAM being marginalized or not recognized as a legitimate healthcare option.

4. **Quality of Education and Training**: Ensuring the quality of CAM education and training can be difficult due to a lack of standardized curricula, insufficiently trained faculty, and inadequate infrastructure for teaching and clinical practice. This variability in education quality can lead to discrepancies in the knowledge and skills of CAM practitioners, affecting the overall standard of care provided to patients.

5. **Research and Evidence Base**: Building a robust evidence base for CAM practices through research is a challenge in many middle-income countries, where research funding is often limited. The lack of high-quality research affects the credibility and acceptance of CAM practices within the broader medical community and among the public. It also hinders the development of evidence-based CAM curricula.

6. **Cultural and Social Acceptance**: While CAM practices may be rooted in the traditional knowledge of many middle-income countries, there can still be challenges related to cultural and social acceptance. This includes balancing traditional beliefs with the need for evidence-based practice, addressing skepticism among certain segments of the population, and overcoming the stigma associated with certain CAM practices.

7. **Access to CAM Therapies and Products**: Access to quality CAM therapies and products can be inconsistent in middle-income countries. Issues with the supply chain, affordability of treatments, and the availability of quality-controlled herbs and supplements can limit patient access to safe and effective CAM options.

8. **Public Awareness and Education**: There is often a need for greater public awareness and education about CAM practices, including their benefits, limitations, and safe use. Misinformation and lack of knowledge can lead to inappropriate use of CAM therapies, potentially endangering public health.

9. **Professional Development and Continuing Education**: Opportunities for professional development and continuing education for CAM practitioners can be limited in middle-income countries. This can hinder practitioners' ability to stay updated on the latest research, advancements, and best practices in CAM, affecting the quality of care they provide.

10. **Healthcare Infrastructure and Support**: The lack of healthcare infrastructure and support for CAM practices, including logistical issues such as the availability of treatment spaces and the integration of CAM services into primary healthcare facilities, can be significant barriers to the widespread acceptance and use of CAM.

11. **Cross-Cultural Competency**: Many middle-income countries are culturally diverse, making it essential for CAM education to incorporate cross-cultural competencies that respect and integrate various traditional health practices. However, developing curricula that are inclusive of diverse traditions while ensuring scientific rigor and evidence-based

practice can be challenging. This diversity also necessitates training practitioners who are sensitive to cultural nuances in health beliefs and practices.

12. **Digital Divide**: While digital technology offers opportunities for expanding access to CAM education through online platforms and telehealth services, the digital divide between urban and rural areas, or across different socioeconomic groups, can limit these opportunities. In some regions, limited internet access and a lack of digital literacy among both practitioners and patients can hinder the adoption of technology in CAM education and practice.

13. **Intellectual Property Rights**: Protecting the intellectual property rights of traditional and indigenous CAM practices poses a significant challenge. There is a need for legal frameworks that safeguard traditional knowledge against exploitation while allowing for its integration into formal CAM education and practice. However, establishing such frameworks that balance protection with sharing and innovation is complex and often contentious.

14. **Intersectoral Collaboration**: Effective integration of CAM into national health systems requires collaboration across various sectors, including health, education, finance, and environment. However, fostering intersectoral collaboration can be difficult due to differing priorities, bureaucratic hurdles, and a lack of a common framework for CAM integration.

15. **Sustainability of CAM Resources**: Many CAM practices rely on natural resources, such as medicinal plants and herbs. Overharvesting, environmental degradation, and climate change pose significant threats to the sustainability of these resources. Educating CAM practitioners and students about sustainable practices and the conservation of medicinal plants is essential but challenging in the face of broader environmental issues.

16. **Ethical Practice and Patient Safety**: Ensuring ethical practice and patient safety within CAM education and practice is a critical challenge. This includes issues such as informed consent, understanding and managing potential interactions between CAM therapies and conventional medicines, and recognizing the limits of CAM practices and when to refer patients to conventional healthcare providers.

17. **Inequities in Healthcare Access**: In many middle-income countries, inequities in healthcare access can be exacerbated by the integration of CAM if not carefully managed. Ensuring that CAM services are accessible and affordable to all segments of the population, including marginalized and underserved communities, is a challenge that requires thoughtful policy and funding strategies.

18. **Faculty Development and Support**: Developing and supporting faculty who are knowledgeable and skilled in both CAM practices and pedagogical strategies is a significant challenge. There may be a shortage of qualified educators who can effectively bridge traditional knowledge with modern scientific approaches in CAM education.

19. **Quality Assurance and Accreditation**: Establishing and maintaining quality assurance and accreditation mechanisms for CAM educational programs and institutions is a complex process. Without universally recognized standards and accreditation bodies, it can be difficult to ensure the quality and credibility of CAM education and practitioners.

20. **Promoting Interdisciplinary Research**: Encouraging and facilitating interdisciplinary research that includes CAM is crucial for building the evidence base and advancing the field. However, securing funding for research that crosses traditional disciplinary boundaries, and fostering collaborations between CAM practitioners, biomedical researchers, and social scientists, can be challenging.

21. **Adapting to Global Health Trends**: Adapting CAM education to align with global health trends and priorities, such as non-communicable diseases (NCDs), mental health, and aging populations, requires continuous curriculum innovation and the ability to integrate CAM practices into broader health strategies.

22. **Language and Communication Barriers**: CAM practices often have roots in local cultures and languages, which can present barriers when trying to standardize education and integrate practices into broader healthcare settings that operate in different or multiple languages. Developing bilingual or multilingual educational materials and ensuring practitioners are proficient in the dominant languages of their practice areas are crucial but challenging tasks.

23. **Navigating Changes in Public Perception**: Public perception of CAM can vary widely, influenced by cultural traditions, colonial histories, and the influence of Western medicine. Educating the public to navigate these perceptions, discern reputable CAM practices from unproven ones, and understand the role of CAM in a comprehensive healthcare system requires strategic communication and community engagement efforts.

24. **Data Privacy and Security in CAM Practice**: As CAM practitioners increasingly use digital tools for patient records, telehealth, and other services, ensuring data privacy and security becomes a significant challenge. Many middle-income countries may lack the robust infrastructure and regulations to protect patient data, posing risks to patient privacy and the credibility of CAM practices.

25. **Access to Scientific Literature and Resources**: For CAM educators and practitioners in middle-income countries, accessing the latest scientific literature and educational resources can be difficult due to high subscription costs and limited library resources. This can hinder the development of evidence-based CAM curricula and the continuous professional development of practitioners.

26. **Balancing Global Influences with Local Needs**: CAM education in middle-income countries must balance the influence of global CAM trends and practices with local health needs and cultural practices. Adapting global knowledge to local contexts in a way that respects cultural heritage while ensuring scientific rigor and relevance to local health challenges requires nuanced understanding and flexibility.

27. **Professional Isolation and Networking**: CAM practitioners and educators in middle-income countries may face professional isolation, particularly if they are working in areas where CAM is not widely practiced or recognized. Building professional networks and finding mentorship opportunities can be difficult, impacting professional development and the exchange of knowledge and best practices.

28. **Environmental Health and CAM**: Many CAM practices emphasize the importance of living in harmony with the natural environment. However, addressing environmental health issues, such as pollution and climate change, within CAM education and practice poses challenges but is increasingly important for ensuring the holistic well-being of communities.

29. **Adapting to Patient Mobility and Migration**: Patient mobility and migration, both within and between countries, can introduce new health challenges and demands on CAM practitioners, including dealing with a range of health beliefs and practices and communicable diseases. Preparing CAM practitioners to work effectively with diverse and transient populations is essential but requires resources and flexibility.

30. **Sociopolitical Factors Influencing CAM**: The sociopolitical environment in middle-income countries can significantly impact CAM education and practice. Issues such as political instability, regulatory changes, and shifts in health policy priorities can affect funding, legal recognition, and the operational environment for CAM institutions and practitioners.

31. **Ethical Considerations in Global Health**: Ethical considerations, including respect for indigenous knowledge, consent in the use of traditional practices, and equity in health access, are critical in CAM education. Navigating these ethical considerations while promoting global health and respecting local cultures requires a careful and informed approach.

32. **Interdisciplinary Education Models**: Developing and implementing interdisciplinary education models that include CAM alongside conventional medical training poses logistical and philosophical challenges. These models require collaboration across disciplines, which can be hindered by differing terminologies, methodologies, and epistemologies.

33. **Clinical Placement and Practical Training Opportunities**: Finding adequate clinical placement and practical training opportunities for CAM students can be challenging, especially in regions where CAM is not fully integrated into the healthcare system. Establishing partnerships with healthcare providers and CAM practitioners to offer these opportunities is essential for comprehensive training.

34. **Responding to Health Emergencies and Pandemics**: The COVID-19 pandemic highlighted the importance of integrating CAM into responses to health emergencies. Preparing CAM practitioners and educators to contribute effectively to pandemic preparedness and response, including understanding public health measures and vaccine education, presents new challenges and opportunities for CAM education.

Addressing these challenges in middle-income countries requires targeted efforts to strengthen financial resources, develop regulatory frameworks, improve education and training quality, foster research, and promote public awareness and acceptance of CAM. Collaborative efforts between governments, educational institutions, healthcare providers, and international organizations can help to overcome these hurdles, ensuring that CAM contributes positively to holistic, accessible, and culturally appropriate healthcare systems.

COMPLEMENTARY AND ALTERNATIVE MEDICINE (CAM) EDUCATION: CHALLENGES IN THE LOW-INCOME COUNTRIES

CAM education in low-income countries faces a unique set of challenges that reflect the broader socioeconomic and healthcare context of these regions. While CAM may be deeply integrated into the cultural and traditional practices of many low-income countries, formalizing and integrating these practices into the healthcare system and education framework presents several hurdles:

1. **Resource Constraints**: The most significant challenge is the severe resource constraints faced by low-income countries. Limited financial, infrastructural, and human resources can hinder the development and delivery of formal CAM education programs. This scarcity of resources affects not only the establishment of educational institutions but also the quality and accessibility of CAM education.

2. **Lack of Formal Recognition and Regulation**: In many low-income countries, there is a lack of formal recognition and regulation of CAM practices and education. This absence of a regulatory framework can lead to variability in the quality of CAM training and services, making it difficult to integrate CAM practitioners into the formal healthcare system and to ensure patient safety and efficacy of treatments.

3. **Integration into National Healthcare Systems**: Integrating CAM into national healthcare systems is a challenge, given the dominance of Western medical models and the potential resistance from conventional medical communities. There may also be a lack of policies supporting the integration of CAM education and practices into primary healthcare services, limiting the role of CAM in public health strategies.

4. **Access to and Preservation of Traditional Knowledge**: In low-income countries, much of the knowledge about CAM practices is passed down through generations and may not be formally documented. There are challenges related to accessing this knowledge, ensuring its accurate transmission through formal education, and protecting it from exploitation.

Additionally, the erosion of traditional knowledge due to globalization and cultural shifts poses a threat to the sustainability of CAM practices.

5. **Research and Evidence-Based Practice**: Conducting research to build an evidence base for CAM practices is particularly challenging in low-income countries due to limited funding, research infrastructure, and capacity. This lack of evidence hampers the legitimacy and acceptance of CAM practices both within the country and in the global health community.

6. **Educational Infrastructure and Faculty Development**: Developing the necessary educational infrastructure, including facilities, learning materials, and qualified faculty, is a significant challenge. There may be a shortage of educators who are trained in both CAM practices and pedagogical methods, affecting the quality of CAM education.

7. **Cultural and Social Acceptance**: While CAM practices may be culturally significant, there can be challenges related to their social acceptance, especially when they conflict with prevailing health beliefs or when there is pressure to conform to Western medical practices. Educating the public and healthcare professionals about the value and potential benefits of CAM in a culturally sensitive manner is crucial.

8. **Quality Control and Standardization**: Ensuring the quality and standardization of CAM practices and products, such as herbal medicines, is challenging. There may be issues related to the adulteration, contamination, and inconsistent potency of herbal medicines, posing risks to patient safety.

9. **Affordability and Access to CAM Therapies**: Despite being in low-income settings, CAM therapies may not always be affordable or accessible to all segments of the population. The cost of training, treatments, and herbal medicines can be prohibitive for many, limiting the reach and impact of CAM.

10. **Professional Development and Continuing Education**: Opportunities for professional development and continuing education for CAM practitioners are limited. This restricts the ability of practitioners to update their knowledge and skills, impacting the quality of care provided.

11. **Public Health Integration**: Integrating CAM into public health initiatives and strategies for disease prevention and health promotion is a challenge. There is a need for CAM practices to be recognized as valuable tools in addressing public health challenges, such as malnutrition, infectious diseases, and chronic conditions.

12. **Intellectual Property and Biopiracy Concerns**: Protecting the intellectual property rights of traditional and indigenous CAM knowledge against biopiracy is a significant challenge. There's a need for legal frameworks that recognize and protect traditional knowledge while allowing for its ethical use and application in CAM education and practice. However, establishing such frameworks and ensuring they are respected internationally can be difficult.

13. **Interdisciplinary Collaboration**: Encouraging and facilitating interdisciplinary collaboration between CAM practitioners and conventional healthcare providers is challenging due to differences in training, language, and professional culture. Such collaboration is essential for integrating CAM into holistic healthcare approaches but requires overcoming mutual skepticism and building respect and understanding across disciplines.

14. **Environmental Sustainability**: Many CAM practices rely on natural resources, which in low-income countries may be under threat from environmental degradation, climate change, and unsustainable harvesting practices. Educating CAM practitioners about sustainable practices and the importance of biodiversity conservation is crucial but challenging amidst broader environmental and economic pressures.

15. **Digital Divide and Technological Access**: While digital health technologies offer the potential to expand access to CAM education and services, the digital divide limits these opportunities in many low-income countries. Lack of access to reliable internet and digital

devices, as well as limited digital literacy among both practitioners and patients, can hinder the integration of technology in CAM practice and education.

16. **Health Literacy and Community Engagement**: Engaging communities in understanding and valuing evidence-based CAM practices requires overcoming barriers related to health literacy. Misconceptions about CAM and traditional practices can persist, and there's a need for educational initiatives that respect cultural beliefs while promoting informed healthcare choices.

17. **Funding for CAM Education and Research**: Securing funding for CAM education programs and research is particularly challenging in low-income countries, where health and education budgets are already stretched thin. International funding and partnerships can play a role, but they must be carefully managed to ensure alignment with local needs and priorities.

18. **Scalability and Replicability of Successful Models**: Identifying and scaling up successful models of CAM education and integration into healthcare systems is challenging. Solutions that work in one context may not be directly transferable to another, requiring adaptations that consider local cultures, healthcare infrastructures, and resource availability.

19. **Ethical Practice and Regulation**: Ensuring ethical practice among CAM practitioners and protecting patients from potential harm is crucial. This includes addressing issues related to informed consent, confidentiality, and the ethical use of traditional medicines and practices. Developing and enforcing ethical guidelines and regulatory standards in environments where CAM is not formally recognized is challenging.

20. **Disaster and Emergency Response**: Integrating CAM into disaster and emergency response efforts in low-income countries presents unique challenges. Traditional and CAM practices can play a vital role in these contexts, providing culturally appropriate care and supporting community resilience. However, formal recognition and integration of CAM into emergency health response plans are often lacking.

21. **Migration and Cultural Exchange**: Migration, both within countries and across borders, can introduce new CAM practices and knowledge but can also lead to the dilution or loss of traditional knowledge. Educating CAM practitioners to navigate these dynamics, and respecting the diversity of practices while ensuring consistent standards of care, is a complex challenge.

22. **Non-communicable Diseases (NCDs)**: With the rising prevalence of NCDs in low-income countries, integrating CAM into prevention and management strategies is increasingly important. However, adapting CAM education to address NCDs, alongside infectious diseases and maternal-child health, requires broadening the scope of CAM curricula and practice.

23. **Conflict and Political Instability**: In low-income countries affected by conflict and political instability, CAM education and practice face disruptions and challenges related to security, infrastructure damage, and displacement of populations. Ensuring the continuity of CAM education and services in these contexts requires resilience and innovative approaches.

24. **Networking and Professional Support**: Building networks and professional support systems for CAM practitioners in low-income countries can be difficult due to geographical isolation, limited professional organizations, and lack of recognition of CAM practices. Establishing these networks is essential for professional development, mentorship, and the advancement of CAM.

25. **Supply Chain and Quality Assurance for CAM Products**: Ensuring a reliable supply chain and quality assurance for CAM products, such as herbal medicines, is a significant challenge. In low-income countries, where regulatory oversight may be minimal, there's a risk of adulteration, contamination, or misidentification of medicinal plants and products. Establishing quality control mechanisms and supply chain standards is crucial but challenging without adequate resources and regulatory infrastructure.

26. **Navigating Global Health Priorities**: Aligning CAM education and practices with global health priorities, such as antimicrobial resistance, HIV/AIDS, and maternal and child health, requires not only a deep understanding of these issues but also an ability to integrate CAM approaches in a way that complements conventional health strategies. This necessitates a broadening of CAM curricula to include global health perspectives and evidence-based integration strategies.

27. **Cultural Preservation vs. Modernization**: Balancing the preservation of traditional CAM practices with the pressures of modernization and globalization presents a delicate challenge. There's a need to respect and preserve cultural heritage while also ensuring that CAM practices are adapted to meet current health needs and standards. This balance involves promoting the documentation and formal education of traditional practices while fostering innovation and evidence-based development.

28. **Healthcare Workforce Development**: Developing a healthcare workforce that is knowledgeable and skilled in both CAM and conventional medicine is essential for integrated healthcare delivery. However, in low-income countries, there may be a shortage of trained healthcare professionals, including CAM practitioners. Expanding CAM education to produce a competent workforce involves addressing broader challenges in healthcare education, recruitment, and retention.

29. **Language and Knowledge Translation**: Translating CAM knowledge, which is often embedded in local languages and cultural contexts, into forms that can be integrated into formal education and healthcare delivery, poses a linguistic and epistemological challenge. Developing bilingual or multilingual educational resources and ensuring that CAM practitioners are proficient in relevant languages for their practice areas are essential but complex tasks.

30. **Infrastructure for Clinical Training and Practice**: The lack of infrastructure for clinical training and practice in CAM is a significant barrier. This includes not only physical facilities but also access to clinical materials, patient populations for training purposes, and opportunities for hands-on clinical experience. Building or enhancing infrastructure requires substantial investment and strategic planning.

31. **Fostering Local and International Partnerships**: Developing local and international partnerships can help address many of the challenges faced by CAM education in low-income countries, including resource constraints, research and development, and integration into healthcare systems. However, fostering effective partnerships requires navigating differences in priorities, expectations, and cultural understandings, as well as ensuring equitable collaboration.

32. **Socioeconomic Determinants of Health**: Addressing the broader socio-economic determinants of health within CAM education and practice is crucial, as these factors significantly impact health outcomes in low-income countries. Integrating an understanding of social determinants into CAM education helps prepare practitioners to address health inequities and advocate for systemic changes.

33. **Ethical Harvesting and Biodiversity Conservation**: The ethical harvesting of medicinal plants and conservation of biodiversity are critical issues in CAM practice, especially in low-income countries rich in biodiversity but facing environmental pressures. Educating CAM practitioners about sustainable and ethical harvesting practices and the importance of biodiversity conservation is essential but requires integration with environmental education and policy.

34. **Adapting to Climate Change**: Climate change poses a significant threat to health and well-being, impacting the availability of medicinal plants and the burden of disease. Integrating climate change adaptation into CAM education and practice involves understanding the impacts of climate change on health and developing strategies to mitigate these impacts through CAM practices and public health interventions.

35. **Emergency and Humanitarian Response**: Enhancing the role of CAM in emergency and humanitarian responses, particularly in low-income countries that may be more vulnerable to natural disasters, conflict, and health emergencies, presents challenges in terms of logistics, integration into emergency health services, and cultural sensitivity. Training CAM practitioners to participate effectively in emergency responses requires a broad skill set and understanding of emergency health systems.

Addressing these challenges in low-income countries requires a multifaceted approach that includes strengthening the educational and healthcare infrastructure, developing regulatory frameworks, enhancing research capacity, and promoting the integration of CAM into national healthcare systems. Collaborative efforts between governments, international organizations, and local communities are essential to leverage the potential of CAM to improve health outcomes and preserve cultural heritage.

COMPLEMENTARY AND ALTERNATIVE MEDICINE (CAM) EDUCATION: RECOMMENDATIONS

Addressing and overcoming the challenges faced by CAM education, especially within diverse economic contexts across high-income, middle-income, and low-income countries, requires a multifaceted and strategic approach. Here are comprehensive recommendations to improve CAM education globally (Al-Worafi, 2020a,b, 2022a,b, 2023a–n, 2024a–c; Hasan et al., 2019):

1. **Enhance Regulatory Frameworks**
   - **Establish clear regulatory standards** for CAM practices and education, ensuring quality control and safety.
   - **Develop accreditation processes** for CAM educational institutions and programs to standardize education and training.
2. **Integrate CAM into Healthcare Systems**
   - **Promote the integration of CAM** within national healthcare systems through policies that recognize the value of CAM practices alongside conventional medicine.
   - **Encourage interdisciplinary collaboration** between CAM practitioners and conventional healthcare providers to offer integrated patient care.
3. **Increase Research and Evidence-Based Practice**
   - **Invest in CAM research** to build a robust evidence base for the efficacy and safety of CAM modalities.
   - **Incorporate evidence-based practice** into CAM education, teaching students to critically evaluate and apply research findings.
4. **Improve Access and Affordability**
   - **Subsidize CAM education** and training for students from diverse backgrounds to enhance accessibility.
   - **Work with insurance companies** to include CAM treatments in coverage plans, making CAM more accessible to patients.
5. **Develop Quality Educational Resources**
   - **Create standardized, high-quality educational materials** and curricula that reflect current research and practices in CAM.
   - **Utilize digital platforms** to extend the reach of CAM education, making it accessible to students and practitioners in remote areas.
6. **Foster Cultural Competence and Preservation**
   - **Incorporate cultural competence training** into CAM education, respecting the diverse origins and traditions of CAM practices.

- **Document and preserve traditional knowledge** in partnership with indigenous and local communities, ensuring ethical use and protection of intellectual property.

7. **Support Professional Development and Networking**
   - **Offer continuing education opportunities** for CAM practitioners to update their knowledge and skills.
   - **Build professional networks** and associations for CAM practitioners to share knowledge, collaborate, and advocate for the recognition of CAM.

8. **Strengthen Intersectoral Collaboration**
   - **Engage in intersectoral collaborations** between health, education, environment, and other relevant sectors to address multifaceted challenges in CAM education and practice.
   - **Promote public-private partnerships** to leverage resources for the development and delivery of CAM education.

9. **Advocate for Public Awareness and Education**
   - **Conduct public awareness campaigns** to educate the public about the benefits and limitations of CAM, promoting informed health choices.
   - **Develop community outreach programs** to demonstrate the practical benefits of CAM in improving health and wellness.

10. **Address Global Health Priorities**
    - **Align CAM education and research** with global health priorities, such as NCDs, mental health, and climate change, to ensure CAM contributes effectively to global health challenges.
    - **Incorporate global health perspectives** into CAM curricula, preparing practitioners to work in diverse health contexts.

11. **Ensure Sustainability and Environmental Responsibility**
    - **Teach sustainable practices** within CAM education, emphasizing the responsible use of natural resources and conservation of medicinal plants.
    - **Promote ethical harvesting and biodiversity conservation** through partnerships with environmental organizations and adherence to international guidelines.

12. **Leverage Technology and Innovation**
    - **Incorporate digital health technologies** in CAM education and practice, including telehealth, mobile health apps, and digital diagnostics.
    - **Explore innovative teaching methods,** such as simulation-based learning and virtual reality, to enhance the learning experience.

13. **Adapt to Local Needs and Contexts**
    - **Tailor CAM education programs** to meet local health needs and cultural contexts, ensuring the relevance and effectiveness of CAM practices.
    - **Engage local communities** in the development and delivery of CAM education to ensure it is grounded in local knowledge and practices.

14. **Enhance Disaster and Emergency Preparedness**
    - **Train CAM practitioners in emergency and disaster response**, integrating CAM into holistic approaches to emergency healthcare.
    - **Develop guidelines for the inclusion of CAM in emergency health services**, ensuring CAM practitioners can contribute effectively in crises.

15. **Develop Specialized Training Centers**
    - **Establish specialized CAM training centers** that serve as hubs for education, research, and clinical practice, providing students with hands-on learning experiences.
    - These centers can also act as resources for continuing education and professional development for existing practitioners.

16. **Promote Ethical International Collaboration**
   - **Foster ethical international collaborations** that respect and protect traditional knowledge while facilitating cross-cultural exchanges of CAM practices and education.
   - Ensure that collaborations are mutually beneficial, with a focus on building capacity in low- and middle-income countries.

17. **Implement Mentorship Programs**
   - **Create mentorship programs** that connect CAM students and early-career practitioners with experienced professionals in their field.
   - Mentorship can provide guidance, enhance clinical skills, and support professional development, particularly in areas with fewer CAM practitioners.

18. **Encourage Student Research and Innovation**
   - **Incentivize research and innovation** among CAM students by offering research grants, awards, and opportunities to present their work at conferences and in academic journals.
   - This encourages a culture of inquiry and evidence-based practice from the outset of their careers.

19. **Strengthen Language and Communication Skills**
   - **Emphasize language and communication skills** in CAM education to ensure practitioners can effectively engage with patients from diverse backgrounds and participate in interdisciplinary healthcare teams.
   - Include training in medical terminology, patient counseling, and cultural sensitivity.

20. **Enhance Access to Scientific Literature**
   - **Improve access to scientific literature and resources** for CAM educators and practitioners, especially in low-resource settings, through partnerships with academic institutions, libraries, and publishers.
   - Utilize open-access journals and digital libraries to reduce barriers to information.

21. **Address Socioeconomic Determinants of Health**
   - **Integrate an understanding of socioeconomic determinants of health** into CAM education, preparing practitioners to address broader factors that impact patient health and wellness.
   - This approach encourages holistic care that considers the patient's environment, lifestyle, and social context.

22. **Utilize Blended Learning Models**
   - **Adopt blended learning models** that combine online education with in-person training to increase the flexibility and accessibility of CAM education.
   - This can make education more accessible to students in remote areas or those with limited mobility.

23. **Standardize Clinical Practicum Requirements**
   - **Standardize clinical practicum requirements** across CAM education programs to ensure that students receive consistent and comprehensive practical experience.
   - Establish partnerships with healthcare facilities to provide diverse clinical placement opportunities.

24. **Promote Patient Safety and Quality Care**
   - **Emphasize patient safety and quality care** in CAM curricula, teaching students about potential interactions with conventional treatments, contraindications, and the importance of referral to other healthcare providers when necessary.
   - Develop guidelines and protocols for safe CAM practice.

25. **Foster Community-Based Learning**
   - **Implement community-based learning projects** where CAM students can engage with local communities, understand their health needs, and apply CAM solutions in real-world settings.

- This approach enhances students' learning experiences and benefits communities by providing access to CAM services.

26. **Advocate for Inclusive Health Policies**
    - **Engage in advocacy for inclusive health policies** that recognize and support the role of CAM in the healthcare system.
    - Work with policymakers to ensure that CAM is included in health planning, funding, and delivery at national and local levels.

Implementing these recommendations requires coordinated efforts from governments, educational institutions, healthcare providers, professional associations, and international organizations. By addressing these challenges comprehensively, CAM education can be improved, making it a valuable component of a holistic and integrative approach to health and wellness globally.

CONCLUSION

The landscape of CAM education faces a multitude of challenges across different economic contexts, from high-income to low-income countries. These challenges range from regulatory and standardization issues, and integration into healthcare systems, to ensuring the quality and accessibility of education and practice. However, with these challenges come opportunities for growth, innovation, and integration that can significantly enhance healthcare delivery and patient outcomes globally. The recommendations provided offer a blueprint for overcoming these challenges. By enhancing regulatory frameworks, integrating CAM into healthcare systems, increasing research and evidence-based practice, and improving access and affordability, CAM education can be significantly improved. Furthermore, developing quality educational resources, fostering cultural competence, supporting professional development, and addressing global health priorities are essential steps toward a more holistic and integrated approach to healthcare. Implementing these strategies requires a collaborative effort among governments, educational institutions, healthcare providers, professional associations, and international organizations. It also necessitates a commitment to respecting cultural traditions while embracing scientific rigor and evidence-based practices. The ultimate goal is to ensure that CAM education contributes to a healthcare landscape that is inclusive, sustainable, and responsive to the diverse needs of populations worldwide. By addressing the outlined challenges and adopting the recommended strategies, there is a tremendous opportunity to leverage the unique strengths of CAM. This approach not only enhances the diversity and richness of healthcare options available but also promotes a more holistic, patient-centered approach to health and wellness. In conclusion, improving CAM education is not just about preserving traditional knowledge or expanding healthcare choices; it's about enriching the healthcare ecosystem with diverse, evidence-based, and culturally sensitive approaches to healing and wellness. As the global healthcare landscape evolves, CAM education and practice stand poised to play a pivotal role in shaping a more integrative, effective, and compassionate approach to healthcare.

REFERENCES

Al-Worafi, Y.M. (Ed.). (2020a). *Drug Safety in Developing Countries: Achievements and Challenges.* Academic Press.

Al-Worafi, Y.M. (2020b). Herbal medicines safety issues. In: Al-Worafi, Y.M. (ed), *Drug Safety in Developing Countries* (pp. 163–178). Academic Press.

Al-Worafi, Y.M. (2022a). *A Guide to Online Pharmacy Education: Teaching Strategies and Assessment Methods.* CRC Press.

Al-Worafi, Y.M. (2022b). Competencies and learning outcomes. In: Al-Worafi, Y.M. (ed), *A Guide to Online Pharmacy Education: Teaching Strategies and Assessment Methods.* CRC Press.

Al-Worafi, Y.M. (2023a). *Patient Safety in Developing Countries: Education, Research, Case Studies.* CRC Press.

Al-Worafi, Y.M. (2023b). *Technology for Drug Safety: Current Status and Future Developments.* Springer Nature.

Al-Worafi, Y.M. (2023c). Patient safety-related issues: Patient care errors and related problems. In: Al-Worafi, Y.M. (ed), *Patient Safety in Developing Countries: Education, Research, Case Studies.* CRC Press.

Al-Worafi, Y.M. (2023d). Patient care errors and related problems: Preventive medicine errors & related problems. In: Al-Worafi, Y.M. (ed), *Patient Safety in Developing Countries: Education, Research, Case Studies.* CRC Press.

Al-Worafi, Y.M. (2023e). Patient care errors and related problems: Patient assessment and diagnostic errors & related problems. In: Al-Worafi, Y.M. (ed), *Patient Safety in Developing Countries: Education, Research, Case Studies.* CRC Press.

Al-Worafi, Y.M. (2023f). Patient care errors and related problems: Non-pharmacological errors & related problems. In: Al-Worafi, Y.M. (ed), *Patient Safety in Developing Countries: Education, Research, Case Studies.* CRC Press.

Al-Worafi, Y.M. (2023g). Patient care errors and related problems: Medical errors & related problems. In: Al-Worafi, Y.M. (ed), *Patient Safety in Developing Countries: Education, Research, Case Studies.* CRC Press.

Al-Worafi, Y.M. (2023h). Patient care errors and related problems: Monitoring errors & related problems. In: Al-Worafi, Y.M. (ed), *Patient Safety in Developing Countries: Education, Research, Case Studies.* CRC Press.

Al-Worafi, Y.M. (2023i). Patient care errors and related problems: Patient education and counselling errors and related problems. In: Al-Worafi, Y.M. (ed), *Patient Safety in Developing Countries: Education, Research, Case Studies.* CRC Press.

Al-Worafi, Y.M. (2023j). Patient safety-related issues: Other medication safety issues. In: Al-Worafi, Y.M. (ed), *Patient Safety in Developing Countries: Education, Research, Case Studies.* CRC Press.

Al-Worafi, Y.M. (2023k). Patient safety culture. In: Al-Worafi, Y.M. (ed), *Patient Safety in Developing Countries: Education, Research, Case Studies.* CRC Press.

Al-Worafi, Y.M. (2023l). Patient safety education: Competencies and learning outcomes. In: Al-Worafi, Y.M. (ed), *Patient Safety in Developing Countries: Education, Research, Case Studies.* CRC Press.

Al-Worafi, Y.M. (Ed.). (2023m). *Clinical Case Studies on Medication Safety.* Academic Press.

Al-Worafi, Y.M. (Ed.). (2023n). *Comprehensive Healthcare Simulation: Pharmacy Education, Practice and Research.* Springer Nature.

Al-Worafi, Y.M. (Ed.). (2024a). *Handbook of Medical and Health Sciences in Developing Countries.* Springer, Cham.

Al-Worafi, Y.M. (2024b). Complementary and alternative medicine (CAM) in developing countries. In: Al-Worafi, Y.M. (ed), *Handbook of Medical and Health Sciences in Developing Countries.* Springer, Cham. https://doi.org/10.1007/978-3-030-74786-2_301-1

Al-Worafi, Y.M. (2024c). Access/equitable access to medical and health sciences in developing countries. In: Al-Worafi, Y.M. (ed), *Handbook of Medical and Health Sciences in Developing Countries.* Springer, Cham. https://doi.org/10.1007/978-3-030-74786-2_147-1

Hasan, S., Al-Omar, M.J., AlZubaidy, H., and Al-Worafi, Y.M. (2019). Use of medications in Arab countries. In: Laher, I. (ed), *Handbook of Healthcare in the Arab World* (p. 42). Springer, Cham.

20 Future of Complementary and Alternative Medicine (CAM) Education

INTRODUCTION

Complementary and alternative medicine (CAM) education is experiencing significant and dynamic shifts. With technological advancements, changing healthcare needs, and a deeper comprehension of human well-being, the approach to teaching future CAM practitioners is evolving.

I. **Technology and Interactive Learning in CAM Education**: The integration of technology into CAM education is set to transform student learning and practice. Virtual reality (VR), augmented reality (AR), and artificial intelligence (AI) are being incorporated into CAM curricula to enrich education in areas like herbal medicine, acupuncture, and mind-body therapies. These innovations support immersive learning experiences, allowing students to explore complex CAM practices in safe, simulated environments.
  1. *Virtual Reality (VR)*: VR enables students to engage deeply with holistic health scenarios, enhancing their understanding and skills in CAM practices without real-world risks.
  2. *Augmented Reality (AR)*: AR helps in visualizing energy meridians, acupuncture points, or the intricate anatomy involved in various CAM therapies, providing a blend of traditional knowledge and modern technology.
  3. *Artificial Intelligence (AI)*: AI applications tailor the educational journey to suit individual learner's needs in CAM, promoting an understanding of diverse treatments and patient responses.
II. **Interdisciplinary Learning and Collaboration**: Future CAM education emphasizes interdisciplinary learning, mirroring the collaborative nature of holistic healthcare. This approach prepares students to work alongside a variety of health professionals, understanding a wide range of therapeutic options and patient care strategies.
  1. *Interdisciplinary Engagement*: Students from CAM disciplines join peers from conventional medicine, nursing, and other health sciences in shared learning experiences, fostering respect and integrated care perspectives.
  2. *Team-Based Simulations*: Through simulated scenarios, CAM students collaborate with other health students, reflecting the collaborative essence of patient-centered care in holistic health settings.
III. **Personalized and Adaptive Learning**: Recognizing the uniqueness of each student, CAM education is moving toward more personalized and adaptive learning strategies.
  1. *Customized Learning Paths*: Leveraging data analytics, educators can tailor learning experiences to address students' specific strengths and areas for growth in CAM practices.
  2. *Adaptive Learning Technologies*: AI-driven platforms adjust in real-time to students' learning pace and comprehension in subjects like herbal pharmacology or nutritional therapy, ensuring optimized learning outcomes.

DOI: 10.1201/9781003327202-20

IV. **Global Awareness and Cultural Competency**: As CAM practices are deeply rooted in various cultures, education in this field is expanding to include global perspectives and cultural sensitivity.
1. *International Learning Opportunities*: CAM students may participate in study abroad programs, gaining exposure to traditional healing systems from around the world.
2. *Cultural Competency*: Courses on cultural diversity, ethics, and communication skills are crucial, enabling practitioners to deliver culturally sensitive and inclusive care.

V. **Focus on Wellness and Preventive Care**: Reflecting the core principles of CAM, education is increasingly emphasizing wellness, preventive care, and public health.
1. *Integrative Health Concepts*: CAM curricula incorporate holistic approaches to disease prevention, health promotion, and wellness, aligning with public health principles.
2. *Preparedness for Health Crises*: Training includes understanding the role of CAM in managing and preventing diseases, including pandemic response with a focus on boosting immunity and resilience.

VI. **Ethical Practice and Social Equity**: With the advancement of CAM, ethical practice and social equity become more crucial in education, preparing students to navigate ethical dilemmas and contribute to reducing health disparities.
1. *Ethics and Equity*: Courses on ethics, social determinants of health, and equity are integral, equipping students to practice with integrity and advocate for accessible, holistic healthcare.
2. *Policy and Advocacy*: CAM education fosters skills in policy engagement and advocacy, encouraging practitioners to support health equity and holistic care access.

VII. **Continuing Education and Lifelong Learning**: Acknowledging the evolving nature of CAM, the education system instills the importance of ongoing learning and professional development.
1. *Lifelong Learning Culture*: Continuous education in emerging CAM practices and research is emphasized; ensuring practitioners remain at the forefront of holistic healthcare.
2. *Flexible Learning Platforms*: Technology facilitates accessible continuing education opportunities, supporting practitioners in their commitment to lifelong learning.

VIII. **Addressing Workforce Needs in CAM**: To meet the growing demand for CAM practitioners, educational strategies are adapting to efficiently prepare skilled professionals without sacrificing quality.
1. *Accelerated and Flexible Programs*: Innovative program formats aim to quickly and effectively train CAM professionals, responding to public interest and healthcare needs.
2. *Training in Remote CAM Services*: With the rise of telehealth, CAM education includes remote consultation skills, and expanding access to holistic therapies.

IX. **Supporting Practitioner Well-being**: The well-being of CAM practitioners is prioritized, with education incorporating mental health, resilience, and self-care strategies.
1. *Resilience and Self-Care*: Students learn techniques to manage stress and maintain well-being, crucial for sustaining a long, fulfilling career in CAM.
2. *Mental Health Resources*: Educational institutions offer support services, fostering a culture of care and openness around mental health.

X. **Encouraging Research and Innovation**: CAM education encourages a culture of inquiry and innovation, driving forward the integration and acceptance of CAM practices in broader healthcare.
1. *Research Opportunities*: Students are engaged in research early, exploring the efficacy, mechanisms, and innovations in CAM.
2. *Entrepreneurship in CAM*: Courses on entrepreneurship inspire students to develop new CAM products and services, contributing to the field's growth and diversity.

As CAM continues to gain recognition and integration within the healthcare system, CAM education evolves to prepare practitioners who are innovative, compassionate, and culturally competent. Embracing technology, fostering interdisciplinary collaboration, and focusing on ethical and equitable care are pivotal in shaping the future of CAM practitioners, ensuring they are ready to meet the holistic health needs of diverse populations.

THE FUTURE OF COMPLEMENTARY AND ALTERNATIVE MEDICINE (CAM) EDUCATION IN DEVELOPED COUNTRIES: OPPORTUNITIES

The future of CAM education in developed countries presents a landscape rich with opportunities. As societal interest in holistic and preventative healthcare approaches grows, the field of CAM is set to play an increasingly significant role in the broader healthcare system. This evolution offers several key opportunities for CAM education:

1. **Integration with Conventional Medical Curricula**: There's a growing opportunity for CAM disciplines to be integrated into the curricula of traditional medical schools and healthcare training programs. This integration can foster a more holistic approach to healthcare, where future practitioners are equipped with a broad toolkit of complementary therapies alongside conventional medicine, enhancing patient care.

2. **Technological Advancements in CAM Education**: The advent of sophisticated technologies such as VR, AR, and AI provides unprecedented opportunities for CAM education. These technologies can simulate real-life patient interactions and complex CAM procedures, offering students immersive and interactive learning experiences. For instance, VR can be used to teach acupuncture or yoga therapy in a highly visual and engaging manner, while AI can customize learning paths for students based on their proficiency and learning styles.

3. **Increased Demand for Personalized Medicine**: As patients increasingly seek personalized healthcare experiences that align with their preferences, lifestyles, and genetic backgrounds, CAM education can expand to include genomics, nutrition, and lifestyle medicine. Educators have the opportunity to develop curricula that emphasize the personalization of CAM therapies, preparing practitioners to offer tailored wellness plans that integrate conventional and alternative approaches.

4. **Globalization of CAM Practices**: The globalization of healthcare provides a unique opportunity for CAM education to incorporate diverse healing traditions from around the world. By exposing students to a wide range of global CAM practices, educational institutions can prepare practitioners with a comprehensive understanding of various cultural approaches to health and wellness, enhancing their ability to serve diverse populations.

5. **Research and Evidence-Based Practice**: There is a growing need for rigorous research in the field of CAM to validate practices and inform clinical use. CAM education programs can seize this opportunity by incorporating research methodology and evidence-based practice into their curricula. Encouraging students to engage in research and contribute to the scientific understanding of CAM modalities can elevate the field's credibility and integration into mainstream healthcare.

6. **Interprofessional Education and Collaborative Practice**: The future of healthcare is inherently interdisciplinary, with a strong emphasis on collaborative practice models. CAM education can embrace this trend by offering interprofessional education opportunities where CAM students learn alongside their peers in conventional medical, nursing, pharmacy, and health sciences programs. This approach can foster mutual respect, enhance communication skills, and prepare students for collaborative patient care environments.

7. **Online and Distance Learning**: The expansion of online and distance learning platforms presents an opportunity for CAM education to reach a wider audience. By developing high-quality online courses and programs, CAM educational institutions can make their offerings more accessible to students across the globe, breaking down geographical barriers and expanding the reach of CAM practices.

8. **Regulatory Recognition and Professionalization**: As CAM practices gain popularity and acceptance, there is an opportunity for further regulatory recognition and professionalization of CAM disciplines. CAM education programs can play a crucial role in this process by ensuring high standards of education, promoting licensure and certification, and advocating for the integration of CAM practices into the healthcare system.

9. **Focus on Prevention and Wellness**: With a shift toward preventive care and wellness in healthcare, CAM education has the opportunity to lead in these areas. CAM curricula that emphasize wellness, nutrition, stress management, and preventive practices can prepare practitioners to meet the growing demand for services that promote health and prevent disease.

10. **Sustainability and Environmental Health**: The intersection of environmental health and CAM offers a broad avenue for expanding CAM education. With increasing awareness of the impact of environmental factors on health, CAM programs have the opportunity to integrate sustainability practices, herbal conservation, and eco-friendly healthcare solutions into their curricula. This approach not only aligns with the holistic principles of CAM but also prepares practitioners to consider the environmental impact of their practices and advocate for sustainable health solutions.

11. **Mental Health and Emotional Wellbeing**: The rise in mental health challenges globally presents an opportunity for CAM education to incorporate comprehensive training in mental and emotional well-being. By including practices such as mindfulness, meditation, and holistic psychotherapy techniques, CAM programs can equip practitioners with the tools to support mental health and resilience in their clients, addressing an urgent and growing area of healthcare need.

12. **Healthcare Policy and Advocacy**: As CAM continues to grow in popularity, there is a significant opportunity for CAM education to include components of healthcare policy, legislation, and advocacy. Educating future CAM practitioners about the regulatory landscape, insurance coverage, and advocacy for CAM practices can empower them to become leaders in shaping healthcare policy that supports the integration of CAM into mainstream healthcare systems.

13. **Community Health and Outreach**: CAM education can expand to emphasize community health, outreach, and public education programs. By training students to design and implement CAM-based community health initiatives, educational programs can play a crucial role in improving public health, reducing healthcare disparities, and increasing access to CAM therapies in underserved populations.

14. **Technological Innovation in CAM Practices**: Beyond the use of technology in education, there is an opportunity for CAM programs to foster innovation in the development of new CAM technologies. Educating students on the latest advancements in health technology, such as wearable devices for monitoring health metrics or apps for stress management, can inspire the creation of new CAM tools and therapies that blend traditional practices with modern technology.

15. **Ethical Considerations in CAM**: With the expanding scope of CAM practices, ethical considerations become increasingly important. CAM education has the opportunity to lead in the development of ethical guidelines for CAM practices, ensuring that students are trained to navigate complex ethical issues, respect patient autonomy, and promote informed consent, particularly in areas where evidence and traditional practices intersect.

16. **Lifelong Learning and Professional Development**: Recognizing the rapid evolution of healthcare and CAM practices, CAM education can prioritize lifelong learning and professional development. By offering advanced courses, continuing education programs, and professional development workshops, CAM educational institutions can support the ongoing growth and skill enhancement of CAM practitioners, ensuring they remain at the cutting edge of healthcare practice.

17. **Interdisciplinary Research Collaborations**: There is a vast opportunity for CAM education to promote interdisciplinary research collaborations between CAM and conventional medical researchers. Such partnerships can explore the efficacy and mechanisms of CAM therapies, contributing to a more integrated understanding of health and healing. Facilitating these collaborations can also open up new funding opportunities for CAM research and increase the evidence base for CAM practices.

18. **Personalized and Genomic Medicine**: The field of genomics offers exciting possibilities for personalized medicine, including in the realm of CAM. CAM education programs can incorporate genomic science to teach students how genetic variations affect individuals' responses to various CAM therapies, such as herbal medicines and nutritional interventions. This knowledge can enable practitioners to tailor CAM treatments to individual genetic profiles, enhancing efficacy and minimizing risks.

19. **Digital Health Integration**: The integration of digital health technologies with CAM practices offers a significant opportunity for CAM education. Programs can teach students how to use health information technology, including electronic health records (EHRs), telehealth platforms, and mobile health apps, to enhance the delivery of CAM services. This knowledge prepares practitioners to operate efficiently within modern healthcare systems and to reach patients in remote or underserved areas, expanding the accessibility of CAM therapies.

20. **Global Health Literacy**: As CAM practices often draw from global traditions, there's a unique opportunity to deepen students' global health literacy. CAM education can include studies on how different cultures approach healing and wellness, providing future practitioners with a broader perspective on health. This global outlook can enhance the adaptability and sensitivity of CAM practitioners working in diverse communities or participating in international health initiatives.

21. **Aging and Geriatric Care**: With aging populations in many developed countries, CAM education can address the specific health needs of older adults. Including specialized training in geriatric care within CAM curricula can prepare practitioners to support the health, mobility, and quality of life of the elderly through non-invasive, complementary therapies tailored to this demographic, such as gentle yoga, tai chi, and nutritional counseling.

22. **Pediatric CAM Practices**: Similarly, focusing on pediatric CAM practices offers an opportunity to fill a niche area of demand. CAM education programs can develop specialized tracks or courses focused on the safe and effective use of CAM therapies in children, addressing a growing interest among parents in alternative approaches to pediatric care, from nutritional strategies to manage Attention-deficit/hyperactivity disorder (ADHD) to acupuncture for pediatric pain management.

23. **Regenerative Medicine and CAM**: The field of regenerative medicine, including therapies like stem cell therapy and platelet-rich plasma (PRP) treatments, presents an intersection with CAM practices focused on healing and restoration. CAM education can explore these cutting-edge treatments, preparing practitioners to offer or integrate regenerative medicine options within their practice, aligning with holistic health principles.

24. **Crisis and Disaster Response**: Training CAM practitioners in crisis and disaster response, including stress management, trauma relief acupuncture, and herbal support for immune function, can prepare them to contribute effectively in emergency situations. CAM education programs can include courses on disaster medicine, psychological first aid, and the

logistics of participating in emergency response teams, equipping practitioners with the skills to provide holistic care under challenging conditions.

25. **Ethnobotany and Herbal Medicine Sustainability**: As interest in herbal medicine continues to grow, there's an opportunity for CAM education to delve deeper into ethnobotany—the study of the relationship between people and plants. This includes sustainable cultivation, ethical sourcing, and conservation of medicinal plants. Educating students on these aspects can ensure the ethical and sustainable use of herbal remedies, preserving these resources for future generations.

26. **Health Coaching and Patient Empowerment**: Health coaching is a rapidly expanding field within CAM, focusing on empowering individuals to make informed health choices and to implement sustainable lifestyle changes. CAM education programs can offer training in health coaching techniques, motivational interviewing, and patient education strategies, preparing practitioners to support patients in achieving their health and wellness goals.

27. **Nutritional Genomics**: The emerging field of nutritional genomics—or nutrigenomics— explores how food interacts with our genes. CAM education can incorporate this field to teach practitioners how to provide personalized nutritional advice based on genetic makeup, enhancing the effectiveness of dietary interventions and supporting holistic health from a genomic perspective.

28. **Neuroscience and Mind-Body Practices**: The intersection of neuroscience and CAM provides a fascinating avenue for expanding CAM education. By incorporating the latest neuroscience research into CAM curricula, programs can offer students deeper insights into how mind-body practices such as meditation, yoga, and biofeedback affect the brain and nervous system. This knowledge can enhance the effectiveness of CAM therapies aimed at mental health, stress reduction, and neurological conditions, grounding them in scientific understanding.

29. **Collaboration with Pharmaceutical Sciences**: As the use of herbal medicines and supplements grows, there's an opportunity for CAM education to foster collaboration with pharmaceutical sciences. This can include training on the pharmacodynamics of herbal compounds, interactions between herbal and pharmaceutical medications, and the development of new integrative therapies. Such collaborations can also lead to the development of standardized, safe, and effective herbal products, bridging the gap between traditional CAM practices and modern pharmacology.

30. **Health Informatics in CAM**: Leveraging health informatics can transform CAM practice by enabling practitioners to analyze health data, track outcomes, and personalize care. CAM education can integrate training on health informatics tools and data analysis, preparing students to use data-driven approaches in their practice. This can improve patient outcomes, contribute to the evidence base for CAM therapies, and facilitate integration into mainstream healthcare systems.

31. **Ethical Harvesting and Biodiversity**: With a growing emphasis on natural products in CAM, education programs have the opportunity to lead in the ethical harvesting and conservation of medicinal plants. This includes teaching sustainable practices that protect biodiversity and respect indigenous knowledge. By instilling these values in students, CAM programs can contribute to the ethical use of natural resources and the preservation of medicinal plant species.

32. **Leadership and Healthcare Management**: As CAM continues to integrate into the healthcare system, there's a need for leaders who can navigate the complexities of healthcare delivery, policy, and administration. CAM education programs can offer leadership and healthcare management tracks, preparing students to take on roles as clinic directors, policy advisors, and advocates for integrative health within healthcare institutions and government bodies.

33. **Cultural Sensitivity and Indigenous Healing Practices**: Recognizing the rich heritage of healing practices among indigenous and traditional communities, CAM education can provide deeper training in cultural sensitivity and the ethical integration of indigenous healing practices. This includes respecting intellectual property rights and ensuring that such practices are used in a way that honors their origins and contributions to the field of health and wellness.

34. **Innovative Delivery Models for CAM Therapies**: As healthcare delivery evolves, there's an opportunity for CAM education to explore innovative models for delivering CAM therapies. This could include mobile health clinics, community wellness programs, and integrative health partnerships with hospitals and primary care settings. Training students in these models can expand access to CAM therapies and integrate holistic approaches into a variety of healthcare settings.

35. **Veterinary CAM**: Interest in CAM is not limited to human health; there's a growing demand for CAM approaches in veterinary medicine as well. CAM education programs can offer tracks or courses in veterinary CAM, covering topics such as acupuncture, herbal medicine, and nutritional therapy for animals. This expands the scope of CAM practice and meets the growing demand for holistic animal care.

36. **Quality Control and Safety Standards**: With the use of CAM products and therapies increasing, ensuring their quality and safety is paramount. CAM education can include training on quality control measures, safety standards, and regulatory compliance for CAM products and practices. This ensures that CAM practitioners are knowledgeable about the products they recommend and can assure patients of their safety and efficacy.

37. **Emergency and Critical Care CAM Applications**: Exploring the role of CAM in emergency and critical care settings offers a novel area for CAM education. Training can cover the use of specific CAM interventions that can be safely and effectively applied in acute care settings, such as stress-reduction techniques for trauma patients or acupuncture for pain management in post-operative care.

38. **Precision Medicine Integration**: Precision medicine, which tailors healthcare to individual differences in genes, environment, and lifestyle, presents a unique convergence opportunity for CAM education. By incorporating principles of precision medicine, CAM programs can teach students to consider individual genetic markers, environmental exposures, and personal health histories in their approach to CAM therapies, thereby enhancing the personalization and efficacy of care.

39. **Environmental Health and CAM**: As environmental factors increasingly impact public health, CAM education has the opportunity to integrate environmental health into its curriculum. This could include training on how environmental toxins affect health and how CAM practices can mitigate these effects, preparing practitioners to address health issues from a holistic perspective that includes environmental wellness.

40. **Social Determinants of Health in CAM**: Understanding the social determinants of health—such as socioeconomic status, education, and community context—can enhance CAM practices. CAM education programs can incorporate these concepts to prepare students to consider the broader context of their patients' lives in treatment plans, promoting health equity and addressing disparities through tailored CAM interventions.

41. **Innovations in CAM Education Methodologies**: The future of CAM education also lies in pedagogical innovation. Exploring and implementing novel educational methodologies, such as flipped classrooms, problem-based learning, and gamification, can enhance student engagement and mastery of complex CAM concepts, preparing them for a dynamic healthcare environment.

42. **Public Health and Epidemiology in CAM**: With an increasing focus on public health, CAM education can include epidemiological principles and public health strategies. This enables CAM practitioners to contribute to health promotion, disease prevention, and

wellness at the community and population levels, integrating CAM practices into broader public health initiatives.

43. **Collaborative Patient-Centered Care Models**: Training CAM practitioners in collaborative, patient-centered care models prepares them to work alongside other healthcare providers in integrated healthcare teams. CAM education can focus on communication skills, understanding of other healthcare disciplines, and shared decision-making processes, ensuring CAM practitioners are valuable team members in interdisciplinary care settings.

44. **Financial Management and Entrepreneurship in CAM**: As many CAM practitioners operate or aspire to operate their own practices, CAM education can benefit from incorporating business skills, financial management, and entrepreneurship training. This prepares students not just to be skilled practitioners but also successful business owners who can sustainably manage and grow their practices.

45. **Advanced Imaging and Diagnostic Tools in CAM**: The use of advanced imaging and diagnostic tools can be integrated into CAM education, enabling practitioners to utilize these technologies in their assessments and treatments. Understanding how to interpret results from MRI, CT scans, and other imaging modalities can enhance the diagnostic precision of CAM practitioners and foster a more integrated approach to patient care.

46. **Global Regulatory and Legal Frameworks for CAM**: As CAM practices and products become more globalized, understanding international regulatory and legal frameworks becomes crucial. CAM education programs can offer courses on international health law, regulation of CAM products, and cross-border practice standards, preparing practitioners for a global practice environment.

47. **Disability and Rehabilitation in CAM**: CAM education can address the needs of individuals with disabilities or those undergoing rehabilitation, offering specialized training in adaptive CAM therapies that support mobility, reduce pain, and enhance quality of life. This specialization fills a critical gap in healthcare, providing holistic options for populations with specific needs.

48. **Nutrition Science and Dietary Therapies**: Deepening the focus on nutrition science and dietary therapies within CAM education responds to the critical role of diet in health and disease. Advanced training in nutritional biochemistry, therapeutic diets, and food as medicine can equip CAM practitioners with the knowledge to prescribe dietary interventions that support holistic health outcomes.

49. **Mental Health First Aid and Crisis Intervention**: Given the rising mental health crisis, CAM education can incorporate mental health first aid and crisis intervention training, preparing practitioners to recognize signs of mental health issues, provide initial support, and refer patients to appropriate mental health services.

50. **Fostering a Culture of Continuous Improvement and Innovation**: Lastly, cultivating a culture of continuous improvement and innovation within CAM education ensures that programs remain responsive to the evolving landscape of healthcare and societal needs. This involves regular curriculum updates, incorporation of cutting-edge research, and fostering an environment where students and faculty are encouraged to innovate and explore new frontiers in CAM.

The future of CAM education in developed countries is poised for significant growth and evolution. By leveraging these opportunities, CAM education can contribute to a more holistic, patient-centered, and integrated healthcare system, ultimately improving health outcomes and enhancing the well-being of communities.

THE FUTURE OF COMPLEMENTARY AND ALTERNATIVE MEDICINE (CAM) EDUCATION IN DEVELOPED COUNTRIES: CHALLENGES

The future of CAM education in developed countries, while bright with opportunities, also faces a range of challenges. Addressing these challenges is crucial for the advancement and integration of CAM into the broader healthcare system. Here are some of the key challenges:

1. **Standardization and Accreditation**: One of the major challenges is the lack of standardized curricula and accreditation processes across CAM disciplines. This variability can lead to inconsistencies in the quality of education and practice, making it difficult to ensure that all CAM practitioners meet a high standard of competency and safety.

2. **Evidence-Based Practice and Research**: Although there is a growing body of research supporting the efficacy of many CAM therapies, significant gaps remain. The challenge lies in integrating evidence-based practice into CAM education and expanding research to cover a wider range of CAM modalities. This is crucial for gaining acceptance within the wider medical community and for ensuring that CAM practices are based on solid evidence.

3. **Integration with Conventional Medicine**: Despite increasing interest in integrative medicine, CAM still faces challenges in being fully accepted and integrated into conventional healthcare systems. CAM education must navigate biases and skepticism from within the conventional medical community, and work toward fostering a collaborative environment where CAM and conventional medicine can coexist and complement each other.

4. **Regulatory and Legal Recognition**: CAM disciplines face varying degrees of regulatory and legal recognition across different countries and regions. This inconsistency can complicate the professionalization of CAM practices, affect insurance coverage for CAM therapies, and limit the ability of CAM practitioners to work within the healthcare system. CAM education programs must prepare students to navigate these complexities while advocating for clearer and more supportive regulatory frameworks.

5. **Cultural Competence and Globalization**: As CAM practices often originate from specific cultural or traditional backgrounds, there is a challenge in ensuring that CAM education respects and accurately represents these traditions while also making them relevant to a global audience. Additionally, the globalization of CAM practices raises questions about cultural appropriation and the ethical use of traditional knowledge.

6. **Technological Integration**: While technology offers exciting opportunities for CAM education, integrating these tools effectively into the curriculum poses challenges. There is a need for significant investment in technological resources, training for educators and students, and the development of pedagogical approaches that effectively leverage technology to enhance learning.

7. **Public Perception and Awareness**: Despite growing interest, there remains a degree of skepticism and misunderstanding about CAM among the general public and healthcare professionals. CAM education faces the challenge of improving public awareness and perception of CAM, demonstrating its value and efficacy, and dispelling myths and misinformation.

8. **Financial Support and Resources**: CAM education programs often struggle with limited financial support and resources compared to conventional medical programs. This can impact the quality of education, research opportunities, and the ability to attract and retain qualified faculty. Securing funding and resources is a critical challenge for the sustainability and growth of CAM education.

9. **Interprofessional Education and Practice**: Developing effective interprofessional education (IPE) models that include CAM and conventional healthcare students is challenging. It requires overcoming cultural and professional barriers, developing shared curricula that

respect and value all perspectives, and creating practical opportunities for interprofessional collaboration.

10. **Adapting to Rapid Changes in Healthcare**: The healthcare landscape is evolving rapidly, with advances in science, technology, and patient care models. CAM education must remain adaptable and responsive to these changes, ensuring that CAM practitioners are prepared to meet the emerging health needs of the population.

11. **Ethical and Sustainable Practice**: As the demand for natural and herbal products increases, ensuring sustainable sourcing and ethical practice in CAM is becoming increasingly challenging. CAM education must address these issues, teaching future practitioners about the importance of sustainability, ethical sourcing, and environmental stewardship in their practices.

12. **Access and Equity**: Ensuring equitable access to CAM education and careers for students from diverse backgrounds remains a challenge. CAM programs need to address barriers to entry, such as high tuition costs and lack of diversity in faculty and student bodies, to create a more inclusive and representative CAM workforce.

13. **Clinical Training and Practicum Opportunities**: A significant challenge in CAM education is providing ample clinical training and practicum opportunities for students. Unlike conventional medical education, which has well-established partnerships with hospitals and clinics for student rotations, CAM programs often struggle to secure sufficient and diverse clinical placements. This limits students' exposure to real-world practice and the variety of conditions that CAM therapies can address.

14. **Quality Control and Safety Standards for CAM Products**: The variability in quality control and safety standards for CAM products, such as herbal supplements and homeopathic remedies, poses a challenge for CAM education. Educators must navigate this landscape to teach students about the importance of product quality, and safety, and how to advise patients on selecting reputable products, amid a market that lacks uniform regulation and oversight.

15. **Balancing Traditional Knowledge with Modern Science**: CAM education must strike a delicate balance between respecting traditional knowledge and practices and integrating modern scientific principles and evidence. This challenge involves preserving the integrity and cultural significance of traditional CAM modalities while also subjecting them to scientific scrutiny and evidence-based standards, which can sometimes lead to tensions within the CAM community.

16. **Adapting to Patient Expectations and Demands**: As patients become more informed and empowered in their healthcare decisions, CAM practitioners face the challenge of meeting evolving patient expectations and demands. CAM education must prepare students to navigate these expectations, communicate effectively about the limitations and strengths of CAM therapies, and integrate patient preferences into holistic care plans.

17. **Professional Identity and Role Definition**: CAM practitioners often face challenges related to professional identity and role definition within the broader healthcare ecosystem. CAM education programs need to address these issues, helping students to understand and articulate their unique contributions to patient care, and to navigate interdisciplinary healthcare environments where their role may be less clearly defined.

18. **Innovative Business Models for CAM Practice**: The economic sustainability of CAM practices is another challenge, particularly for new practitioners entering the field. CAM education programs can respond by teaching entrepreneurial skills, innovative business models, and strategies for integrating CAM services into conventional healthcare settings or offering them directly to the public in a financially viable manner.

19. **Licensing and Credentialing Variances**: The variances in licensing and credentialing requirements for CAM practitioners across different jurisdictions create a complex landscape for CAM education. Programs must prepare students to navigate these regulatory

environments, ensuring they meet the necessary qualifications to practice legally and ethically in their chosen locality.

20. **Health Insurance and Reimbursement Issues**: The lack of consistent health insurance coverage and reimbursement for CAM services is a significant barrier. CAM education must equip future practitioners with the knowledge and skills to advocate for broader recognition and integration of CAM therapies into health insurance plans and to navigate the financial aspects of providing CAM services.

21. **Addressing Scope of Practice Limitations**: CAM practitioners often face limitations in their scope of practice, which can restrict the range of services they can offer and their ability to fully integrate into healthcare teams. CAM education needs to prepare students to understand these limitations, work within legal frameworks, and advocate for scope of practice expansions where appropriate.

22. **Maintaining Cultural Humility and Ethical Use of Traditional Practices**: As CAM draws from a wealth of global traditional practices, maintaining cultural humility and ensuring the ethical use of these traditions is paramount. Educators face the challenge of instilling these values in students while navigating the complex issues of cultural appropriation and the commercialization of traditional knowledge.

23. **Responding to Rapid Technological Advancements**: The rapid pace of technological advancement presents a challenge for CAM education to keep curricula up-to-date and to integrate emerging technologies that can enhance CAM practice, such as digital health apps, wearables for health monitoring, and telehealth platforms.

24. **Building Interdisciplinary Research Competencies**: Finally, fostering a research culture and building competencies in interdisciplinary research among CAM students and practitioners remains a challenge. CAM education must emphasize the importance of research, provide training in research methodologies that are appropriate for CAM studies, and encourage collaboration with researchers in conventional medicine and other disciplines to build a robust evidence base for CAM practices.

Addressing these challenges requires concerted efforts from educators, practitioners, researchers, policymakers, and the broader healthcare community. By tackling these issues, CAM education in developed countries can continue to evolve, contributing to a more holistic, integrated, and patient-centered approach to health and wellness.

THE FUTURE OF COMPLEMENTARY AND ALTERNATIVE MEDICINE (CAM) EDUCATION IN DEVELOPING COUNTRIES: OPPORTUNITIES

The future of CAM education in developing countries presents a unique set of opportunities that can significantly contribute to healthcare delivery, public health, and economic development. These opportunities capitalize on the rich cultural heritage, biodiversity, and growing interest in holistic and integrative health practices within these regions. Here are some key opportunities for CAM education in developing countries:

1. **Leveraging Traditional Knowledge**: Developing countries often have a rich history of traditional medicine that is deeply embedded in their culture. CAM education can formalize and preserve this indigenous knowledge by integrating it into academic curricula, ensuring that valuable traditional practices are not lost but rather validated, improved, and passed down to future generations.

2. **Addressing Public Health Challenges**: CAM can play a vital role in addressing public health challenges in developing countries, especially where access to conventional healthcare is limited. Education programs in CAM can focus on preventive care, nutritional

interventions, and low-cost, accessible treatments for common and chronic conditions, thereby improving public health outcomes.

3. **Promoting Biodiversity and Sustainable Use of Resources**: Developing countries are often rich in biodiversity, providing a vast array of medicinal plants and natural resources. CAM education can promote the sustainable use of these resources, teaching students about ethical harvesting, conservation, and the development of herbal medicines, which can support both health and environmental conservation efforts.

4. **Enhancing Healthcare Accessibility**: By training more CAM practitioners, developing countries can enhance the accessibility of healthcare services, especially in rural or underserved areas. CAM therapies often require fewer resources and infrastructure than conventional medicine, making them a feasible option for expanding healthcare access in resource-limited settings.

5. **Economic Development through CAM Industries**: The global demand for CAM products and therapies provides an economic opportunity for developing countries. CAM education can foster entrepreneurship and support the development of local CAM industries, including the cultivation of medicinal plants, the production of herbal medicines, and wellness tourism, contributing to economic development and job creation.

6. **Integrating CAM into Primary Healthcare**: CAM education can support the integration of CAM practices into primary healthcare systems in developing countries, offering a more holistic and person-centered approach to healthcare. This integration can enhance the quality of care and patient satisfaction, especially for chronic conditions where CAM therapies can complement conventional treatments.

7. **Research and Innovation in CAM**: There is significant potential for research and innovation in CAM within developing countries, particularly in exploring the efficacy and mechanisms of traditional remedies. CAM education can encourage a culture of research and innovation, fostering collaborations with international research institutions and contributing to the global knowledge base on CAM.

8. **Global Partnerships and Knowledge Exchange**: Developing countries can engage in global partnerships and knowledge exchange programs in CAM education, benefiting from international expertise while also sharing their rich traditional medicine practices with the world. These partnerships can enhance the quality of CAM education and research, promoting cross-cultural understanding and cooperation in the field of integrative health.

9. **Strengthening Regulatory Frameworks and Quality Assurance**: CAM education in developing countries can play a role in strengthening regulatory frameworks and quality assurance mechanisms for CAM practices and products. This includes developing standards for education, practice, and products, ensuring safety, efficacy, and quality in the CAM sector.

10. **Empowering Communities with Health Literacy**: CAM education can empower communities with health literacy, teaching individuals about preventive health, nutrition, and self-care practices. This empowerment can lead to healthier lifestyles, reduced disease burden, and improved overall well-being in communities.

11. **Innovative Delivery Models for CAM Education**: Developing countries have the opportunity to explore innovative delivery models for CAM education, such as online learning platforms, mobile health education, and community-based training programs. These models can make CAM education more accessible and scalable, reaching a wider audience of future practitioners.

12. **Fostering Local and Global Collaborations for CAM Research**: Developing countries can leverage their unique CAM resources by fostering local and global research collaborations. This not only elevates the scientific understanding and global appreciation of their indigenous CAM practices but also attracts funding and resources for further development. Collaborative research projects can help standardize methodologies, validate traditional

knowledge through scientific inquiry, and integrate CAM more effectively into global healthcare practices.

13. **Utilizing Digital Health Technologies**: The rapid expansion of digital health technologies offers significant opportunities for CAM education in developing countries. By incorporating digital tools, such as telemedicine, mobile health apps, and online educational resources, CAM education can overcome geographical barriers, enhance learning experiences, and provide wider access to CAM services. This digital approach can also facilitate the collection and analysis of data on CAM efficacy and safety, contributing to evidence-based CAM practices.

14. **Building Capacity for Local CAM Production**: Developing countries can harness their natural and human resources to build capacity for local production of CAM products. CAM education programs that include pharmacognosy, herbal medicine production, and quality control can support the development of local industries, reduce dependency on imported health products, and ensure the availability of high-quality, affordable CAM options for the local population.

15. **Enhancing Interdisciplinary Healthcare Education**: Integrating CAM education into the broader interdisciplinary healthcare education system in developing countries can foster a more holistic approach to health and wellness. This integration encourages mutual respect and understanding among future healthcare professionals from various disciplines, promoting collaborative care models that leverage the best of CAM and conventional medicine to meet diverse patient needs.

16. **Empowering Women and Marginalized Groups**: In many developing countries, women and marginalized groups often serve as primary caregivers and keepers of traditional health knowledge. CAM education offers an opportunity to empower these groups by formalizing their knowledge, providing professional training, and creating pathways for economic empowerment through participation in the CAM sector.

17. **Addressing Non-Communicable Diseases (NCDs)**: With the rising burden of NCDs in developing countries, CAM education can play a pivotal role in addressing these challenges. Training in lifestyle medicine, nutritional therapy, and stress reduction techniques, for example, can equip CAM practitioners with tools to support patients in managing conditions such as diabetes, hypertension, and heart disease through holistic approaches.

18. **Cultural Sensitivity and Ethical Practice**: CAM education in developing countries has the opportunity to model cultural sensitivity and ethical practice in healthcare. By respecting and incorporating local health traditions and ensuring that CAM practices are conducted with ethical considerations for patients and communities, CAM education can set a high standard for healthcare delivery that honors cultural diversity and patient autonomy.

19. **Promoting Environmental Health and One Health Concepts**: Given the close relationship between environmental health and many CAM practices, CAM education can incorporate One Health concepts, which recognize the interconnectedness of human, animal, and environmental health. This approach can promote sustainable practices, encourage the use of ethically sourced and environmentally friendly CAM products, and contribute to the overall health of the planet.

20. **Leveraging CAM for Health Tourism**: Developing countries with rich traditions in CAM have the opportunity to leverage this asset for health tourism, attracting visitors seeking wellness and traditional healing experiences. CAM education can support this industry by ensuring high standards of practice, creating certification programs for practitioners, and promoting the integration of CAM into wellness tourism offerings.

21. **Enhancing Global Health Security**: By integrating CAM education and practices into public health strategies, developing countries can enhance their contributions to global health security. Training in traditional and alternative methods for disease prevention, outbreak response, and health promotion can complement conventional public health

measures, offering diverse tools for addressing health crises and improving community resilience.

22. **Creating Inclusive Health Policies**: Finally, CAM education in developing countries can inform and influence health policy development, advocating for the inclusion of CAM in national health systems and insurance schemes. This requires engaging policymakers, demonstrating the value and efficacy of CAM practices, and ensuring that CAM is recognized as a valuable component of comprehensive health care.

23. **Strengthening Mental Health Services**: CAM education in developing countries presents an opportunity to strengthen mental health services, which are often underfunded and understaffed. By incorporating CAM approaches such as mindfulness, meditation, and herbal treatments into mental health care, practitioners can offer low-cost, accessible interventions that complement conventional mental health therapies. This holistic approach can help address the growing mental health crisis by providing a range of options that cater to individual needs and cultural preferences.

24. **Enhancing Nutritional Health and Food Security**: Education in CAM can also focus on the crucial role of nutrition in maintaining health and preventing disease. By leveraging local knowledge and resources, CAM programs can teach sustainable farming practices, the nutritional value of indigenous foods, and how to use food as medicine. This focus not only promotes better health outcomes but also supports food security and the local economy.

25. **Disaster Preparedness and Response**: CAM education can include training in disaster preparedness and response, focusing on how CAM practices can be integrated into emergency health care. This includes using CAM for stress relief, trauma recovery, and the management of acute conditions when conventional medical resources are scarce. By preparing CAM practitioners to participate in disaster response efforts, developing countries can enhance their resilience to natural disasters and other emergencies.

26. **Promoting Healthy Aging**: With global populations aging, CAM education can address the needs of older adults, promoting healthy aging and improving the quality of life for this demographic. Training can cover age-appropriate CAM therapies that focus on mobility, pain management, cognitive health, and chronic disease management, offering non-invasive, affordable alternatives or complements to conventional geriatric care.

27. **Developing CAM Informatics**: The field of CAM informatics can be developed as part of CAM education, focusing on the use of information technology to manage patient information, conduct research, and improve the quality of CAM practice. This includes developing databases of traditional medicine practices, creating digital platforms for patient education, and using data analytics to study CAM efficacy and safety.

28. **Cultivating a Global CAM Community**: By connecting CAM practitioners, educators, and students across developing countries through digital platforms and networks, CAM education can foster a global community of practice. This community can share knowledge, collaborate on research, and advocate for the integration of CAM into global health strategies. Such collaboration can amplify the voice of CAM practitioners and ensure that CAM is represented in international health dialogues.

29. **Innovating in CAM Delivery Models**: Developing countries have the opportunity to innovate in how CAM education and services are delivered. This could include mobile clinics to reach remote areas, community health worker training programs that include CAM practices, and the integration of CAM services into existing healthcare facilities to offer a more holistic approach to patient care.

30. **Addressing Climate Change and Health**: CAM education can incorporate training on the health impacts of climate change and how CAM practices can mitigate these effects. This includes understanding the environmental determinants of health, using natural

resources sustainably, and promoting practices that reduce the health impacts of pollution and climate-related diseases.

31. **Leveraging Microfinance and Social Entrepreneurship for CAM Development**: Developing countries can leverage microfinance and social entrepreneurship models to support the development of CAM practices and products. CAM education programs can teach business skills, ethical marketing, and social entrepreneurship, empowering practitioners to start and sustain CAM businesses that address local health needs while also supporting economic development.

32. **Fostering Ethical Leadership in CAM**: Finally, CAM education in developing countries can emphasize the development of ethical leadership within the CAM profession. This involves training practitioners to advocate for patient rights, navigate ethical dilemmas in healthcare, and lead efforts to integrate CAM into health systems in a way that is equitable, sustainable, and respectful of cultural traditions.

33. **Building Resilience to Antimicrobial Resistance (AMR)**: As the world grapples with the growing challenge of antimicrobial resistance, CAM education in developing countries can contribute to resilience against AMR. By incorporating training on the judicious use of antimicrobial herbs alongside conventional antibiotics, and promoting natural immunity-boosting practices, CAM practitioners can offer complementary strategies to mitigate the impact of AMR. This approach also includes educating communities about the importance of antibiotic stewardship and the role of CAM in maintaining overall health without overreliance on antibiotics.

34. **Enhancing Maternal and Child Health**: CAM education can play a significant role in improving maternal and child health in developing countries by integrating traditional and modern CAM practices focused on women and children. Training in safe, culturally sensitive CAM therapies for prenatal care, childbirth, postnatal care, and common childhood ailments can provide accessible, low-cost options to complement conventional care and improve health outcomes for mothers and children.

35. **Developing Specialized CAM Research Centers**: Establishing specialized CAM research centers within developing countries can drive innovation and evidence-based practice in CAM. These centers can focus on rigorous scientific research to validate the efficacy and safety of traditional remedies, explore new CAM therapies, and contribute to the global CAM knowledge base. By partnering with academic institutions, governments, and international health organizations, these research centers can attract funding, enhance the credibility of CAM practices, and influence health policy and practice.

36. **Promoting Mental Resilience and Community Well-being**: CAM education can emphasize the importance of mental resilience and community well-being, especially in regions facing socioeconomic challenges, conflict, or natural disasters. Training in community-based CAM approaches, such as group meditation, community gardens for medicinal plants, and traditional healing ceremonies, can foster a sense of community, resilience, and collective well-being.

37. **Integrating CAM with Non-Governmental Organization (NGO) Health Initiatives**: CAM education can synergize with health initiatives led by NGOs in developing countries. By training CAM practitioners to work within NGO frameworks, CAM can be integrated into broader health programs, including those focusing on HIV/AIDS, malaria, tuberculosis, and chronic diseases. This integration can expand the reach of CAM services and contribute to holistic, community-centered health initiatives.

38. **Leveraging Traditional Birth Attendants (TBAs) in CAM Education**: In many developing countries, traditional birth attendants play a crucial role in maternal and child health. CAM education programs can include specialized training for TBAs in CAM practices that are safe and beneficial during pregnancy and childbirth, enhancing their skills and integrating their invaluable work into the formal healthcare system.

39. **Addressing Urban Health Challenges with CAM**: As urbanization accelerates in developing countries, CAM education can address the unique health challenges faced by urban populations, including stress, pollution-related health issues, and lifestyle diseases. Urban-specific CAM programs can focus on therapies suited to urban living conditions, such as indoor herbal gardens, urban beekeeping for medicinal honey, and mindfulness practices for stress reduction.

40. **Fostering Health Equity through CAM Education**: By making CAM education accessible to a diverse range of students, including those from underprivileged backgrounds, developing countries can foster health equity. Providing scholarships, remote learning options, and community-based training programs can help break down barriers to CAM education, ensuring that a diverse group of practitioners can serve their communities.

41. **Utilizing CAM for Veterinary Health**: Expanding CAM education to include veterinary applications can address the need for affordable, accessible animal health care in developing countries. Training in veterinary herbal medicine, acupuncture, and other CAM therapies can provide farmers and pet owners with alternatives to conventional veterinary treatments, contributing to the health of livestock and pets and supporting livelihoods.

42. **Promoting Sustainable Health Tourism**: Developing countries with rich CAM traditions can promote sustainable health tourism, offering authentic CAM experiences that attract international visitors while respecting and preserving local cultures and environments. CAM education can include training in the management of health tourism enterprises, ensuring that practitioners can provide high-quality, ethical CAM services to tourists.

43. **Incorporating One Health Approaches in CAM Education**: Recognizing the interconnectedness of human, animal, and environmental health, CAM education in developing countries can adopt One Health approaches. This interdisciplinary strategy can teach students about the impact of environmental changes on health and how CAM practices can contribute to the sustainability and balance of ecosystems, leading to healthier communities and a healthier planet.

44. **Strengthening Disease Surveillance and Public Health through CAM**: CAM education can play a role in strengthening disease surveillance and public health systems in developing countries. By training CAM practitioners in basic public health principles and surveillance techniques, they can become valuable contributors to early warning systems for outbreaks and play an active role in community health monitoring and education.

45. **Utilizing CAM for Non-Communicable Disease (NCD) Prevention**: With the rise of NCDs in developing countries, CAM education can focus on lifestyle interventions, dietary practices, and traditional exercises that have been shown to prevent or manage conditions such as diabetes, hypertension, and obesity. This preventive focus not only addresses the root causes of NCDs but also reduces the long-term healthcare costs associated with these diseases.

46. **Expanding Access to Mental Health Care through CAM**: CAM education can contribute to expanding access to mental health care by incorporating traditional and holistic approaches to mental well-being. Training in CAM modalities that support mental health, such as herbal medicine, meditation, and therapeutic massage, can provide practitioners with additional tools to address the mental health needs of their communities.

47. **Developing Local CAM Guidelines and Protocols**: To ensure the safe and effective use of CAM in developing countries, there is a need for localized CAM guidelines and treatment protocols that consider the specific health needs, available resources, and cultural contexts of these regions. CAM education can include the development and dissemination of these guidelines, ensuring that CAM practices are standardized and evidence-based.

48. **Enhancing Access to Quality CAM Education through Technology**: Leveraging technology to enhance access to quality CAM education can address geographical and logistical barriers. Online courses, virtual reality simulations, and mobile apps can deliver CAM

education to remote or underserved areas, ensuring that a broader audience can benefit from CAM training.

49. **Promoting Gender Equality in CAM Professions**: CAM education can also serve as a platform for promoting gender equality in healthcare professions. By ensuring equal access to CAM education for women and addressing gender-specific health issues within CAM curricula, developing countries can empower women as healthcare providers and leaders in their communities.

50. **Fostering Innovation in CAM Practice and Product Development**: Encouraging innovation within CAM education can lead to the development of new CAM practices and products that are tailored to the specific needs of populations in developing countries. This includes innovating in herbal medicine formulations, therapeutic devices, and holistic health apps that can improve health outcomes and contribute to economic development.

51. **Building Resilience in Healthcare Systems through CAM Integration**: Integrating CAM into healthcare systems can build resilience, particularly in resource-constrained settings. CAM education that focuses on system integration can prepare practitioners to work within multidisciplinary healthcare teams, contributing to a more robust, flexible, and responsive healthcare system.

52. **Advocating for Inclusive Health Policies that Recognize CAM**: CAM education stakeholders can advocate for health policies that recognize and integrate CAM practices. By demonstrating the value of CAM through research and community health outcomes, CAM educators and practitioners can influence policy changes that support the inclusion of CAM in national health systems and insurance schemes.

53. **Enhancing Disaster Resilience with CAM Practices**: CAM education can equip practitioners with knowledge and skills to use CAM therapies as part of disaster resilience and recovery efforts. Training in stress-reducing techniques, herbal remedies for common ailments, and energy healing can provide communities with additional tools to cope with the aftermath of natural disasters, potentially reducing reliance on overstretched conventional medical resources.

54. **Incorporating CAM into School Health Programs**: Integrating CAM education into school health programs in developing countries offers an opportunity to instill healthy habits from a young age. Teaching children about the principles of nutrition, herbal medicine, and mindfulness can foster a culture of wellness and preventive care, laying the foundation for healthier future generations.

55. **Leveraging Traditional Healers in Public Health Initiatives**: Traditional healers hold a respected position in many communities within developing countries. CAM education initiatives that involve traditional healers can bridge the gap between modern healthcare systems and community-based practices, leveraging their influence to support public health campaigns, such as vaccination drives or disease prevention efforts.

56. **Promoting Ecotourism and Cultural Preservation through CAM**: Developing countries can use CAM education to promote ecotourism and cultural preservation, highlighting traditional healing practices as part of their cultural heritage. This not only supports economic development through tourism but also encourages the conservation of medicinal plants and traditional knowledge.

57. **Developing Community-Based CAM Programs**: Community-based CAM education programs can empower local populations with knowledge and skills to address their health needs, promoting self-reliance and reducing healthcare costs. These programs can focus on home remedies, basic herbal medicine, and preventive health practices tailored to the specific needs and resources of the community.

58. **Addressing Occupational Health with CAM**: CAM education can address occupational health challenges faced by workers in developing countries, particularly in the agriculture, mining, and manufacturing sectors. Training in ergonomic practices, natural pain

management, and stress reduction can improve worker health and productivity, contributing to economic development.

59. **Enhancing Palliative Care with CAM**: Integrating CAM into palliative care education can provide healthcare practitioners with additional tools to improve the quality of life for patients with life-limiting illnesses. CAM practices such as massage, aromatherapy, and meditation can complement conventional palliative care by addressing physical symptoms, emotional distress, and spiritual needs.

60. **Fostering Global Networks for CAM Education**: Developing a global network for CAM education can facilitate the exchange of knowledge, resources, and best practices between developing and developed countries. This network can support curriculum development, faculty exchange programs, and collaborative research projects, enhancing the quality and impact of CAM education worldwide.

61. **Integrating CAM with Veterinary Practices**: Expanding CAM education to include veterinary applications can address the health needs of livestock and pets in developing countries, providing cost-effective and accessible treatment options. This can improve animal welfare, support livelihoods dependent on agriculture, and reduce the risk of zoonotic diseases.

62. **Utilizing CAM for Environmental Health Challenges**: CAM education can address environmental health challenges such as pollution and toxic exposures. Training in detoxification practices, environmental health assessments, and the use of medicinal plants to mitigate environmental health risks can contribute to healthier communities and ecosystems.

63. **Promoting Self-Care and Wellness in the Workplace**: CAM education can extend into the workplace, promoting self-care and wellness practices among employees. This can include training in stress management techniques, ergonomic practices, and nutritional counseling, contributing to a healthier, more productive workforce.

64. **Supporting Mental Health in Conflict and Post-Conflict Settings**: In areas affected by conflict, CAM education can offer tools for mental health support and trauma recovery. Practices such as yoga, meditation, and traditional healing ceremonies can provide individuals and communities with nonverbal means of processing trauma, fostering resilience and social cohesion in post-conflict recovery.

By seizing these opportunities, CAM education in developing countries can contribute significantly to health system strengthening, economic development, and the preservation of cultural heritage. It requires strategic investment, international support, and a commitment to integrating CAM into the broader healthcare and educational systems, ensuring that CAM practices are used safely, effectively, and sustainably for the benefit of all.

THE FUTURE OF COMPLEMENTARY AND ALTERNATIVE MEDICINE (CAM) EDUCATION IN DEVELOPING COUNTRIES: CHALLENGES

The future of CAM education in developing countries, while rich in opportunities, faces several significant challenges that could impede its development and integration into the broader healthcare landscape. Addressing these challenges is crucial for the effective utilization of CAM in improving health outcomes. Here are some key challenges:

1. **Limited Recognition and Regulation**: One of the primary challenges is the limited recognition and regulation of CAM practices and education within the healthcare systems of developing countries. This lack of formal recognition can lead to challenges in standardizing curricula, ensuring quality education, and integrating CAM practitioners into the mainstream healthcare workforce.

2. **Resource Constraints**: Developing countries often face significant resource constraints, including limited funding for healthcare education and services. This can affect the availability and quality of CAM education programs, access to educational materials, and the development of research infrastructure to support evidence-based CAM practices.

3. **Quality of Education and Training**: Ensuring high-quality CAM education and training is a challenge, particularly in settings where there are no standardized curricula or accreditation processes for CAM programs. This can lead to variability in the competence of CAM practitioners and potentially impact the safety and efficacy of CAM treatments offered to patients.

4. **Integration into Healthcare Systems**: The integration of CAM into existing healthcare systems poses significant challenges, including resistance from conventional healthcare providers, lack of policy support, and difficulties in establishing effective collaboration and referral systems between CAM practitioners and other healthcare professionals.

5. **Research and Evidence Base**: Developing a robust evidence base for CAM practices is a major challenge in developing countries, where research funding is limited, and the capacity for conducting high-quality clinical research may be lacking. This impacts the acceptance of CAM practices by conventional medicine and limits the ability to inform policy and practice with solid evidence.

6. **Cultural and Traditional Beliefs**: While the rich cultural and traditional heritage of CAM in developing countries is an asset, it can also pose challenges in terms of standardization and scientific validation. Balancing respect for traditional knowledge with the need for evidence-based practice requires sensitive handling to avoid cultural insensitivity or appropriation.

7. **Access to CAM Education**: Accessibility to CAM education can be limited by geographical, financial, and societal barriers. This includes the rural-urban divide, where rural areas may have a greater need for CAM practitioners but fewer educational opportunities, and gender or socioeconomic factors that limit access to education for certain groups.

8. **Ethical and Sustainable Practice**: Ensuring ethical and sustainable practices in CAM, particularly concerning the use of medicinal plants and other natural resources, presents challenges. Overharvesting, habitat destruction, and loss of biodiversity can threaten the sustainability of CAM practices that rely on these resources.

9. **Public Awareness and Perception**: Public awareness and perception of CAM can vary widely, with some communities highly valuing traditional practices, while others may be skeptical or unaware of the benefits and limitations of CAM. Educating the public about safe and effective CAM practices is essential for its acceptance and integration into health and wellness regimes.

10. **Interprofessional Education and Collaboration**: Developing effective models for interprofessional education and collaboration that include CAM alongside conventional healthcare disciplines is challenging but essential for integrated care. Overcoming professional biases and establishing equal partnerships requires concerted efforts in educational reform and practice.

11. **Technological Advancements**: Keeping pace with technological advancements in healthcare education and practice, including telemedicine, digital health records, and online learning platforms, can be challenging for CAM education programs, particularly in resource-limited settings.

12. **Regulatory and Legal Frameworks**: The development of supportive regulatory and legal frameworks for CAM practices and products is a significant challenge. This includes issues related to licensure, scope of practice, product regulation, and the integration of CAM into national health insurance schemes.

13. **Quality and Safety Standards**: Ensuring the quality and safety of CAM therapies and products is a persistent challenge. In developing countries, the absence of stringent quality

control and safety standards can lead to the availability of substandard or adulterated products, posing health risks to consumers. Developing and enforcing strict quality and safety guidelines for CAM practices and products is crucial for patient safety and the credibility of CAM as a whole.

14. **Intellectual Property Rights and Traditional Knowledge**: Protecting the intellectual property rights of traditional knowledge holders without hindering access to traditional medicines for those who need them is a complex challenge. There's a delicate balance between commercializing traditional remedies to support economic development and ensuring that such practices do not exploit or disenfranchise the indigenous and local communities that have developed and maintained these practices over generations.

15. **Education and Training of Trainers**: A significant challenge is the lack of qualified educators and trainers in CAM disciplines. Developing a cadre of highly skilled, knowledgeable trainers who can impart high-quality CAM education is essential for the growth and standardization of CAM practices. This requires investment in higher education and training programs, as well as incentives to attract talented professionals into CAM education roles.

16. **Scalability of CAM Education Programs**: Expanding CAM education to meet the growing interest and demand while maintaining the quality of education is challenging, especially in resource-constrained environments. Scalability issues can limit the reach and impact of CAM education, hindering the development of a sufficiently large workforce of CAM practitioners to meet the healthcare needs of the population.

17. **Cultural Competence in CAM Curriculum**: Incorporating cultural competence into the CAM curriculum is crucial to ensure that CAM practitioners are sensitive to the diverse cultural backgrounds of patients. Developing culturally competent CAM practitioners who can navigate the nuances of various cultural beliefs and practices related to health and healing is a challenge that requires careful curriculum design and training.

18. **International Recognition and Mobility**: The international recognition of CAM qualifications from developing countries is a challenge for CAM practitioners seeking to practice or continue their education abroad. Establishing international accreditation and recognition agreements can help improve the mobility of CAM practitioners and facilitate the exchange of knowledge and skills across borders.

19. **Access to CAM Therapies for Vulnerable Populations**: Ensuring equitable access to CAM therapies for all segments of the population, especially vulnerable groups such as the poor, elderly, and those living in remote areas, is a significant challenge. Developing strategies to make CAM therapies more accessible and affordable is essential for achieving health equity.

20. **Integration of CAM into National Health Policies**: The lack of integration of CAM into national health policies and programs is a significant barrier to its wider acceptance and use. Advocacy and dialogue with policymakers are needed to highlight the potential benefits of CAM and to ensure that CAM is included in health policy planning and implementation.

21. **Sustainability of Medicinal Plant Resources**: Overreliance on wild medicinal plants without adequate conservation and sustainable harvesting practices can lead to the depletion of these valuable resources. Implementing sustainable practices and promoting the cultivation of medicinal plants are essential for the long-term viability of many CAM therapies.

22. **Digital Divide**: The digital divide between urban and rural areas can limit access to online CAM education and resources. Bridging this divide by expanding internet access and developing offline educational materials is crucial for the widespread dissemination of CAM knowledge.

23. **Navigating Changes in Healthcare Demand**: As populations in developing countries undergo demographic and epidemiological transitions, the demand for healthcare services, including CAM, changes. Adapting CAM education and services to address emerging health issues, such as chronic diseases and mental health disorders, is a challenge that requires ongoing curriculum updates and practitioner training.

24. **Financial Sustainability of CAM Practices**: Many CAM practitioners in developing countries face financial sustainability challenges, particularly those operating in rural or underserved areas where patients may have limited ability to pay for services. Developing business models that allow for the sustainable practice of CAM, possibly through integration with public health systems or innovative financing mechanisms like microinsurance, is crucial for ensuring that CAM services remain accessible and practitioners are adequately compensated.

25. **Data Privacy and Security in CAM Practices**: As CAM practices increasingly incorporate digital health technologies, ensuring data privacy and security becomes a challenge. Developing and enforcing robust data protection regulations that cover the unique aspects of CAM practice, including the handling of sensitive patient information related to traditional and alternative therapies, is essential for maintaining patient trust and compliance with international standards.

26. **Climate Change Impacts on Medicinal Plant Availability**: Climate change poses a significant threat to the availability of medicinal plants, many of which are sensitive to changes in temperature, precipitation, and extreme weather events. Developing adaptive strategies to protect these resources, such as climate-resilient agriculture practices and the establishment of medicinal plant reserves is essential for the sustainability of CAM practices that rely on herbal medicines.

27. **Bridging the Gap between Traditional Healers and Formal CAM Education**: In many developing countries, there is a gap between traditional healers, who often possess generations of empirical knowledge, and formal CAM education systems. Bridging this gap through programs that recognize and certify the knowledge of traditional healers, and integrating this knowledge into formal CAM curricula, can help preserve traditional practices while ensuring they meet modern standards of efficacy and safety.

28. **Combating Misinformation and Pseudoscience**: The proliferation of misinformation and pseudoscience related to CAM poses a significant challenge, particularly in the digital age where false health claims can spread rapidly. Developing critical thinking and evidence appraisal skills within CAM education, and creating public awareness campaigns to combat misinformation, are essential for protecting public health.

29. **Ensuring Inclusivity in CAM Education**: Ensuring that CAM education is inclusive and accessible to students from diverse backgrounds, including ethnic minorities, women, and individuals from low-income families, is a challenge. Implementing policies and programs that promote diversity and inclusivity within CAM educational institutions is crucial for creating a CAM workforce that reflects the diversity of the populations they serve.

30. **Adapting to Technological Advancements in Healthcare**: Keeping pace with rapid technological advancements in healthcare, such as artificial intelligence, machine learning, and advanced diagnostics, and integrating these technologies into CAM practices and education is a significant challenge. Continuous professional development and the inclusion of technological training in CAM curricula are essential for preparing CAM practitioners for the future healthcare landscape.

31. **Professional Isolation of CAM Practitioners**: CAM practitioners in developing countries may experience professional isolation, particularly if they are operating in areas where CAM is not widely recognized or integrated into the healthcare system. Creating professional networks and associations for CAM practitioners can provide support, facilitate

knowledge exchange, and strengthen the professional identity and advocacy efforts of CAM practitioners.

32. **Maintaining the Balance between Globalization and Localization**: As CAM practices gain popularity worldwide, maintaining the balance between globalization—adopting and adapting CAM practices from around the world—and localization—preserving local traditions and practices—becomes a challenge. Ensuring that CAM education respects and preserves local cultural and medicinal practices while also embracing beneficial global perspectives is essential for the holistic development of CAM.

Addressing these challenges requires a multifaceted approach involving government policy changes, investment in CAM education and research, collaboration between CAM and conventional healthcare sectors, and efforts to increase public awareness and acceptance of CAM. Overcoming these hurdles is essential for fully realizing the potential of CAM to contribute to health and wellness in developing countries.

THE FUTURE OF COMPLEMENTARY AND ALTERNATIVE MEDICINE (CAM) EDUCATION: RECOMMENDATIONS

To address the opportunities and challenges facing CAM education, especially in the context of both developed and developing countries, a comprehensive set of recommendations is proposed (Al-Worafi, 2020a,b, 2022a,b, 2023a–n, 2024a–c; Hasan et al., 2019). These recommendations aim to enhance the quality, accessibility, integration, and impact of CAM education, ensuring it contributes effectively to global health and wellness.

GLOBAL STANDARDIZATION AND ACCREDITATION

1. **Develop International Standards**: Establish international accreditation standards for CAM education programs to ensure consistency in training quality.
2. **Create a Global CAM Education Framework**: Develop a global framework for CAM curricula that respects cultural diversity in healing practices while ensuring scientific rigor and safety.

INTEGRATION INTO HEALTHCARE SYSTEMS

3. **Foster Interprofessional Education**: Promote interprofessional education programs that include CAM and conventional medicine students to foster mutual respect and collaboration.
4. **Advocate for Policy Support**: Work with healthcare policymakers to recognize and integrate CAM into national healthcare systems, including insurance coverage.

ENHANCING RESEARCH AND EVIDENCE BASE

5. **Increase Funding for CAM Research**: Advocate for increased funding for research on CAM modalities to strengthen the evidence base and inform practice and policy.
6. **Promote Collaborative Research Initiatives**: Encourage partnerships between CAM institutions and conventional research universities to undertake interdisciplinary research projects.

QUALITY OF EDUCATION AND TRAINING

7. **Enhance Educator Training**: Invest in the training of CAM educators to ensure high-quality teaching and mentoring for the next generation of CAM practitioners.
8. **Incorporate Technological Advances**: Utilize digital technologies, such as online learning platforms and virtual reality, to enhance CAM education and training.

CULTURAL AND TRADITIONAL KNOWLEDGE PRESERVATION

9. **Document and Validate Traditional Knowledge**: Support initiatives to document, validate, and integrate traditional healing practices into CAM education, respecting intellectual property rights and cultural heritage.
10. **Promote Cultural Competence**: Include cultural competence training in CAM education programs to prepare practitioners to work effectively with diverse populations.

PUBLIC AWARENESS AND ACCEPTANCE

11. **Launch Public Education Campaigns**: Conduct campaigns to raise public awareness about the benefits and limitations of CAM, aiming to inform and empower healthcare consumers.
12. **Counter Misinformation**: Develop strategies to counter misinformation and pseudoscience related to CAM practices through evidence-based public education.

REGULATORY AND LEGAL FRAMEWORKS

13. **Develop Supportive Legal Frameworks**: Work with legal experts and policymakers to create supportive regulatory frameworks for CAM practice and education.
14. **Standardize Licensing and Practice Standards**: Advocate for the standardization of licensing and practice standards for CAM practitioners across jurisdictions.

ACCESS AND EQUITY

15. **Expand Access to CAM Education**: Implement policies and programs to make CAM education more accessible to underrepresented and economically disadvantaged students.
16. **Ensure Equitable Access to CAM Therapies**: Develop community-based CAM programs to extend the reach of CAM services to underserved and rural populations.

SUSTAINABILITY AND ENVIRONMENTAL HEALTH

17. **Promote Sustainable Practices**: Educate CAM practitioners and students about sustainable harvesting, cultivation, and use of medicinal plants.
18. **Integrate One Health Concepts**: Integrate One Health concepts into CAM education to highlight the interconnections between human health, animal health, and environmental sustainability.

PROFESSIONAL DEVELOPMENT AND LIFELONG LEARNING

19. **Encourage Continuous Professional Development**: Develop continuing education programs for CAM practitioners to keep pace with advances in healthcare and CAM practices.
20. **Build Professional Networks**: Support the creation of professional associations and networks for CAM practitioners to facilitate knowledge exchange and professional support.

Implementing these recommendations requires collaboration among educators, practitioners, policymakers, researchers, and the public. By addressing these key areas, the future of CAM education can be shaped to meet the evolving health and wellness needs of populations globally, ensuring that CAM contributes effectively to comprehensive, culturally sensitive, and person-centered healthcare.

CONCLUSION

The future of CAM education represents a dynamic and promising frontier in the global healthcare landscape. As societies continue to embrace a more holistic and integrative approach to health and wellness, the role of CAM education becomes increasingly pivotal. The recommendations outlined above aim to address the multifaceted challenges and harness the vast opportunities within CAM education, from enhancing the quality and accessibility of education to integrating CAM practices into mainstream healthcare systems and ensuring the sustainability of traditional knowledge and practices. Implementing these recommendations will require concerted efforts from a broad spectrum of stakeholders, including educational institutions, healthcare providers, policymakers, researchers, and the community at large. Collaboration across disciplines and borders will be key to advancing the CAM field, promoting evidence-based practice, and ensuring that CAM contributes meaningfully to the health and well-being of individuals and communities worldwide. The evolution of CAM education also hinges on its ability to adapt to emerging healthcare challenges, technological advancements, and changing societal needs. By fostering an environment of continuous learning, innovation, and respect for diverse healing traditions, CAM education can contribute to a more inclusive, effective, and holistic healthcare system. As we look to the future, it is clear that CAM education holds immense potential to impact global health positively. By embracing these recommendations, stakeholders can work together to realize this potential, ensuring that CAM continues to play a vital role in promoting health, preventing illness, and enhancing the quality of life for people around the world. The journey ahead is both exciting and challenging, but with shared commitment and vision, the future of CAM education and practice can be bright and transformative, contributing to a healthier and more harmonious world.

REFERENCES

Al-Worafi, Y.M. (Ed.). (2020a). *Drug Safety in Developing Countries: Achievements and Challenges.* Academic Press.

Al-Worafi, Y.M. (2020b). Herbal medicines safety issues. In: Al-Worafi, Y.M. (ed), *Drug Safety in Developing Countries* (pp. 163–178). Academic Press.

Al-Worafi, Y.M. (2022a). *A Guide to Online Pharmacy Education: Teaching Strategies and Assessment Methods.* CRC Press.

Al-Worafi, Y.M. (2022b). Competencies and learning outcomes. In: Al-Worafi, Y.M. (ed), *A Guide to Online Pharmacy Education: Teaching Strategies and Assessment Methods.* CRC Press.

Al-Worafi, Y.M. (2023a). *Patient Safety in Developing Countries: Education, Research, Case Studies.* CRC Press.

Al-Worafi, Y.M. (2023b). *Technology for Drug Safety: Current Status and Future Developments.* Springer Nature.

Al-Worafi, Y.M. (2023c). Patient safety-related issues: Patient care errors and related problems. In: Al-Worafi, Y.M. (ed), *Patient Safety in Developing Countries: Education, Research, Case Studies.* CRC Press.

Al-Worafi, Y.M. (2023d). Patient care errors and related problems: Preventive medicine errors & related problems. In: Al-Worafi, Y.M. (ed), *Patient Safety in Developing Countries: Education, Research, Case Studies.* CRC Press.

Al-Worafi, Y.M. (2023e). Patient care errors and related problems: Patient assessment and diagnostic errors & related problems. In: Al-Worafi, Y.M. (ed), *Patient Safety in Developing Countries: Education, Research, Case Studies.* CRC Press.

Al-Worafi, Y.M. (2023f). Patient care errors and related problems: Non-pharmacological errors & related problems. In: Al-Worafi, Y.M. (ed), *Patient Safety in Developing Countries: Education, Research, Case Studies*. CRC Press.

Al-Worafi, Y.M. (2023g). Patient care errors and related problems: Medical errors & related problems. In: Al-Worafi, Y.M. (ed), *Patient Safety in Developing Countries: Education, Research, Case Studies*. CRC Press.

Al-Worafi, Y.M. (2023h). Patient care errors and related problems: Monitoring errors & related problems. In: Al-Worafi, Y.M. (ed), *Patient Safety in Developing Countries: Education, Research, Case Studies*. CRC Press.

Al-Worafi, Y.M. (2023i). Patient care errors and related problems: Patient education and counselling errors and related problems. In: Al-Worafi, Y.M. (ed), *Patient Safety in Developing Countries: Education, Research, Case Studies*. CRC Press.

Al-Worafi, Y.M. (2023j). Patient safety-related issues: Other medication safety issues. In: Al-Worafi, Y.M. (ed), *Patient Safety in Developing Countries: Education, Research, Case Studies*. CRC Press.

Al-Worafi, Y.M. (2023k). Patient safety culture. In: Al-Worafi, Y.M. (ed), *Patient Safety in Developing Countries: Education, Research, Case Studies*. CRC Press.

Al-Worafi, Y.M. (2023l). Patient safety education: Competencies and learning outcomes. In: Al-Worafi, Y.M. (ed), *Patient Safety in Developing Countries: Education, Research, Case Studies*. CRC Press.

Al-Worafi, Y.M. (Ed.). (2023m). *Clinical Case Studies on Medication Safety*. Academic Press.

Al-Worafi, Y.M. (Ed.). (2023n). *Comprehensive Healthcare Simulation: Pharmacy Education, Practice and Research*. Springer Nature.

Al-Worafi, Y.M. (Ed.). (2024a). *Handbook of Medical and Health Sciences in Developing Countries*. Springer, Cham.

Al-Worafi, Y.M. (2024b). Complementary and alternative medicine (CAM) in developing countries. In: Al-Worafi, Y.M. (ed), *Handbook of Medical and Health Sciences in Developing Countries*. Springer, Cham. https://doi.org/10.1007/978-3-030-74786-2_301-1

Al-Worafi, Y.M. (2024c). Access/equitable access to medical and health sciences in developing countries. In: Al-Worafi, Y.M. (ed), *Handbook of Medical and Health Sciences in Developing Countries*. Springer, Cham. https://doi.org/10.1007/978-3-030-74786-2_147-1

Hasan, S., Al-Omar, M.J., AlZubaidy, H., and Al-Worafi, Y.M. (2019). Use of medications in Arab countries. In: Laher, I. (ed.), *Handbook of Healthcare in the Arab World* (p. 42). Springer, Cham.

Index

Printed in the United States
by Baker & Taylor Publisher Services